8th International Conference on Advanced Infocomm Technology (ICAIT 2015)

Journal of Physics: Conference Series Volume 679

Hangzhou, China
25 – 27 October 2015

Editors:

Min Qiu
Wenhua Zhao
Perry Ping Shum

ISBN: 978-1-5108-2234-4
ISSN: 1742-6588

Printed from e-media with permission by:

Curran Associates, Inc.
57 Morehouse Lane
Red Hook, NY 12571

Some format issues inherent in the e-media version may also appear in this print version.

Printed by Curran Associates, Inc. (2016)

For permission requests, please contact the Institute of Physics at the address below.

Institute of Physics
Dirac House, Temple Back
Bristol BS1 6BE UK

Phone: 44 1 17 929 7481
Fax: 44 1 17 920 0979

techtracking@iop.org

Additional copies of this publication are available from:

Curran Associates, Inc.
57 Morehouse Lane
Red Hook, NY 12571 USA
Phone: 845-758-0400
Fax: 845-758-2633
Email: curran@proceedings.com
Web: www.proceedings.com

Table of contents

Volume 679

8th International Conference on Advanced Infocomm Technology (ICAIT 2015)
25–27 October 2015, Hangzhou, China

Accepted papers received: 10 December 2015
Published online: 29 February 2016

Preface

Papers

Journal of Physics: Conference Series **679** (2016) 011001 doi:10.1088/1742-6596/679/1/011001

PREFACE

The 2015 IEEE 8th International Conference on Advanced Infocomm Technology (ICAIT 2015) was held in Hangzhou, China, during 25-27, October 2015, following the successes of previous events held in Shenzhan, Xi'an, Haikou, Wuhan, Paris, Hsinchu, and Fuzhou. This year the ICAIT 2015 aimed to bring together researchers, developers, and users in both industry and academia in the world for sharing state-of-art results, for exploring new areas of research and development, and to discuss emerging issues on advanced infocomm technology.

The conference was hosted by Zhejiang University and China Satellite Maritime Tracking and Control Department. It was organized by the State Ley Laboratory of Modern Optical Instrumentation of Zhejiang University, in collaboration with the Joint International Research Laboratory of Photonics of Zhejiang University. More than 150 international participants from 9 foreign countries attended the conference.

The ICAIT 2015 was featured with 4 plenary lectures (by Xiaoyi Bao, Benjamin J. Eggleton, Min Gu, and Chinlon Lin, respectively), and 40 invited talks, in which a wide range of topics were covered and the most recent significant results were presented. Including oral and poster presentations, 138 abstracts were presented in the conference, some of which were selected to publish in full papers in this edition of Journal of Physics: Conference Series.

With the excellent quality of the presentations, the ICAIT 2015 was a success. We also wish to thank the sponsors of the conference, and particularly the technical program committee and the local organizing committee.

Min Qiu, *Zhejiang University*
Wenhua Zhao, *China Satellite Maritime Tracking and Control Department*
Perry Ping Shum, *Nanyang Technological University, Singapore*

Journal of Physics: Conference Series **679** (2016) 011001 doi:10.1088/1742-6596/679/1/011001

Journal of Physics: Conference Series **679** (2016) 011001 doi:10.1088/1742-6596/679/1/011001

Techninal Program Committee

Kamal Alameh	Edith Cowan University, Australia
Xiaoyi Bao	University of Ottawa, Canada
Jiahong Chen	China Satellite Maritime Tracking and Control Department
Nan-Kuang Chen	National United University, Taiwan
Yu Chen	University of Maryland, USA
Bo Cong	China Satellite Maritime Tracking and Control Department
Ken Grattan	City University London, UK
Hongwei Jiao	China Satellite Maritime Tracking and Control Department
Raman Kashyap	Ecole Polytechnique de Montre, Canada
Byoung Yoon Kim	Korea Advanced Institute of Science and Technology, Korea
Shen Lianfeng	Southeast University, China
Chinlon Lin	Lightel, USA
Bing Liu	China Satellite Maritime Tracking and Control Department
Xu Liu	Zhejiang Universiy, China
John Marsh	University of Glasgow, UK
Ci-Ling Pan	National Tsing Hua University, Taiwan
Junle Qu	Shenzhen University, China
Lars Thyen	Royal Institute of Technology, Sweden
Limin Tong	Zhejiang University
Vien Van	University of Alberta, Canada

Organizing Committee

Co-Chairs

Min Qiu	Zhejiang University
Wenhua Zhao	China Satellite Maritime Tracking and Control Department
Perry Ping Shum	Nanyang Technological University, Singapore

Local Organizing committee: Xu Liu, Liming Tong, Cuifang Kuang, Jun Lu, Liying Chen, Yanfang Lu

Plenary Speakers

Xiaoyi Bao	University of Ottawa, Canada
Benjamin J. Eggleton	University of Sydney, Australia
Min Gu	Swinburne University of Technology, Australia
Chinlon Lin	Lightel, USA

Journal of Physics: Conference Series **679** (2016) 011002 doi:10.1088/1742-6596/679/1/011002

Peer review statement

All papers published in this volume of *Journal of Physics: Conference Series* have been peer reviewed through processes administered by the proceedings Editors. Reviews were conducted by expert referees to the professional and scientific standards expected of a proceedings journal published by IOP Publishing.

Journal of Physics: Conference Series **679** (2016) 012001 doi:10.1088/1742-6596/679/1/012001

Reliability Assessment and Robustness Study for Key Navigation Components using Belief Rule Based System

Yuan You[1], Liuying Wang[1], Leilei Chang [1], and Xiaodong Ling [2] ,Nan Sun [3]

[1]High-Tech Institute of Xi'an, Xi'an, Shaanxi 710025, P. R. China
[2] China Satellite Maritime Tracking and Control Laboratory, Wuxi 214400, P. R. China
[3]The military representative in Wuxi of the military representative bureau in Nanjing, P. R. China

Email: you_xtz@163.com

Abstract. The gyro device is the key navigation component for maritime tracking and control, and gyro shift is the key factor which influences the performance of the gyro device, which makes conducting the reliability analysis on the gyro device very important. For the gyro device reliability analysis, the residual life probability prediction plays an essential role although it requires a complex process adapted by existed studies. In this study the Belief Rule Base (BRB) system is applied to model the relationship between the time as the input and the residual life probability as the output. Two scenarios are designed to study the robustness of the proposed BRB prediction model. The comparative results show that the BRB prediction model performs better in Scenario II when new the referenced values are predictable.

1. Introduction

As is known that the gyro device is a critical component for maritime navigation, and its residual life probability prediction is the main subject of its reliability assessment. Traditional residual life probability follows similar means[1]. First, suppose that certain distribution, such as the Weibull distribution, is followed by the degradation process. Second, estimate the parameters to model the residual life prediction model of the gyro device, in which the Wiener process is mostly applied. Third, update the parameters for the prediction model proposed in the second step. Fourth, use the Expectation-Maximization (EM) algorithm as well as the prior degradation data to estimate the value of the parameters.

The biggest challenge is that the gyro device has a very high reliability degree, which makes it almost impossible to collect enough data to conduct aforementioned reliability analysis. With this, the Belief Rule Base (BRB) expert system is applied in this study to model the residual life probability prediction process since BRB can model complex system behaviors under uncertainty[2] [3].

In order to improve the prediction precision, a corresponding BRB parameter learning approach is applied, which determines the best fit parameters of the BRB prediction model. In BRB parameter learning, the Evidential Reasoning (ER) algorithm is applied as the inference engine and the Differential Evolutionary (DE) algorithm is applied as the optimization means. Due to its excellent performance versus the more expensive leave-one-out cross-validation, the ten-folder cross-validation[4] is applied to validate the efficiency and robustness of the proposed parameter learning approach.

The metrics to validate the efficiency and robustness of the proposed parameter learning approach are the average/minimum/variance of the error, the average variance among the results of the ten-folder cross-validation, and the running time. Furthermore, the robustness is studied when the datasets are systematically or randomly sampled.

The remainder of this study is organized as follows. The BRB prediction model is introduced in Section 2. Section 3 and Section 4 give the reliability assessment result and robustness study result of the case study, respectively. This study is concluded in Section 5.

2. Belief Rule Base prediction model

2.1. The Belief Rule Base

The BRB system is comprised of a series of rules in the same belief structure. The kth rule is described as:

$$R_k : if \quad A_1^k \wedge A_2^k \wedge ... \wedge A_M^k, \quad then\{(D_1, \beta_{1,k}), ..., (D_N, \beta_{N,k})\}$$

where $A_i^k (i = 1,..., M; k = 1,...L)$ represents the referenced values of the *i*th attribute, *M* represents the number of the attributes, *N* represents the number of the scales, *L* represents the number of the rules, $\beta_{j,k}$ represents the belief for the *j*th scale, D_j. The *k*th rule is complete when $\sum_{j=1}^{N} \beta_{j,k} = 1$, and the *k*th rule is incomplete when $\sum_{j=1}^{N} \beta_{j,k} < 1$.

2.2. The BRB prediction model

Fig. 1 gives the BRB prediction model.

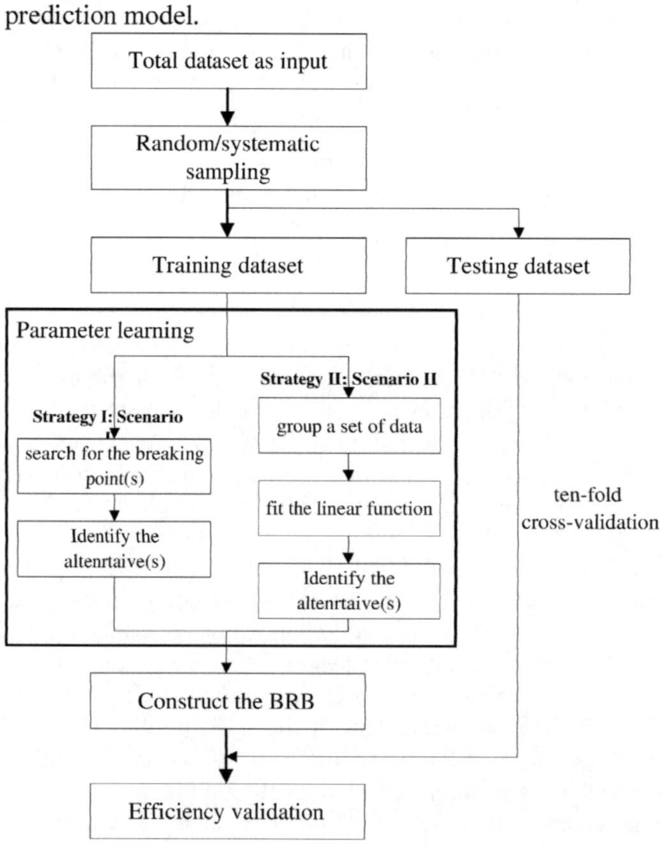

Fig. 1 The BRB prediction model.

Steps are given as following:
Step 1: Random or systematic sample from the original total dataset;
Step 2: Divide the observation data into the training dataset and the testing dataset;
Step 3: Identify key referenced values;
Step 4: Construct the new BRB prediction model;
Step 5: Validation.

For Strategy I, only the data in the training dataset is available and the referenced values are "blindly" selected. After selecting the referenced values, the BRB prediction model is constructed. For the validation part, the testing dataset is compared with the predicted result of the constructed BRB prediction model, and then the errors are summed up.

For Strategy II, new referenced values are allowed to be predicted rather than just those from the training dataset. First, group certain data from the training dataset by the number of the referenced values as previously set. And then, fit the linear equation using the grouped data and calculate the intersecting points. At last, the BRB prediction model is constructed using the intersecting points. For the validation part, the testing dataset is compared with the predicted result of the constructed BRB prediction model, and then the errors are summed up.

Details of the Differential Evolutionary (DE) algorithm and the Evidential Reasoning algorithm is not introduced in this study, readers can refer to[3] [5] and referenced therein for more information.

3. Case study: Reliability Assessment

In this section, the residual life probability prediction of the gyro device is solved by a BRB prediction model using the proposed parameter learning approach. The input is the time and the output is the residual life probability.

The 500 sets of data derived from observation are divided into two dataset. The training dataset consists of 450 sets of data while the testing dataset consists of the left 50 sets of data.

The case study is implemented in Matlab R2010b on Core (TM) i5-2450M CPU 2.50GHz with Windows 7 system. For each instances, 50 independent runs are performed. The population size of the DE algorithm is set to be 40 and the generation number is set to be 100.

3.1. Preparation

In order to effectively construct the BRB prediction model, the starting and ending referenced values are pre-determined. Normally it would be the first and last referenced values in the training dataset. However, there are multiple sets of data close to zero and do not require to be treated differently. For this specific case, the referenced values whose probabilities are the first and last above 0.1E-5, 7100 and 44700, are taken as the two ending referenced values of the BRB prediction model. Any referenced value beyond this range is considered to be zero.

3.2. Scenario I: only the training dataset is available

For Scenario I, it is supposed that only the data in the training dataset are available.

When the number of the referenced values is below five, all the possible combinations of the possible referenced values are enumerated. When the number of the referenced values is beyond five, the running time is too long for a complete enumeration. For Scenario I with any number of referenced values, the DE algorithm is applied and calculated.

As previously discussed, the samples are systematically selected.

The comparison on the minimum/average/maximum error for Scenario I with different number of referenced values ranging from three to seven is shown as in Fig. 2, in which the running time is also presented. Although all the possible combinations of the referenced values are enumerated when there are no more than five referenced values, the corresponding running time is not added in Fig. 2 because it is not in the same order of magnitude with the time consumed by the DE algorithm.

Journal of Physics: Conference Series **679** (2016) 012001 doi:10.1088/1742-6596/679/1/012001

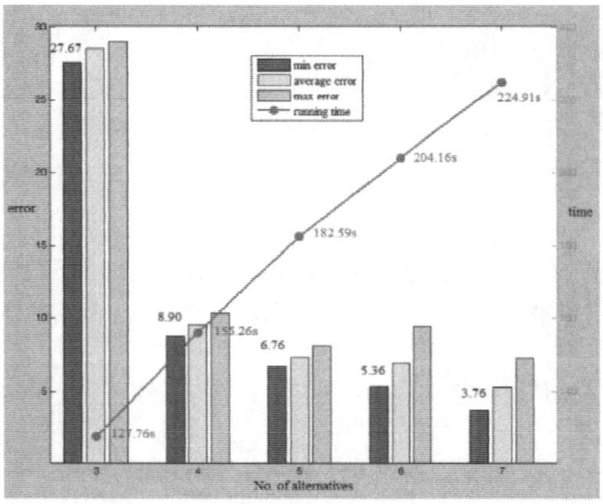

Fig. 2 The comparison of statistics of Scenario I.

Fig. 2 shows that the minimum/average/maximum error decreases with the increase of the number of the referenced values. When there are seven referenced values, the minimum/average error per each set of data in the testing dataset (a total fifty sets of data) is 0.075/0.1 (E-05), which is rather low and acceptable.

As for the running time, it grows linearly with the increase of the number of the referenced values, which is far less time-consuming than that of the enumeration. The running time of the enumeration, being 0.16s (for three referenced values)-17.54s (for four referenced values)-2086.69 (for five referenced values), is not in the same order of magnitude as that using the DE algorithm, being 127.76s (for three referenced values) to 224.91 (for seven referenced values) with 50 runs.

3.3. Scenario II: predicting new referenced values

For the second scenario, new referenced values are allowed to be predicted.

The comparison on the minimum/average/maximum error for Scenario II with different number of referenced values is shown as in Fig. 3, in which the running time is also presented.

Fig. 3 shows that the minimum/average/maximum error decreases with the increase of the number of the referenced values. When there are four referenced values, compared with seven referenced values of Scenario I as shown in Fig. 2, the minimum/average error per each testing data is 0.08/0.104 (E-05), which is rather low and acceptable.

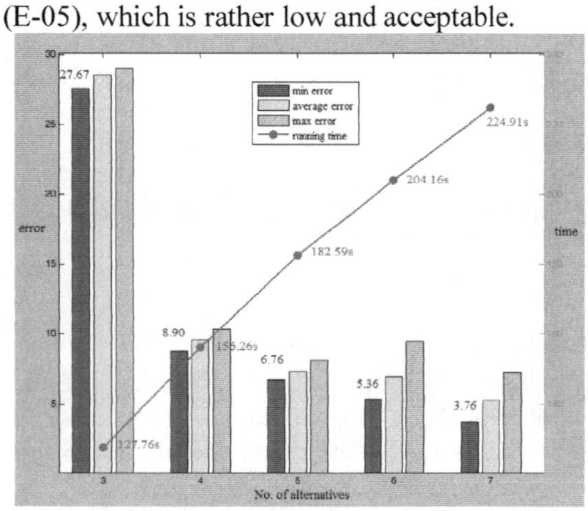

Fig. 3 The comparison of statistics of Scenario II.

As for the running time, it grows linearly with the increase of the number of the referenced values. However it is far more time-consuming than that of Scenario I: The running time of Scenario II, being 2126.47s (for three referenced values) to 5590.36 (for seven referenced values), is almost 20 times to 40 times as long as that of Scenario I, being 127.76s (for three referenced values) to 224.91 (for seven referenced values).

4. Case study: Robustness study

The robustness study is conducted in two stages. First, the average/minimum/variance of the errors in 50 runs are compared in Section 4.1 and some preliminary conclusions are drawn. And then a more comprehensive discussion is held based on more statistics including the average variance of the ten-folder cross-validation and the running time.

4.1. Robustness study on error for fifty runs

Fig. 4 shows the comparison on the error in 50 runs in both scenarios. For the purpose of visualization, the 50 runs are arranged by the values of the metrics in ascending order.

(1) For the average/minimum error, there shows similar trends between systematic and random sampling means in both scenarios. It is clear that when the numbers of referenced values are the same, the values of random sampling and systematic sampling in each scenario are very close. This trend is less clear with the variance of error.

(2) For the variance of error in both scenarios, the biggest variance is shown with random samples and three referenced values. When there are more than three referenced values, the variances in both scenarios are rather low and become acceptable. Taking one step backward, even for the biggest variance in both scenarios, it does not affect the ability of the DE algorithm in identifying the optimal referenced values, which can be proved by the fact that the average/minimum error as shown in Fig. 4(a) are rather stable and remain at low values. This is because the size of the training dataset is rather big and there are always the optimal referenced values in it.

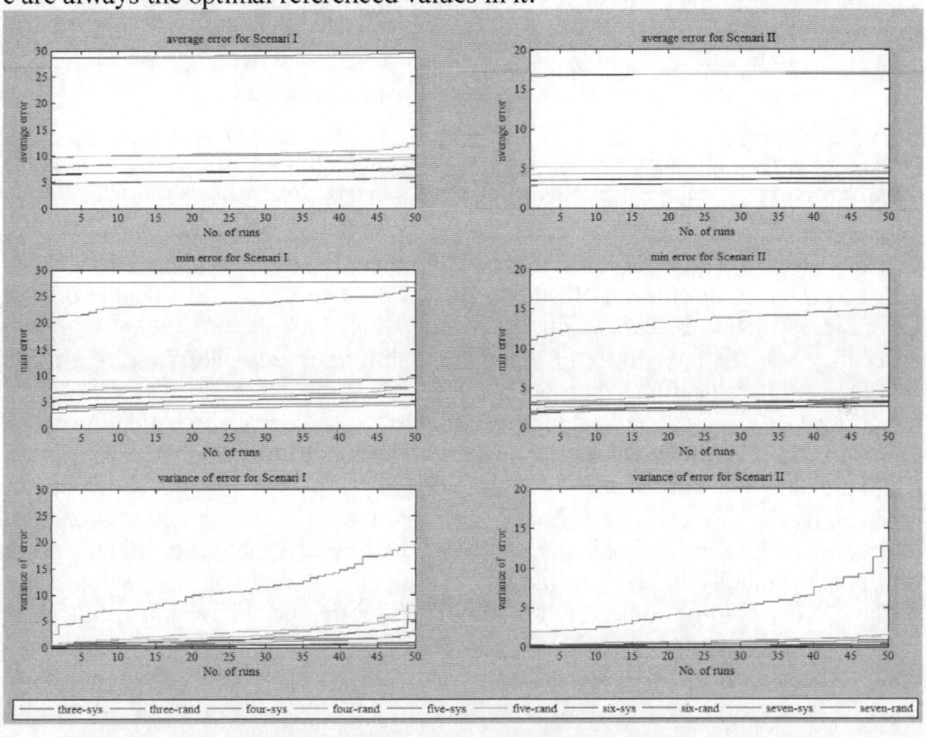

Fig. 4 The comparison of statistics of Scenario II.

(3) For all the metrics, the values of either metric in Scenario II are smaller than those in Scenario I. This is because new referenced values are predicted in Scenario II while only the data in the training dataset is available in Scenario I. New referenced values can be recognized as information extracted from the parameter learning process, and adding new information means less ambiguity and more precision.

4.2. Robustness study on the statistics between Scenarios I/II

In this section, a more comprehensive compassion on multiple statistics is conducted. A graphic demonstration of this comparison is shown as in Fig. 5. In Fig. 5(a-c), the bars represent systematic samples while the dots represent random samples. In Fig. 5(d), the bars represent Scenario I while the dots represent Scenario II.

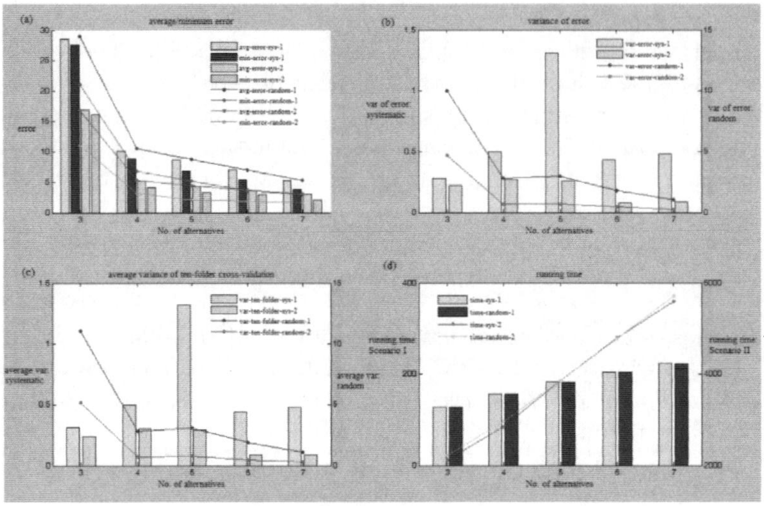

Fig. 5 Further comparison on statistics between.

(1) For the average/minimum error as in Fig. 5(a), it is observed that the results are quite consistent within and between different and different sampling means. Comparatively the average/minimum of the error in Scenario II is smaller than that in Scenario I. This is caused by new referenced values added in Scenario II.

For the values of the average/minimum error, when there are seven referenced values in Scenario I and four referenced values in Scenario II, the values of the two metrics are around 0.075/0.01 (E-05) per each set of data, which is fairly low and acceptable.

The consistency and the low values of the average/minimum error are the direct evidence for the robustness of the parameter learning approach.

(2) The variance of error and the average variance of ten-folder cross-validation as in Fig. 5(b/c) show the same trends. The following is explained from the perspective of different sampling means.

For systematic samples, both metrics remain rather low, below 1.5, yet shows no distinguished trends. Interestingly, there is an abnormal peak when there are five referenced values. This is because there are multiple local optimal combinations of five referenced values and the DE algorithm cannot escape from local optimality in order to identify the global optimal solution. When the samples are systematic, these local optimal solutions "randomly" show up and this would result in a peak in the variance.

For random samples, there is a decreasing trend with the increase of the number of referenced values. When there are over four referenced values, the two metrics would be within 3 and become acceptable. The advantage of random sampling is its ability to demonstrate the improvement of the proposed approach. This is reflected in the decreasing trend of the values of both metrics which is not observed when the samples are systematic.

Note that even for the peak of the systematic samples, the values of the two metrics are far less than those of the random samples. This is because systematic samples are equally distributed and tend to produced smaller errors.

(3) For the running time as shown in Fig. 5(d), the only decisive factor is the number of referenced values while the sampling strategy is not relevant. In both scenarios, the running time grows linearly along with the increase of the number of referenced values while they are not in the same order of magnitude.

5. Conclusion

The residual life probability prediction of the gyro device is crucial for the reliability analysis of the navigation system which is an important device for the thermonuclear fusion. The BRB prediction model is used to model this prediction process since the traditional means and approaches are rather complicated and very much conditional. The parameter learning of the BRB prediction model is proposed with two strategies and each strategy is designed for one specific scenario regarding on whether new referenced values are predictable or not.

Assessment results show that Strategy II outperforms Strategy I while consumes much longer time. Preliminary assessment results are as following:

(1) In Scenario I with no more than five referenced values, all the possible combinations of the referenced values are enumerated. The optimal referenced values of the enumeration are identical with the ones of the DE algorithm.

(2) In both scenarios, the average/minimum error decreases along with the increase of the number of referenced values.

(3) The running time in both Scenario I/II grows linearly along with the increase of the number of referenced values while Scenario II consumes much longer time than Scenario I.

The robustness of the proposed approach is studied as well. Certain conclusions are drawn as following:

(1) For the average/minimum error, the consistency between both systematic/random samples and the two scenarios is observed.

(2) For the variance of the error and the average variance of ten-folder cross-validation, except when there are three referenced values with random samples, the values of both metrics are below 3 in Scenario II and below 1.5 in Scenario I.

Since the residual life probability prediction of the gyro device requires one and only one attribute/scale, it actually simplifies the parameter learning process. However it is believed that the principle of the proposed parameter learning approach is transferable when there are multiple attributes/scales. This leads to future work that needs to be done: the extension of the parameter learning approach of the BRB prediction model in other practical cases studies with multiple attributes/scales.

Acknowledgment
This research is supported by the National Science Foundation of China under Contract no. 61305084 and no. 71201168, the open funding program of Joint Laboratory of Flight Vehicle Ocean-based Measurement and Control under Grant FOM2015OF017 and FOM2014OF14.

References
[1] Wang X L, Guo B and Cheng Z J 2012 Real-time estimation based on bivariate Wiener degradation process with time-scale transformation," *J Statis. Comput. Simul.* 1-19
[2] Yang J B, Singh M G 1994 An evidential reasoning approach for multiple-attribute decision making with uncertainty *IEEE Trans. Syst., Man, Cybern., A, Syst., Humans* **24** 1-18
[3] Yang J B and Xu D L 2013 Evidential Reasoning rule for evidence combination *Artif. Intell.* **205** 1-29
[4] Efron B and Tibshirani R 1993 An introduction to the bootstrap, *Chapman & Hall*
[5] Das S and Suganthan P N 2011 Differential Evolution: A survey of the state-of-the-art *IEEE T. on Evolut. Comput* **15** 4-31.

Journal of Physics: Conference Series **679** (2016) 012002 doi:10.1088/1742-6596/679/1/012002

Refractive Index Sensitivity of Tilted Long Period Fibre Gratings Written in Thinned Cladding Fibre

Yunhe Zhao, Yunqi Liu*, Qiang Guo, Tingyun Wang
Key Lab of Specialty Fiber Optics and Optical Access Networks, School of Communication and Information Engineering, Shanghai University, 149 Yanchang Road, Shanghai 200072, China.

*E-mail: yqliu@shu.edu.cn

Abstract. We demonstrated the fabrication of tilted long period fibre gratings written in the thinned cladding fibre by CO_2 laser. The refractive index characteristics of the gratings with different tilted angles were investigated experimentally. The experimental results show that the grating with larger tilted angle has higher sensitivity to the surrounding refractive index changes.

1. Introduction

Long period fibre gratings (LPFGs) have attracted wide attention in the last few decades. LPFGs have been demonstrated to measure curvature [1], pressure [2], twist [3], temperature [4], strain [4][5], as well as refractive index (RI) [5] in highly corrosive and electromagnetic environments. Due to its extraordinary high sensitivity on SRI, LPFGs have been exploited as a RI sensor in biological and chemical applications. The index sensitivity of the LPFGs can be increased by various methods, such as applying an overlay of high RI [6], changing the tilt angle of LPFGs [7-8] and reducing the cladding diameter [9-11].

Compared with the conventional LPFGs, tilted LPFGs (TLPFGs) make the LP_{01} core mode couple to the forward-propagating cladding modes with dissimilar azimuthal order. Recently, TLPFGs were fabricated by CO_2 laser with a tilt angle up to 80° [7-8]. Due to the increased asymmetry of index modulation of TLPFGs, the sensing characteristics of the gratings may change. Another method to increase the index sensitivity is reducing the fibre cladding including the hydrofluoric (HF) acid etching [9], tapering the fibre [10] and reaction-ion etching (RIE) [11].

The thinned cladding fibre (TCF) used in our experiment has a thin core and cladding diameter compared with the conventional single-mode fibre (SMF). The SMF-TCF-SMF (STCS) structure, which is formed by the TCF spliced between two conventional SMFs, has been verified to be a RI sensor [12]. In this paper, we demonstrate the fabrication of TLPFGs with different period and tilt angles in TCF the using CO_2 laser. It can be found that the writing efficiency of the TLPFGs can be increased for the gratings with large tilt angles. The refractive index characteristics of TLPFGs were investigated experimentally. The RI sensitivity of TCF-TLPFG is found to be higher with the increasing tilt angle of the TLPFGs.

Journal of Physics: Conference Series **679** (2016) 012002 doi:10.1088/1742-6596/679/1/012002

2. Experimental results and discussion

2.1. STCS structure

Figure 1 shows the transverse RI profile of the TCF measured by a RI profiler (S14, Photon Kinetics). The TCF has a core diameter of 5.6 μm and a small cladding diameter of 80 μm. The RI of the core is 0.012 larger than that of the cladding. The TCF was spliced with two conventional SMFs. The schematic diagram of the STCS structure is shown in figure 2(a). A small part of the incident light travels into the fibre cladding of the TCF at the SMF-TCF splicing point. Then the light travels through the core and cladding of the TCF, is combined together at the TCF-SMF splicing point. An in-fibre Mach-Zehnder (M-Z) interferometer is formed in the STCS structure. The wavelength separation of the interference fringes will decrease as the TCF length increases [12]. The contrast of interference fringes is only 1 dB, so the coupling between core mode and cladding mode is weak. Figure 2 (b) shows the picture of the spliced point of SMF and TCF by use of a commercial fusion splicer (Fujikura FSM-100P+).

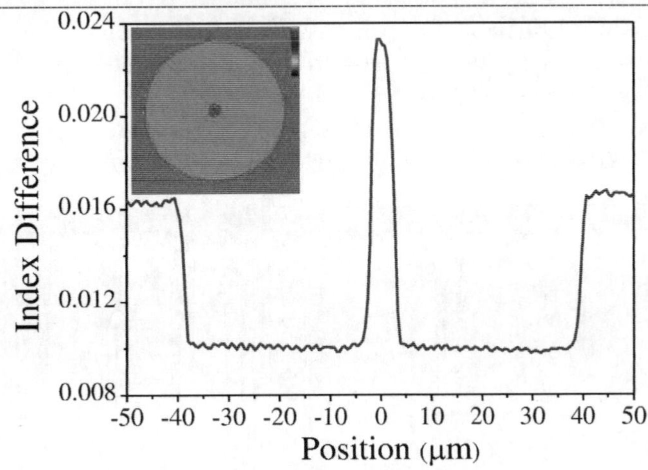

Figure 1. The transverse refractive index profile of the TCF measured by S14.

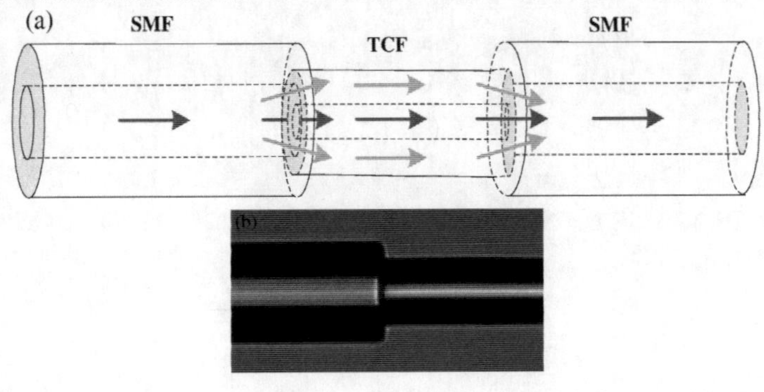

Figure 2. (a) The schematic diagram of the STCS structure, (b) The picture of the TCF spliced with the conventional SMF

2.2. Gratings fabrication

We wrote the TLPFGs in the TCF by irradiating the fibre from one side with a CO_2 laser (CO_2-H10, Han's Laser). The average power of the CO_2 laser and the frequency of the laser pulses were 0.6 W and 5 kHz, respectively. The laser beam was focused on the exposed fibre (with its jacket removed) to a spot with a diameter of ~30 μm, which was computer programmed to scan across the fibre point wise in the transverse direction at a controlled particular speed. Transverse scanning was advanced along the fibre with each step equal to the grating period. The transmission spectra of the LPFGs were measured with a supercontinuum broadband source (NKT Photonics) and an optical spectrum analyzer (OSA, AQ6375, YOKOGAWA). The resonance wavelength and contrast of the LPFGs were measured after every scanning cycle of CO_2 laser. The TLPFG were fabricated with different period Λ (400 μm, 500 μm and 600 μm) and different tilt angle (0°, 10°, 45° and 60°), respectively. For the TLPFG with period of 400 μm, the writing energy density of the CO_2 laser was 4.8615, 4.5837, 3.6114 and 3.0558 J/mm^2 when the tilt angle was 0°, 10°, 45° and 60°, respectively. It is obvious that the writing energy density of the CO_2 laser is much lower than that used for the grating fabrication in conventional SMF and the writing efficiency can be improved with the increased tile angle. The transmission spectra of the TLPFGs are shown in Figure 3. It can be found that the transmission spectrum is more complex when the tilt angle is larger than 45°, because the LP_{01} core mode couple to not only the LP_{0m} (m=2, 3, 4...) cladding modes but also the LP_{1n} (n=1, 2, 3...) cladding modes.

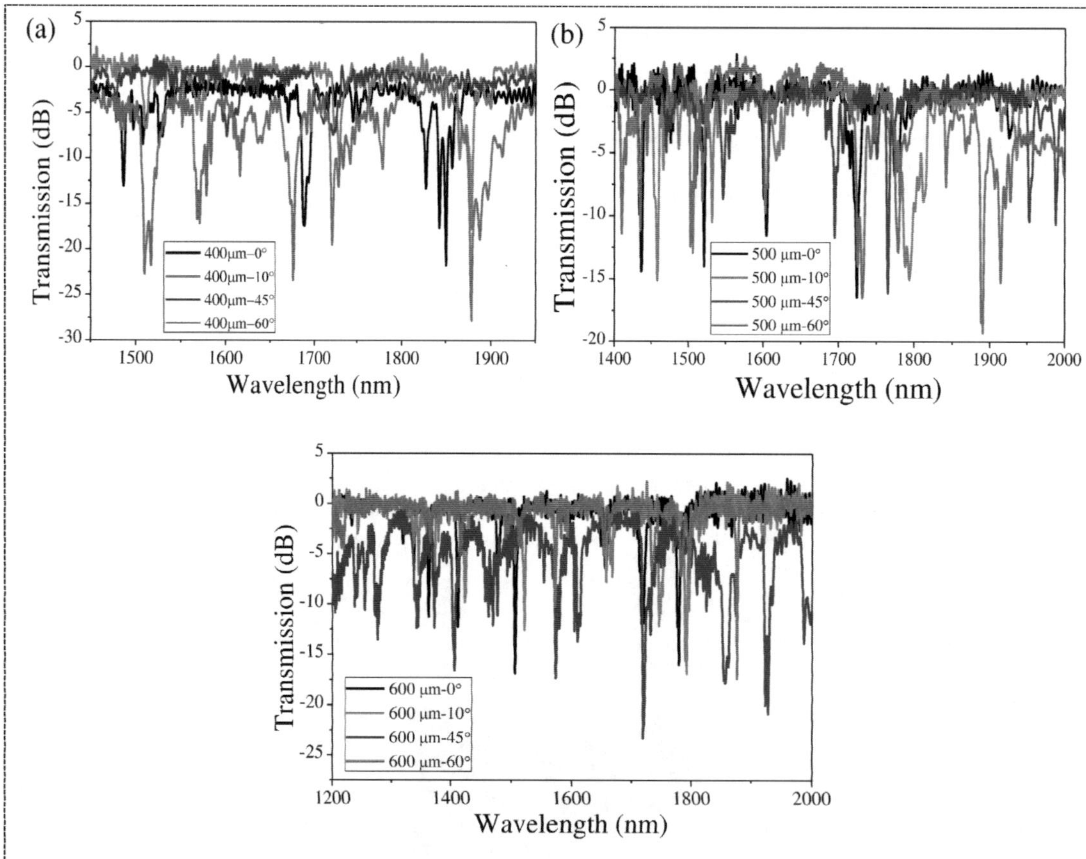

Figure 3. The transmission spectra of TCF-TLPFGs with different periods and tilt angles. (a) Λ = 400 μm, (b) Λ = 500 μm, and (c) Λ = 600 μm.

2.3. RI characteristics of TLPFG

We studied the RI characteristics of the TCF-TLPFG. The index range of the RI liquid used in our experiment is from 1.335 to 1.455. Among the resonance dips of each TLPFG, we selected the dip which is the most sensitive to the surrounding RI (SRI). Figure 4 shows the variations of resonance wavelength on SRI for different TCF-TLPFGs, and it can be found that all the selective resonance dips shift toward shorter wavelengths with the increase of the SRI. The selective resonance wavelength for TCF-TLPFG with period of 400 μm is 1688 nm (0°), 1729.8 nm (10°), 1606.6nm (45°) and 1721 nm (60°), respectively. When the SRI is 1.455, the correspondent wavelength shift is -62, -73, -61 and -67.2 nm, respectively. The selective resonance wavelength for TCF-TLPFG with period of 500 μm is 1724 nm (0°), 1792.8 nm (10°), 1695nm (45°) and 1731 nm (60°), respectively. When the SRI is 1.455, the correspondent wavelength shift is -133.8, -188.4, -36.4 and -84 nm, respectively. The selective resonance wavelength for TCF-TLPFG with period of 600 μm is 1716 nm (0°), 1746 nm (10°), 1853.6nm (45°) and 1732 nm (60°), respectively. When the SRI is 1.455, the correspondent wavelength shift is -69.6, -78.8, -111.8 and -42.6 nm, respectively.

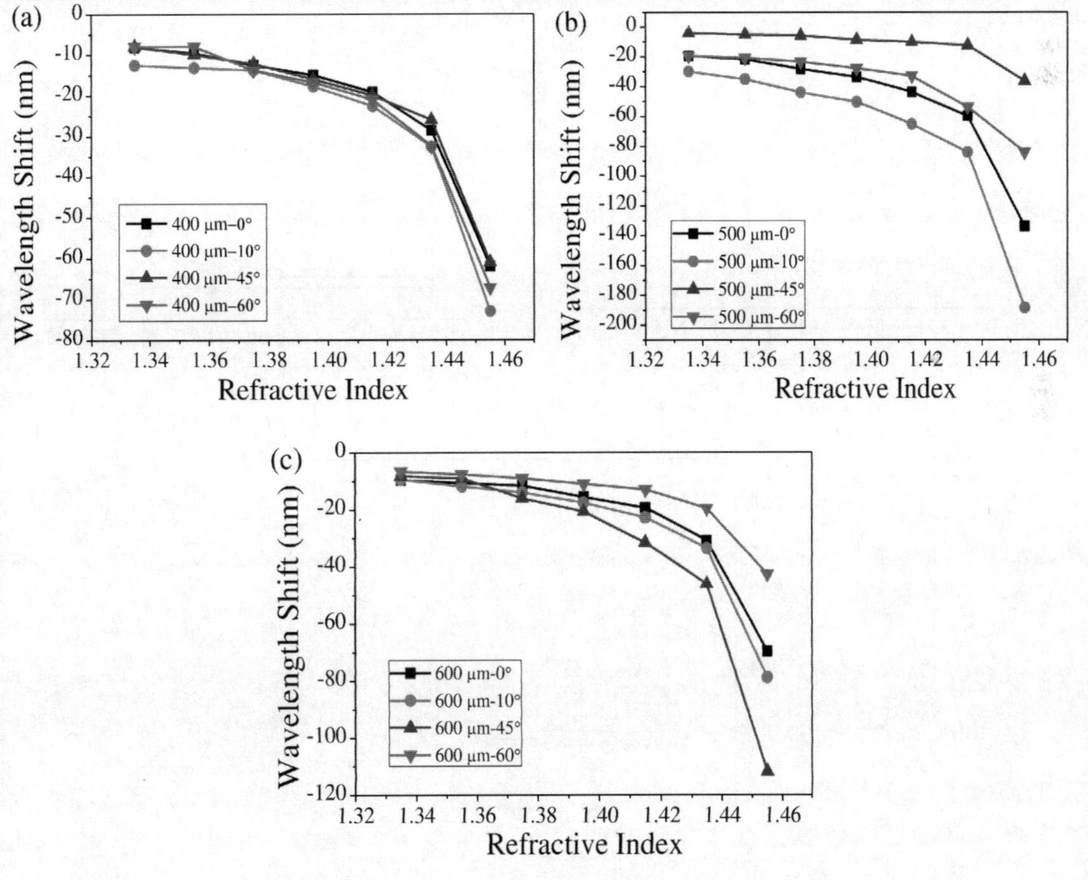

Figure 4. The variations of resonance wavelength on SRI for TCF-TLPFG with different periods and tilt angles. (a) $\Lambda = 400$ μm , (b) $\Lambda = 500$ μm, and (c) $\Lambda = 600$ μm.

Moreover, we can calculate the RI sensitivity which is defined by $S_n = \Delta\lambda/\Delta n$, where $\Delta\lambda$ is the resonance wavelength shift and Δn is the SRI variation. Figure 5 shows the RI sensitivity of different TCF-TLPFGs. By comparison, we can find that the TCF-TLPFGs with the same period and different tilt angles have different RI sensitivity. As shown in figure 5(a), the RI sensitivity of the TCF-TLPFG whose period is 400 μm with the tilt angle of 10° is higher than that of the other gratings. When the SRI ranges from 1.435 to 1.455, the correspondent RI sensitivity of the TLPFG with the tilt angle of 0°, 10°, 45°and 60° is 1670, 2020, 1756 and 1740 nm/RIU, respectively. As shown in Figure 5(b), the RI sensitivity of the TCF-TLPFGs whose period is 500 μm with the tilt angle 0°, 10°, 45°and 60° is 3700, 5220, 1200 and 1520 nm/RIU, respectively, when the SRI ranges from 1.435 to 1.455. The selective resonance wavelength for TCF-TLPFG with the tilted angle of 10° is the most sensitive to the SRI. As shown in Figure 5(c), the RI sensitivity of the TCF-TLPFGs with period of 600 μm with the tilt angle of 45° is much higher than that of the other gratings. When the SRI ranges from 1.435 to 1.455, the correspondent RI sensitivity of the TLPFG with tilt angle of 0°, 10°, 45°and 60° is 1940, 2260, 3280 and 1156 nm/RIU, respectively. Therefore it can be found that the RI sensitivity of TLPFG can be higher with the increasing tilt angle. The RI sensitivity of the TLPFG with the same tilt angles is not the same when the grating periods are different, because the TLPFGs with different tilt angles and grating periods couple the LP_{01} core mode to different order cladding modes.

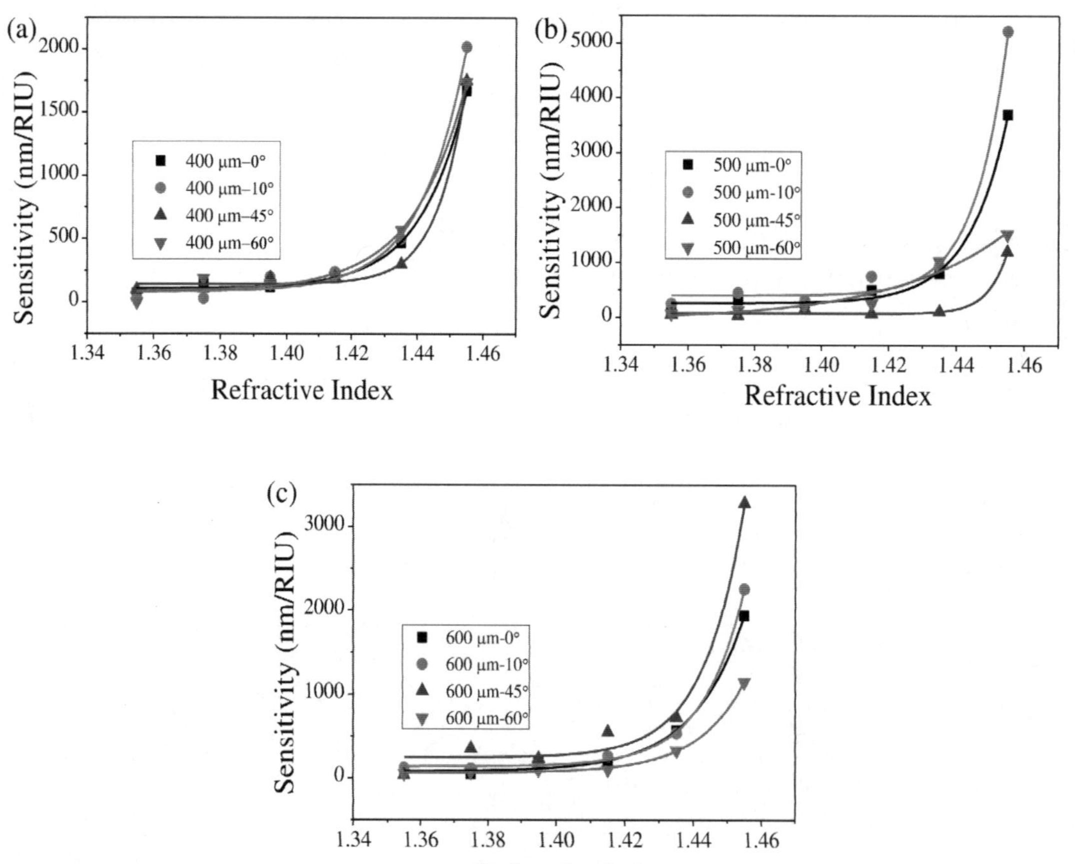

Figure 5. The refractive index sensitivity of TCF-TLPFG with different periods and tilt angles. (a) Λ = 400 μm, (b) Λ = 500 μm, and (c) Λ = 600 μm.

3. Conclusion

In conclusion, we have demonstrated the fabrication of the TLPFGs with different period and tilt angles in the TCF by CO_2 laser. It can be found that the writing efficiency of the TLPFGs can be increased for the gratings with large tilt angle. And the cladding modes of LP_{0m} and LP_{1n} modes are excited by the LP_{01} core mode when the tilt angle is larger than 45°. The index characteristics of the gratings were investigated experimentally. The RI sensitivity of TCF-TLPFG can be higher with the increasing tilt angle.

Acknowledgments
The research was jointly supported by the National Natural Science Foundation of China (61377083, 61077065).

4. References

[1] Saffari P, Allsop T, Adebayo T, Webb D, Haynes R and Roth M M 2014 Long period grating in multicore optical fiber: an ultra-sensitive vector bending sensor for low curvatures *Opt. Lett.* **39** 3508
[2] Smietana M, Bock W J, Mikulic P and Chen J 2012 Tuned pressure sensitivity of dual resonant long-period gratings written in boron co-doped optical fiber *J. Lightw. Technol.* **30** 1080
[3] Gao R, Jiang Y and Jiang L 2014 Multi-phase-shifted helical long period fiber grating based temperature-insensitive optical twist sensor *Opt. Express* **22** 15679
[4] Li W, Zhang W, Wang B, Chen L, Bai Z, Gao S, Li J, Liu Y, Zhang L, Zhou Q and Yan T 2014 Simultaneous strain and temperature measurement by cascading few-mode fiber and single-mode fiber long-period fiber gratings *Appl. Opt.* **53** 7045
[5] Huang Q, Yu Y, Ou Z, Chen X, Wang J, Yan P and Du C 2014 Refractive index and strain sensitivities of a long period fiber grating *Photonic Sensors* **4** 92
[6] Zou F, Liu Y, Deng C, Dong Y, Zhu S, Wang T 2015 Refractive index sensitivity of nano-film coated long-period fiber gratings *Opt. Express* **23** 1114
[7] Wei S, Zhang W, Fan H, Geng P, Shang J, Yin L and Xue X 2011 Study on spectral properties of tilted long-period fiber grating written by high-frequency CO_2 laser pulses *Acta Optica Sinica* **31** 1
[8] Wu R, Liu Y, Zou J, Chen N, Pang F, Wang T 2011 Fabrication of tilted long-period fibre gratings by CO_2 laser *Proceedings of the SPIE in Passive Components and Fibre-Based Devices VIII* (Shanghai, China) p 83072D
[9] Chen H and Gu Z 2013 Design of a gas sensor based on a cladding-reduced long period fiber grating coated with a sensitive film *Optik* **124** 219
[10] Xuan H, Jin W and Zhang M 2009 CO_2 laser induced long period gratings in optical microfibers *Opt. Express* **17** 21882
[11] Kumazaki H, Yamada Y, Oshima T, Inaba S and Hane K 2000 Micromachining of optical fiber using reactive ion etching and its application *Jpn. J. Appl. Phys.* **39** 7142
[12] Li L, Li X, Xie Z and Liu D 2012 All-fiber Mach-Zehnder interferometers for sensing applications *Opt. Express.* **20** 11109

Journal of Physics: Conference Series **679** (2016) 012003 doi:10.1088/1742-6596/679/1/012003

Quantitative phase imaging of Breast cancer cell based on SLIM

Huaqin Wu, Zhifang Li, Hui Li*, Shulian Wu
Key Laboratory of Optoelectronic Science and Technology for Medicine, Ministry of Education, Fujian Normal University, Fuzhou, China

E-mail: hli@fjnu.edu.cn

Abstract We illustrated a novel optical microscopy technique to observe cell dynamics via spatial light interference microscopy (SLIM). SLIM combines Zernike's phase contrast microscopy and Gabor's holography. When the light passes through the transparent specimens, it could render high contrast intensity and record the phase information from the object. We reconstructed the Breast cancer cell phase image by SLIM and the reconstruction algorithm. Our investigation showed that SLIM has the ability to achieve the quantitative phase imaging (QPI).

1 Introduction

Phase contrast, discovered by Zernike in 1935, it provides the ability to observe transparent or phase objects that do not absorb or scatter light, especially living cell. Phase contrast is sensitive to optical path-length changes, so it can reveal inner details of cells without labeling and staining [1]. SLIM is based on the Zernike phase contrast which renders high contrast intensity images and reveals the intrinsic contrast of cell structure. The accuracy of resulting topography is comparable to the atomic force microscopy and the acquisition speed is higher [2]. There are various methods to observe phase objects, such as phase contrast, dark field image, and differential interference contrast. They are widely used in biomedical signal processing. However, they cannot provide quantitative measurement but qualitative measurement.

Quantitative phase imaging (QPI) has become a rapidly growing area of bioresearch which maps the phase distribution of image field[3]. And it has been widely used in imaging red blood cell[4,5,6], cell growth[5,7], cell refractive[8,9], optical properties of tissue[10]. However, QPI illumination often produced speckles and the experimental setup was too complex, which limits the deep observation of biological application, such as living cell imaging. To overcome the shortcomings of QPI mentioned above, a new method, SLIM, was demonstrated in this paper. It mainly consisted of a commercial phase contrast microscopy and a spatial light modulator. And the white light illumination of SLIM diminished speckle effects of traditional QPI method.

SLIM combined with Laplace operator can reveal a highly detailed quantitative phase image in cell structure without gradient artifacts, such as differential interference contrast microscopy or photo-bleaching and photo-toxicity limitation in fluorescence microscopy[11]. Combining SLIM with fluorescence imaging provides a unique method for studying cell cycle-dependent growth[12], and it is an easy method to measure the growth rate of individual adherent cells of various conditions. SLIM can be used for instantaneous spatial light interference microscopy which combines the benefits of white light illumination in Zernike's phase contrast microscopy and phase stability associated diffraction phase microscopy[13]. And static iSLIM, Dynamic iSLIM, RGB iSLIM are all presented in the reference. SLIM could provide clinically relevant parameters for red blood cell analysis with unprecedented detail and sensitivity [14]. SLIM would set the basis for novel high-throughput topography and refractometry of man-made and biological nanostructures[15]. By measuring the system point-spread function and the modeling of coherent image formation in SLIM, it was found that the resolution and contrast of living cell images were significant improved[16].

The SLIM presented in this paper consisted of a commercial phase contrast microscopy and a spatial light modulator. We measured the phase of breast cancer cell to prove that SLIM could achieve quantitative phase imaging, and to lay a foundation for studying breast cancer cell dry mass. Female breast is composed of skin, fibrous tissue, mammary glands and fat composition. In the late 1970s the global breast cancer incidence showed linear growth trend. Breast cancer cells do not have some normal cell characteristics, and the connections between the cells are loose, which make the breast cancer cells easily to shed. Once the cancer cells shed, they can be transferred with the blood or lymph disseminated the whole body. At present, breast cancer has become one of the most severe threats to the health among common tumors. Thus studying breast cancer cells has a significance meaning.

2 SLIM Setup and Principle

2.1Experimental Setup

Fig.1 showed the schematic of instrument setup. The system combined Olympus IX73 commercial phase contrast microscopy with Holoeye spatial light modulator. The basic principle of SLIM is that the scattered light intervenes with unscattered light after the light passes through the sample. A phase ring provides a $\pi/2$ phase shift between scattered light and unscattered light, which is in the back focal plane of the objective. Then a liquid crystal phase modulator on the back focal plane was used to import phase shift. We recorded 4 intensity images with 4 different phases, which are 0, $\pi/2$, π, $3\pi/2$, respectively. Therefore the quantitative phase map of the sample can be uniquely determined by 4 intensity images.

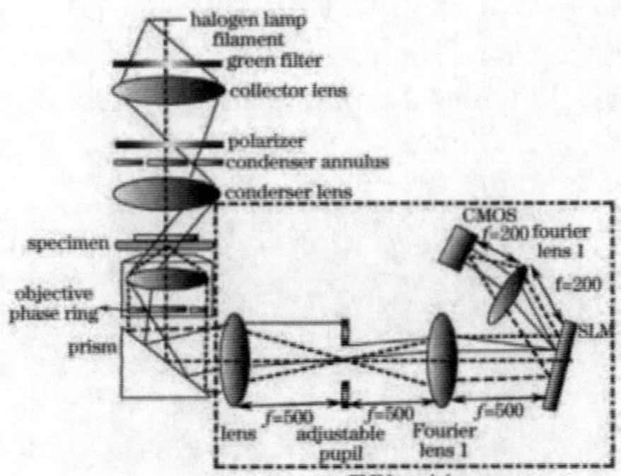

Figure.1 Experimental setup[17]. A lamp filament was projected onto a condenser annulus, which was located in the front of focal plane of the condenser. We added a Olympus color filter (431F550-W45) to eliminate the halo effects and improve the quality of image before the condenser. The central wavelength was 550nm. We used a lens and two 500mm lenses guide the light into SLM to make the phase grating modulate the scattered and unscattered light. A 20 ×/0.45 objective was used here. At last, a 200mm Fourier lens was used to guide the light into CMOS.

2.2 Principle

The principle of SLIM can be found in reference [2], and is summarized as follows. When light passes the sample, one portion of light remains scattered and contains the structure information of sample, and the other portion of light remains unscattered and forms a uniform background of the image. Here $U_1(x, y)$ denotes scattered light, U_0 denotes unscattered light, and $I(x, y; \varphi)$ denotes the light intensity accepted by the CCD, which can be expressed as equation (1).

$$I(x, y; \varphi) = |U_0|^2 + |U_1(x, y)|^2 + 2|U_0||U_1(x, y)|\cos[\Delta\varphi(x, y) + \varphi] \qquad (1)$$

$\Delta\varphi(x, y)$ in equation (1) is the phase difference between$U_1(x, y)$ and U_0, and φ is the additional phase modulation introduced by SLM. $\Delta\varphi(x, y)$ can be expressed as equation (2).

$$\Delta\varphi(x, y) = \tan^{-1}\left[\frac{I(x, y; 3\pi/2) - I(x, y; \pi/2)}{I(x, y; 0) - I(x, y; \pi)}\right] \qquad (2)$$

$\varphi(x, y)$ in equation (1) is the phase delay associated with the auto-correlation function, and can be expressed as equation (3), where $\alpha(x, y)$ is defined as the ratio of $|U_1|$ and $|U_0|$.

$$\varphi(x, y) = \tan^{-1}\left[\frac{\alpha(x, y)\sin(\Delta\varphi(x, y))}{1 + \alpha(x, y)\cos(\Delta\varphi(x, y))}\right] \qquad (3)$$

Thus, together with equation (1) (2) (3), we can get the quantitative phase image associated with sample. Moreover, the local refractive index, used to calculate the sample mass, can be retrieved from equation (4), where n_0-n is the local refractive index contrast between surrounding culture medium and sample, h is the local thickness of the sample, and λ is the central wavelength of the illumination light.

$$\varphi(x, y) = \frac{2\pi}{\lambda}\int_0^{h(x,y)} [n(x, y, z) - n_0]dz \qquad (4)$$

3 Data Analysis

We placed the MCF-7 breast cancer cells with 10% FBS added in DMEM buffer and incubated for three days at 37 °c, 5% carbon dioxide incubator. The average size of breast cancer cells was 13μm. Figure 2 presented the MCF-7 breast cancer cells under the microscope. Figure 3 showed four phase rings, and the corresponding phase were 0, π/2, π, 3π/2. Figure 4 showed the images recorded by CCD corresponding to 0, π/2, π, and 3π/2 phase.

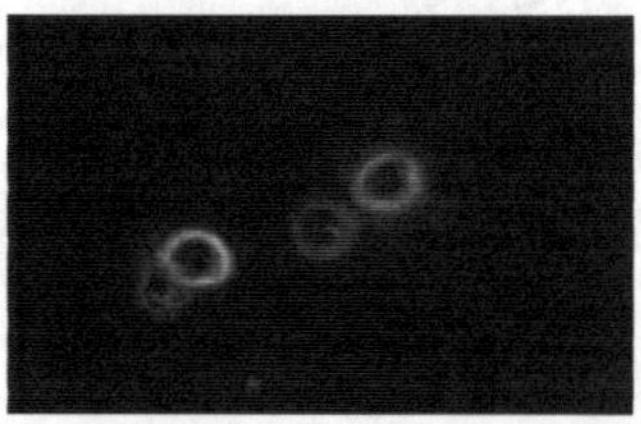

Figure.2 The MCF-7 breast cancer cells under the microscope.

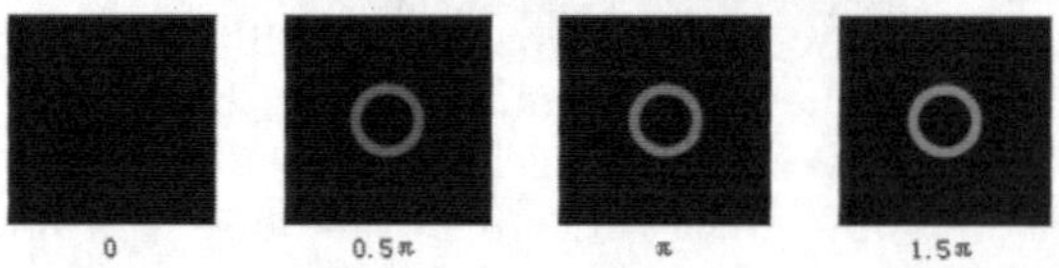

Figure.3 The phase rings respectively were 0, π/2, π, 3π/2.

Journal of Physics: Conference Series **679** (2016) 012003 doi:10.1088/1742-6596/679/1/012003

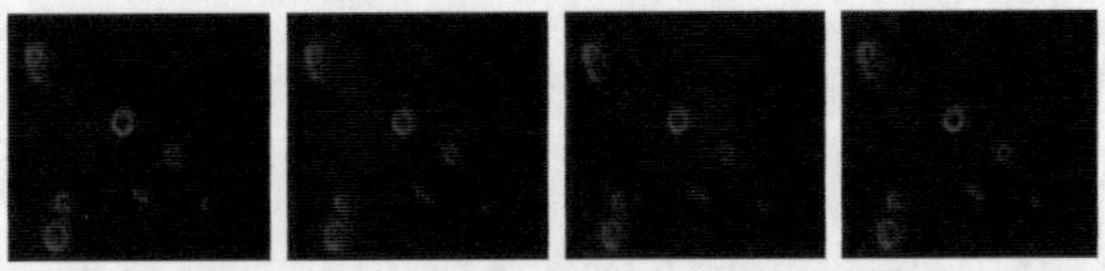

Figure.4 The images recorded by CCD corresponding to $0, \pi/2, \pi, 3\pi/2$ phase.

Then the breast cancer cell phase can be reconstructed using equation (2), and the quantitative phase image of breast cancer cell was shown in Figure 5.

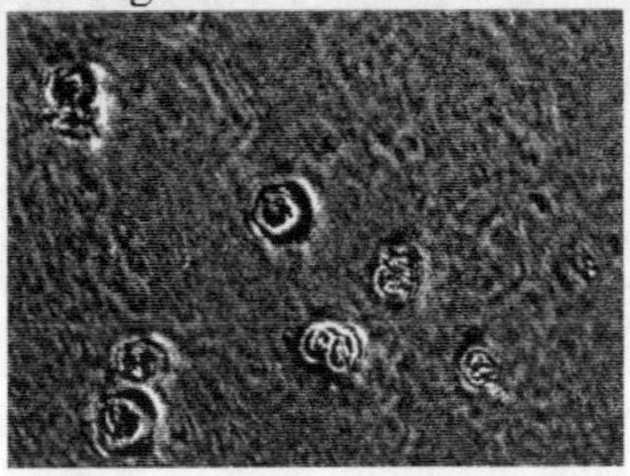

Figure.5 The quantitative phase image of Breast cancer cell.

4 Summary and Outlook

SLIM uses common interferometry, which enables temporally sensitive optical path-length measurement and provides speckle-free images. It will be widely used to extract the quantitative and nanoscale information about cell density, which can be used to demonstrate the physiological activity and status of individual cells, or reveal the heterogeneity between normal Breast cells. Meanwhile, SLIM also plays an important role in the studies on cellular differentiation and pathology, as well as early clinical diagnosis and treatment.

This paper took advantage of the ability of SLIM, which combines commercial microscopy and spatial light modulator, to reconstruct the quantitative phase image of Breast cancer cell, and obtained Breast cancer single-cell biophysical characterization. However, because of the noise introduced by the culture environment, the error of the system was larger than intrinsically allowed by the optical instrument. Thus more effort is needed to diminish the system error and improve the quality of quantitative phase image further.

Acknowledgment

This work was funded by the National Natural Science Foundation of China (No.61178089) and the Natural Science Foundation of Fujian Province (No.2014J01226).

Reference

[1] Zernike F 1955 *Sci.* **121** 345-349
[2] Wang Z, Millet L, Mir M 2011 *Opt. Express.* **19** 1016-1026
[3] Popescu G 2008 *Methods in Nano Cell Biology*. **Chap. 5, Elsevier**
[4] Paganin D and Nugent K A 1998 *Phys. Rev. Lett.* **80** 2586–2589
[5] Zicha D, Dunn GA 1995 *J. Microsc.* **179** 11–21
[6] Yang C H, Wax A, Dasari R R 2001 *Opt.Lett.* **26** 686–688
[7] Popescu G Park Y, Lue N 2008 *Am. J. Physiol.* Cell Physiol. **295** C538–C544
[8] Rappaz B, Marquet P, Cuche E 2005 *Opt. Express.* **13** 9361–9373

[9] Lue N, Popescu G Ikeda T 2006 *Opt.Lett.* **31** 2759–2761
[10] Ding H F, Nguyen F, Boppart S A 2009 *Opt.Lett.* **34** 1372–1374
[11] Wang Z, Millet L, Chan V 2011 *J Biomed Opt.* **16** 026019
[12] Mir M, Wang Z, Shen Z 2011 *PNAS.* **108** 13124-13129
[13] Ding H F and Popescu G 2010 *Opt. Express.* **18** 1569-1575
[14] Mir M, Tangella K, Popescu G 2011 *Bio. Opt.Express.* **2** 3259-3266
[15] Wang Z, Chun I S, Li X L 2010 *Opt.Lett.* **35** 208-210
[16] Haldar J P and Wang Z 2011 *IEEE T BIO-MED ENG* **58** 2489-2497
[17] Chen D Q, Zheng M J, Li Z F,Wu H Q 2014 *Laser & Optoelectronics Progress.* **51** 121204

Journal of Physics: Conference Series **679** (2016) 012004 doi:10.1088/1742-6596/679/1/012004

Macrobending Loss Measurements of G.657 Fiber with Suppression of Ripple Effect Induced by Whispering Gallery Modes

Chunsheng Li[1, 2], Tao Wang[2], Linyin Li[3], and Xiaoguang Zhang[1]

[1]State Key Laboratory of Information Photonics and Optical Communications, Beijing University of Posts and Telecommunications, Beijing, China 100876, xgzhang@bupt.edu.cn

[2]Beijing Blade Science & Tech Co, Ltd., Beijing, China 100876, lwm1975@qq.com

[3] Quality Supervision and Test Center for Cables and Communication Equipments of information industry, Sichuan, China 610062

Abstract During the macrobending loss performance tests of ITU-T G.657 fiber under small bending radius, the test results show big differencesin many tests for the same test samples and conditions. Research shows that the main reason for the difference is Whispering gallery modes phenomenon in small bending radius [1]. The inappropriate test conditions can affect the accuracy of macrobending loss test results. In the test of product validation and field application, single wavelength light source and optical power meter were often used. How to judge whether there is effect existing and how to remove the Whispering gallery modes influence in the testing process has become the key to correctly test macrobending loss by light source and optical power meter. This paper introduces the method of eliminating Whispering gallery modes effect during the macrobending test under small bending radius by single wavelength light source and optical power meter.

1. Introduction

With the fast pace of development of FTTH in China, bending insensitive single mode fibers like G.657 fiber are widely used the FTTH system. Macrobending loss is one of the key issues to be test for G.657 fibers. Theoretically, macrobending loss shows an exponential function curve with respective to the wavelength when the radius of the fiber is fixed. But for practical fiber there is strong ripple effect in the measurement of bending loss induced by so called Whispering Gallery Modes (WG modes) when radius of the fiber is small [1]. It is found that this kind of ripple or oscillation is result of the interference of guided propagating optical signal in the fiber core with radiating modes (WG modes) reflected by the cladding-buffer interface or the buffer-air interface. And oscillating bending loss curve is not repeatable and stable for the reason of the irregular nature of the reflections. The standard IEC 60793-1-47 [2] gives a curve fitting method to smooth the curve, but we need the effective measurement methods to suppress the ripple or oscillation in order to get more accurate measurement.

Journal of Physics: Conference Series **679** (2016) 012004 doi:10.1088/1742-6596/679/1/012004

In this paper, we verify experimentally the oscillating phenomenon of the spectral bending loss curve, and also propose a ripple suppression method by immersing the fiber into some carefully chosen index matching material solution. The experiments show that two proposed methods are very effective.

2. Measurement Experiment and Principle

2.1. The principle of whispering gallery modes

The whispering gallery modes are the result of the interference of guided mode in the core with the radiating modes (WG modes) reflected by the cladding-buffer interface or the buffer-air interface, as shown in Fig. 1.

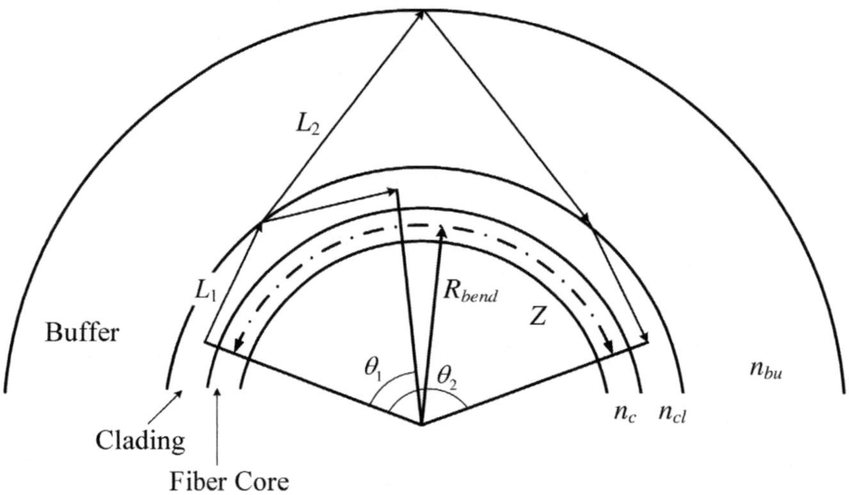

Figure 1. Bend geometry and formation of whispering gallery modes.

The constructive interference condition for interference between the guided mode with ith group of WD modes, corresponding to the maximum of the ripple, is (1).

$$\frac{2\pi}{\lambda} l_i - \beta Z_i + \phi_i = 2\pi N_i \quad (1)$$

where li is the optical path of the ith WD mode, λ is the wavelength, β is the propagation constant of the core guided mode, Z_i is the arc length over the angle θi, ϕ_i is phase shift at the corresponding interface, N_i is the integer[3].

2.2. Measurement setup

The measurement setup for macrobending loss of the fiber under test is shown in Fig. 2. A monochromator filters the monochromatic light working at the wavelength ranging from 1400nm to 1650nm. The filtered monochromatic light is fed into the fiber under test, and output power is measured, where the fiber is arranged at first as the reference power level measurement, and then is bended N turns for macrobending loss measurement [4].

Therefore in the bending loss tests should be arranged with larger radius circle and more turns. But we can see from Figure 3 that even if we make the bending fiber lager circle and many turn wrapping, the macrobending loss curves also show to some extent unrepeated and unstable and make the measurement inaccurate. So the bending loss measurement methods with ripple suppression should be found.

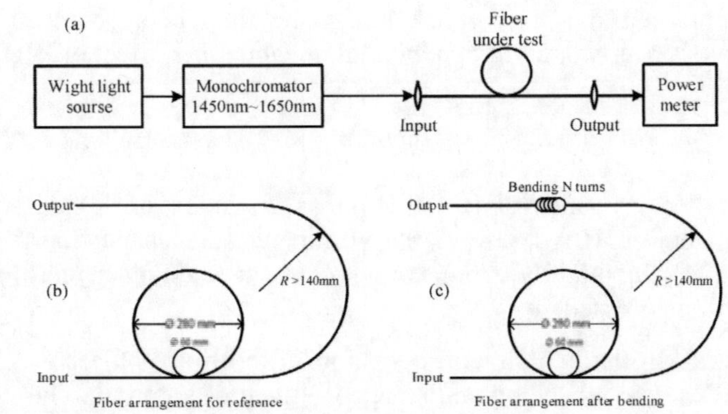

Figure 2. (a) Macrobending loss measurement setup, (b) The fiber arrangement for the measurement of reference power level, (c) The fiber arrangement after bending.

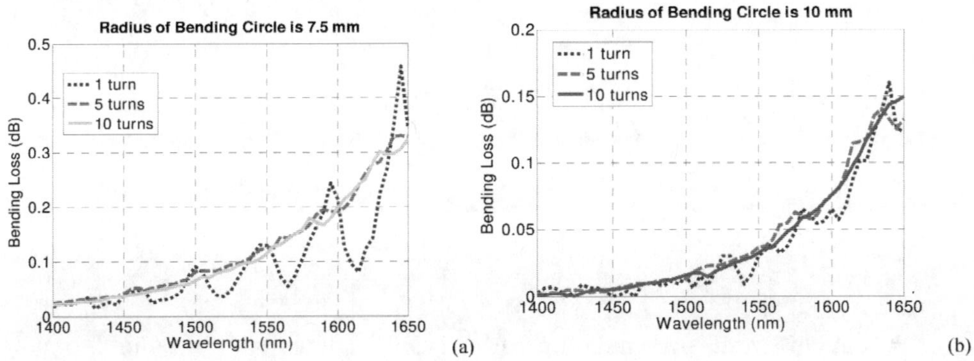

Figure 3. Bending loss results when the fiber is bent in 1, 5, and 10 turns, with radius of bending circle is (a) 7.5 mm and (b) 10 mm respectively.

3. Ripple Suppression by fiber immersed in the Index Matching Solutions

If we found some index matched materials, solid or liquid, revolve around the fiber under test, the reflections from interfaces outside the fiber core [5].

Table 1 lists the indices of the cladding, buffer, and two index matched materials. By using this kind of materials we can greatly suppress or remove the WD mode induced ripple, and will discussed in detailed as follows.

Table 1. The indices of fiber cladding, buffer, and two matched materials

	Cladding	Buffer	$C_{11}H_{12}O_2$	$C_3H_8O_3$
Indices	1.462	1.480	1.558	1.473

3.1. Fiber immersed in ethyl cinnamate solution

We can see from Table 1 that the index of ethyl cinnamate ($C_{11}H_{12}O_2$) is larger than that of buffer. If we let the fiber under test immersed in matching solution of ethyl cinnamate, the total reflection cannot occur at the buffer-matching material interface, leads to effective suppression of the bending loss ripple.

Fig. 4(a) demonstrates the measurement results when the fiber is in 1 turn circle of radius 7.5 mm with or without immersed in the solution of ethyl cinnamate. It is clear that with immersed in the solution, effective suppress is achieved.

Journal of Physics: Conference Series **679** (2016) 012004 doi:10.1088/1742-6596/679/1/012004

If the experiment just mentioned above is repeated with the fiber being bent 10 turns, the ripple is almost absent and test curve is nearly smooth exponential, meaning the effect of WD mode is eliminated (Fig. 4(b)).

3.2. Fiber buffer removed and immersed in glycerol solution

Glycerol which is also called propanetriol (C3H8O3) has an index of 1.473, which is larger than cladding, and smaller than buffer. If we remove the buffer out and let the fiber without buffer being immersed in glycerol solution, the total reflection cannot occur at cladding-glycerol interface. Therefore the ripple effect can also be suppressed.

Fig. 5 shows the experiment. In the experiment, a section of the fiber' buffer is removed, and then is bent to be a circle of radius 7.5 mm being immersed in glycerol solution. The comparison is made between the case mentioned in Section 3.2 and the case here. It is clear that the method of buffer removed combined with immersed in glycerol solution can effectively depress the ripple.

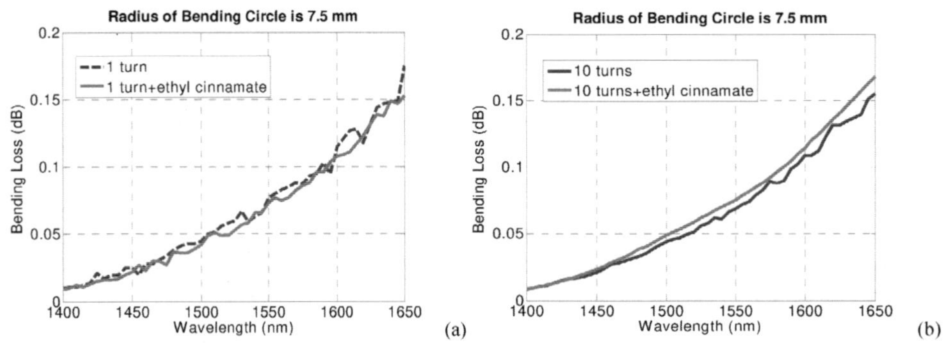

Figure 4. Bending loss results when the fiber is in 1 and 10 turn circles of radius 7.5 mm (a) with or (b) without immersed in the solution of ethyl cinnamate.

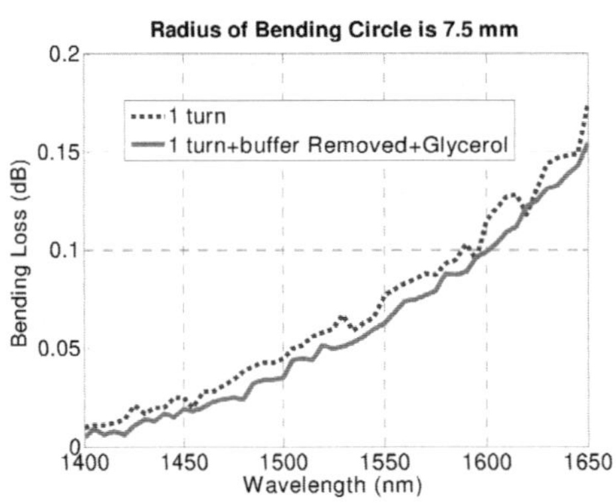

Figure 5. Bending loss results when the removed fiber is in 1 circle of radius 7.5 mm with immersed in the solution of ethyl cinnamate.

3.3. Statistical analysis

Because long wavelength such as 1625nm has more serious oscillation, so select 1625nm as typical wavelength to research this oscillation phenomenon. Using above methods, the single wavelength light source and optical power meter are used for measurement of macrobending loss at 1625nm fifteen times, to test the macrobending loss of the same fiber sample by the methods of 1 loop,10 loops,1 loop+ Ethyl cinnamate and 1loop+strip coating+ glycerin and the average value and standard deviation are calculated respectively. The results are listed in Table 2.

Table 2. The statistical results (15 samples) for different cases mentioned in the text

Statistical item	1 turn	10 turns	1 turn + C11H12O2	1 turn + buffer removed + C3H8O3
Mean value	0.126333	0.130933	0.128133	0.131267
Standard Deviation	0.029250	0.008312	0.010501	0.016342

For N statistical samples, the standard deviation σ reflects to some extent distribution convergence or the accuracy. Therefore the bending loss measurement with 10 turn circles (σ = 0.008312) is more accurate than that with 1 turn (σ = 0.029250) for the reasons mentioned above. The ripple suppression methods are effective. Among them 1 turn + C11H12O2 is better.

4. Conclusion

Obviously ethyl cinnamateor glycerol matching method is not applicable to field optical fiber macrobending loss measurement. Method of multiple coils is the most convenient, and recommended in the field tests. In general, 10 loops are easy to operate andobtain a more stable test results. In the laboratory absorbent matching method should be prior recommended for high test efficiency and accuracy.Combined with the method of multi circle, the method of ethyl cinnamate matching is best choice in laboratory.

5. Acknowledgments

Special thanks to Shixin Yang of Coring Company for his help to this paper. The project was supported by Fund of State Key Laboratory of Information Photonics and Optical Communications (Beijing University of Posts and Telecommunications), P. R. China .

Reference

[1] You Shanhong, Hao Sujun, 2004, Analysis and Test of Bend Loss in Single-mode Fiber, *ACTA PhotonicSinica*,Vol. 32, No. 4, pp. 409-411

[2] IEC 60793-1-47, 2009, Optical fibers-Part 1-47: Measurement methods and test procedures Macrobending loss", *Edition* 3.0 2009-03

[3] R. Morgan, J. D. C. Jones, P. G. Harper and J. S. Barton, 1991, Observation of secondary bend loss oscillations arising from propagation of cladding modes in buffered monomode optical fibers, *Opt. Commun.*, 85(1), pp.17-20

[4] Li Linyin,Gan Lu, Li Chunsheng, Song Zhituo, 2014, Measurement of Macrobend Spectral Loss for Bend Insenstive Single Mode Fiber, *Modern Transmission*, No. 4, pp. 63-65

[5] Siping Ning, and Onno R. Bresser,eds 2014, Use of Weighted Least Square Fitting Methods in Macrobending Loss Tests on Single-mode Fibres, in Proceeding of the 62nd *IWCS*, Providence, Rhode Island, pp. 92-96

Journal of Physics: Conference Series **679** (2016) 012005 doi:10.1088/1742-6596/679/1/012005

Controllable Delay of Optical Pulses for Laser Communication via Stimulated Brillouin Scattering

Haihuan Chen, Huanbi Yang and Wenhan Wu

China Satellite Maritime Tracking and Control Department, China

Email: chenhaihuan2010@sina.com

Abstract. We present and discuss the time delay of optical pulses for all-fiber laser communication at 2 μm band, which is a newly demonstrated region for communication since it's the broadest emission band among those of rare-earth ions. The simulation investigated initial considerations when performing laser communication at 2 μm.

1. Introduction

In recent years, the tremendous increase of information transmission amount has nearly driven the current network based on fiber communication at 1.5 μm to its limit of capacity. Laser sources at 2 μm based on Tm ions can cover ~1700-2200 nm region, which is much broader than that of Er ions at 1.5 μm. Thus, laser communication employing laser sources at 2 μm can improve the capacity of transmission network, which is mainly based on wavelength division multiplexing technique.

Actually, the data transmission at 2 μm has been demonstrated by Petrovich et al. in 2012, where they used a low-loss hollow core photonic bandgap fiber to reduce the background loss of fiber at 2 μm (~4.5 dB/km)[1]. Li et al. also presented Tm-doped fiber amplifier for optical communication at 2 μm in 2013 [2]. Recently, Wang et al. experimentally demonstrated slow light pulses via stimulated Brillouin scattering (SBS) effect at 2 μm, which is a meaningful step to make all-fiber laser communication at 2 μm practical[3]. Due to the high loss of fiber used in experiment, the performance of delayed pulses should be further improved. Here we theoretically investigate the delayed pulses at 2 μm to make referential conclusions for the practical work in the near future.

2. Results and discussion

The theoretical model is set up employing the three-wave coupled wave equations[4], and parallelizable and bidirectional method is used to perform the calculation since it is more precise and efficient compared with the traditional sequential and unidirectional method [5]. The pump light is at 1971 nm and the Stokes light is at the Brillouin shift band of the pump light (about 1971.1 nm). The Brillouin frequency shift is about 8.77 GHz in silica-based fibers. The delay fiber is 450 m passive single mode fiber (SMF) with mode diameter of about 9 μm. The Brillouin gain has a peak value of about 2.31×10^{-11} m/W and has a Lorentzian shape. The input stokes pulse has a width of 40 ns and peak power of 15 mW. The fiber loss is set to be 4.5 dB/km.

Figure 1 shows the reference pulse and the delayed pulses. The time data of x-axis indicate the propagating time after 450 m SMF. One can see that pulse is delayed apparently and broadened to a certain extent, which matches with the experimental results. Furthermore, when the pump power is relatively low, the Stokes pulse is delayed. When the pump power is further increased, the pulse is advanced, i.e., the delay time has negative value, which means the system is in gain-saturation regime. Thus, the characters of time-delay at 2 μm are much like that at 1.5 μm.

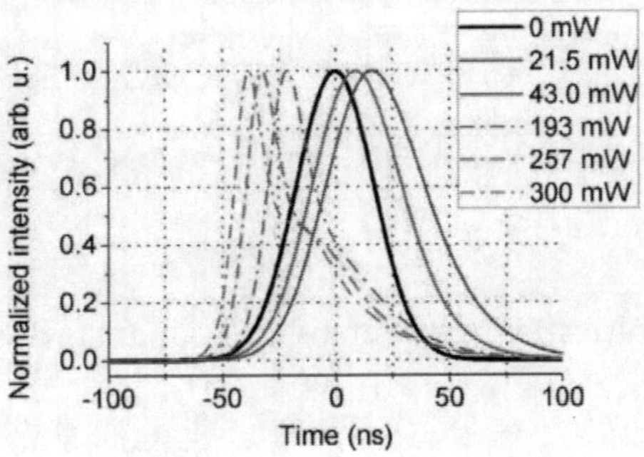

Figure 1. Pulse delay at different pump power levels.

At 2 μm regime, the fiber background loss plays an important role in the operation of pulse delay. Figure 2 shows the influence of fiber loss on the performance of time-delay base on SBS. The fiber loss of 20 dB/km is the typical loss of current commercial SMF at 2 μm, and the 4.5 dB/km is the loss when using low-loss fibers (hollow core bandgap photonic fiber, for example). Apparently, the lower the fiber loss is, the higher time delay can be generated at a fixed pump power (well below the gain-saturation level, 30 mW, for instance). However, we find out that when the fiber loss is higher, the achievable maximum delayed time is longer, which is also accompanied by the more serious pulse broadening.

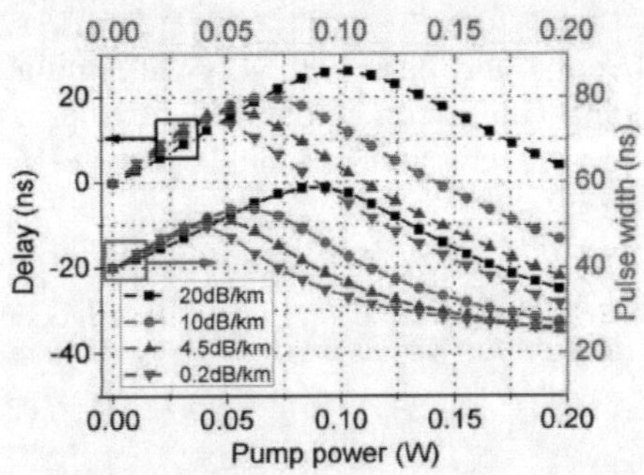

Figure 2. Time delay and pulse width data versus the pump power with different fiber losses.

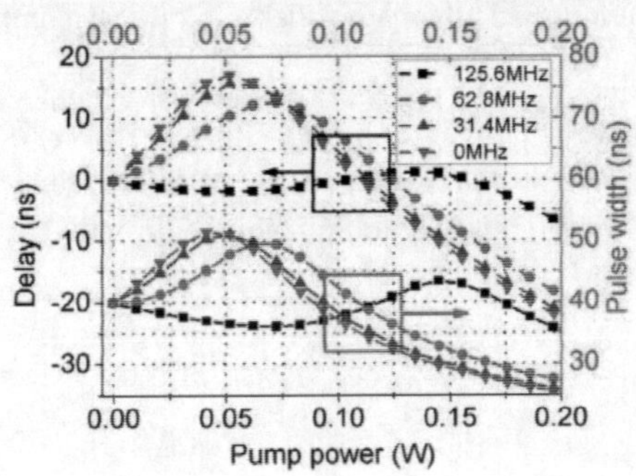

Figure 3. Time delay and pulse width data versus the pump power with different frequency detuning values.

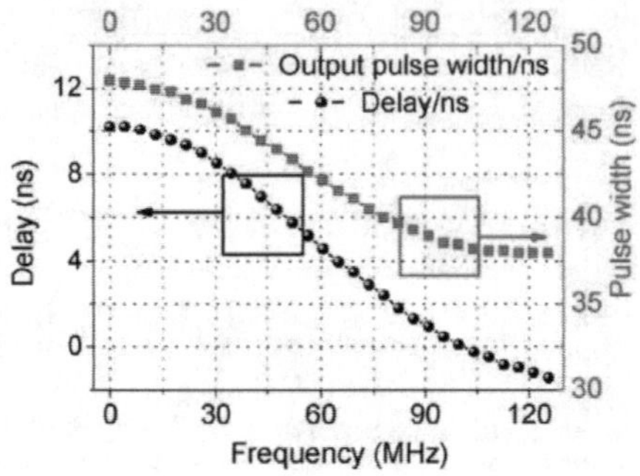

Figure 4. Time delay and pulse width data versus different frequency detuning values at a fixed pump power.

We investigated the influence of frequency detuning of pump light on the performance of the time-delay system, as shown in figure 3. The Brillouin gain reaches the maximum when the pump light and Stokes light have the exact Brillouin frequency shift, which may not always necessarily being guaranteed, however. And when the pump light's frequency is detuned from the center of Brillouin frequency shift, the Brillouin process is degraded. It is shown that even when the detuning frequency reaches 62.8 MHz, one can still obtain time delay effectively, only with lower efficiency. Figure 4 can show the trend more obviously. The pump power is fixed at 25 mW. One can obtain positive and effective time delay as long as the frequency detuning is less than about 105 MHz in this case. In addition, the pulse width broadening can be nearly avoided when the frequency detuning is about 75 MHz.

3. Conclusion

We investigated the time delay at 2 μm based on SBS via simulation method. The proper control of fiber loss and frequency detuning values can be useful when design all-fiber buffer at 2 μm, which is thought to be a vital technique in the next generation of optical communication system.

Acknowledgment

The authors thank Xiong Wang and Haibin Lv from National University of Defense Technology for useful discussion.

References

[1] M. N. Petrovich *et al* 2012 First demonstration of 2 μm data transmission in a low-loss hollow core photonic bandgap fiber *European Conference and Exhibition on Optical Communication* **3** 327
[2] Z. Li *et al* 2013 Thulium-doped fiber amplifier for optical communications at 2 μm *Opt. Express* **21** 9289
[3] X. Wang, P. Zhou, X. Wang, H. Xiao and Z. Liu 2015 Tunable slow light via stimulated Brillouin scattering at 2 μm based on Tm-doped fiber amplifiers *Opt. Lett.* **11** 2584
[4] G. P. Agrawal 2007 *Nonlinear Fiber Optics (Fourth Edition)* (New York, Academic Press) p 529
[5] D. Hollenbeck, C. D. Cantrell 2009 Parallelizable bidirectional method for simulating optical-signal propagation *J. Lightw. Technol.* **12** 2140

A Novel Architecture of Radio Tracking System Based on Photonics

Bo Cong[1], Xiaofeng Jin[2] and Xinfeng Yu[1]

[1]China Satellite Maritime Tracking and Control Department, China
[2]Department of Information & Electronic Engineering, Zhejiang University, China

E-mail:1151625288@qq.com

Abstract. We propose a photonics-based novel architecture of radio tracking system in space applications. This system can operate in multiple frequency bands, implement various functions, and realize integrated, generalized, frequency–independent design. This new scheme can meet the needs of future space missions, and especially suitable for space tracking ship.

1. Introduction

The future space missions require that a single ground-based radio tracking system can support many types of mission, i.e., operate in multiple frequency bands, implement various functions and have large bandwidth[1].

In this background, the traditional electronics methodology faces great challenges. Firstly, each frequency band needs separate RF links and the system will be very bulky and complicated. Rotary joint is needed to transfer many different signals and the reliability will be very low especially in oceanic environment. Secondly, the system performance will be limited by the mutual interference between multiple frequencies and intended interference.

Furthermore, due to the large loss of the cables, there is no much degree of freedom of the processing units' location.

The dynamic allocation of tracking equipment needs large-scale RF switch matrix, but as the operation frequency and the scale of the matrix rises, the insert loss will increase rapidly. The change of temperature will induce instability of phase and group delay. To compensate this loss, multi-stage amplifiers are usually adopted and more instability is induced. Besides, the isolation and interference problems are also significant.

To meet the need of deep space communication, a large G/T ratio is needed. To improve the G/T ratio, we can increase the antenna aperture, however, the effect-cost ratio is quite low. An alternative approach is antenna arraying but the large loss caused by the electrical cable must be overcome.

Microwave photonics is an interdisciplinary area that studies the interaction between microwave and optical signals[2], for applications such as broadband wireless access networks, sensor networks, radar, satellite communications, instrumentation, and warfare systems. In the past few years, there has been an increasing effort in researching new microwave photonics techniques for different applications.

Combining optics and radar together and applying microwave photonic in the development of new generation of radar is now a hot topic and shows great potential advantages[3,4].

Journal of Physics: Conference Series **679** (2016) 012006 doi:10.1088/1742-6596/679/1/012006

2. System Design

In this project, we propose a novel photonics-based system which operates at several frequency band simultaneously, realizes ultra-wide band and integrated design. The diagram of the proposed system is shown in figure 1.

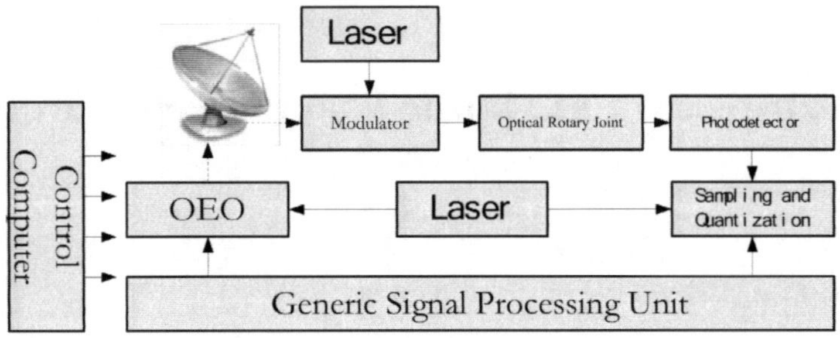

Figure 1. The diagram of photonics-based radio tracking system.

In this scheme, output signals from the LNA(Low-Noise Amplifier) are modulated to optical wave and mixed to a lower frequency, transferred through optical fiber by WDM(Wave Division Multiplex) to a photodetector. After being detected and sampled, the signals are sent to a generic signal processing unit in which the downlink information from the satellite, and the range, range rate, direction of the satellite are extract.

The generic signal processing unit also generate the uplink data and modulates the optic-electronic oscillator. The RF signal is transmitted to the satellite by the antenna.

3. Performance Analysis and Potential Advantages

The proposed scheme can realize frequency-independent and ultra-width design, has high ability of anti-interference and high stability.

Compared with conventional approach, the RF links are significantly simplified. Complicated up-converters/down-converts for each band are replaced by a single optical mixer and the development cost is thus decreased and the reliability is improved.

High loss, phase drifting and amplitude fluctuating caused by electrical cable are overcame and a large-scale low-loss RF switch matrix based on photonics can be realized

In sequential tone ranging, tone dispersion in the channel affects greatly the ranging accuracy. Since the fiber has very large bandwidth, this effect can be eased and the ranging accuracy will be improved. Compared with electronic oscillator, optoelectronic oscillator has very low phase noise and a higher accuracy of range rate can be achieved[1]. Phase noise also increase the bit error rate, so a lower phase noise can also improve the communication/telemetry performance[1].

4. Challenges

To realize a system mentioned above, there are still many challenges.

Firstly, a rotary joint that can transfer optical signal and large-power RF signal simultaneously must be developed. This rotary joint differs from existing ones and might be very difficult to design.

Secondly, the difference signal in the tracking channel is very low, especially when the antenna axis is pointing at the satellite, the difference signal will be approximate zero. How to transfer such a weak signal through fiber is a problem must be solved.

References

[1] Jiaxing Liu 2011 *Spacecraft TT&C and Information Transmission Technology* (Beijing, Defense Industry Press) p 304
[2] Jianping Yao 2009 Microwave Photonics *J. Lightw. Technol.* **27** 1131
[3] Paolo Ghelfi1*et al* 2014 A fully photonics-based coherent radar system *Nature* **507** 341

[4] Jason D. Mckinney 2014 Photonics illuminates the future of radar *Nature* **507** 310

Tracking Ability Evaluation and Calculation of Shipborne Tracking Radar for Space Applications

Yuanxin Qu, Bo Cong

China Satellite Maritime Tracking and Control Department, China,

E-mail: 1151625288@qq.com

Abstract. This paper analyzes the dynamic lag and tracking ability of shipborne tracking radar for space applications. A novel calculation formula of azimuth accuracy is proposed which is more reasonable compared with conventional methods.

1. Introduction

Shipborne tracking radar plays an important role in space flight missions and its accuracy determines the spacecraft's orbiting accuracy, so the dynamic lag of the tracking radar must be quantitatively evaluated and the tracking accuracy also needs to be scrutinized.

2. Dynamic Lag of Shipborne Tracking Radar

Dynamic lag is caused by the relative angle acceleration between the target and the tracking radar, and can be expressed by the input function, derivatives and error constants of the servo system[1],

$$e(t) = \frac{\theta_i}{K_P} + \frac{\dot{\theta}_i}{K_v} + \frac{\ddot{\theta}_i}{K_a} + \frac{\dddot{\theta}_i}{K_3} + \cdots \tag{1}$$

in which, θ_i、$\dot{\theta}_i$、$\ddot{\theta}_i$ ...are the derivatives of the input function, K_P, K_v, K_a ...... are the error constants of position, velocity and acceleration of the servo system.

For the shipborne tracking radar, the servo system is usually designed as second-order system which includes two integrations in the forward channel. The typical open-loop transfer function is:

$$K_a W(s) = \frac{K_a(1+T_2 s)}{s^2(1+T_3 s)} \tag{2}$$

, in which, Ka is the acceleration constant, T2 is the time constant of 1st-order derivation, T3 is the time constant of the inertial part, so the dynamic error of the shipborne tracking radar can be denoted as:

$$e(t) \approx \frac{\ddot{\theta}_i}{K_a} \tag{3}$$

If the dynamic error is larger than the beam width of the antenna, the tracking will be lost.

Take the shipborne unified X-band system as an example, which has a half power beam width of 0.2°, i.e., $\theta_{0.5}$=0.2°. By evaluation, we can get that, in azimuth, $K_a=3/s^2$, in elevation, $K_a=1.75/s^2$. When the radar tracks an aircraft with the highest elevation of 85°, the highest acceleration in azimuth is $a_A=3.7437°/s^2$, in elevation is $a_E=0.9526°/s^2$, then the dynamic lags are:

$$e_A = \frac{3.7437}{3} * \cos 85 = 0.11° > \frac{1}{2}\theta_{0.5} = 0.1° \tag{4}$$

$$e_E = \frac{0.6652}{1.75} = 0.38° > \frac{1}{2}\theta_{0.5} = 0.1° \tag{5}$$

From the results, we can see that if special measures such as object forward control etc. aren't taken to increase the value of K_a, the tracking will be lost with high elevation.

3. Angle Accuracy Analysis of Shipborne Tracking Radar

For a tracking radar with azimuth-elevation antenna, the azimuth servo system need secant compensation. The relationship between the azimuth, sweeping angel and elevation is[2]

$$A = A' \sec E \tag{6}$$

Different both sides and we can get the angle velocity, angle acceleration of azimuth

$$\omega_A = \frac{dA}{dt} = \omega_{A'} \sec E\big|_{E=\text{costant}} \tag{7}$$

$$a_A = \frac{d\omega_A}{dt} = a_{A'} \sec E\big|_{E=\text{constant}} \tag{8}$$

In the technical specifications, the azimuth accuracy means the accuracy of the sweeping angle. However, in the literature[3], the azimuth accuracy is calculated by the azimuth of the antenna. The result is that in flight test, the azimuth error we calculated is much larger than real value. Therefore, to calculate the azimuth accuracy correctly, we need to calculate the sweeping angle,

$$\Delta A' = -[\beta_m \cdot \sin(A - A_m) \cdot \sin E + \delta_m \cdot \sin E + S_b + C_S + \Delta A_Z)] \tag{9}$$

4. Conclusions

(1)The dynamic lag of the shipborne tracking radar is quantitatively evaluated to predict whether or not the tracking will be lost.

(2)The method of calculating the azimuth accuracy is reconsidered and a more reasonable one is proposed.

References

[1] Yuanxin Qu *et al* 2009 *Antenna Stabilization Techniques on Space Tracking Ship*(Beijing, National Defense Press) p 329

[2] Dean Zhong *et al* 2009 *Radar Calibration and Flight Test for Space Tracking Ship*(Beijing, National Defense Press) p 276

[3] Yuxi Lou *et al* 1979 *Radar Accuracy Analysis*(Beijing, National Defense Press) p 174

Research on Shore-Ship Photonic Link Performance for Two-Frequency-Band Signals

Yanqin Zuo, Bo Cong

China Satellite Maritime Tracking and Control Department, China

E-mail:1151625288@qq.com

Abstract. Ka and Ku bands links for shore-ship communications suffer limited bandwidth and high loss. In this paper, photonics-based links are proposed and modeled. The principle of phase modulation (PM) is elaborated and analyzed. It is showed that PM can effectively suppress high-order inter-modulation distortion (IMD), reduce the insert loss and improve the reliability of the system.

1. Introduction

Microwave photonic link can be explored for shore-ship communication of Ka and Ku bands, due to the ad-vantages in terms of small size, light weight, high bandwidth and low loss.

Previously, most researches choose the Mach-Zehnder modulator (MZM) as the optimized way, which requires the DC bias. This would cause high insertion loss and bias drift [1].

In this paper, the Ku and Ka bands links via the phase modulation (PM) are modeled. By the analysis of the PM principle and simulation, it is shown that PM can effectively suppress high-order inter-modulation distortion (IMD), reduce the insert loss and improve the reliability of the system.

2. System Model

Shore-ship photonic link is modeled in figure 1. The signal of Ku and Ka bands is received by the satellite, and then transmitted to the antenna on the ship. After amplified by the optic power amplifier (OPA), it is filtered from background noise by the optics filter (OF). Then, the lightwave signal is processed by a phase modulator (PM) and a tunable band-pass filter (TBF), and divided into two channels after a photo detector by the modems.

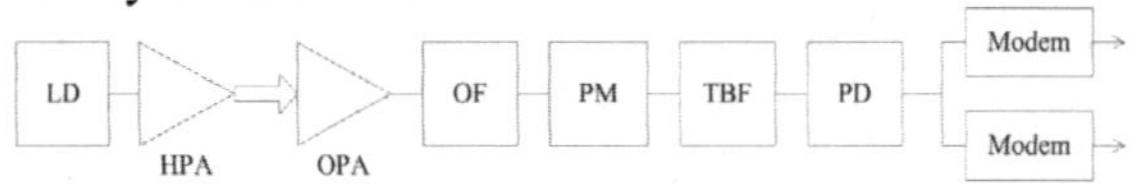

Figure 1. Microwave photonic link model for Ka and Ku bands.

3. Theoretical Analysis

The input signal is $V_{in}(t)$ is assumed to consist of two frequencies ω_1 and ω_2:

$$V_{in}(t) = V_1 \cos(\omega_1 t) + V_2 \cos(\omega_2 t) \tag{1}$$

, and the carrier frequency is ω_0.

3.1. Theory for MZM

The output of the Mach-Zehnder modulator is that:

$$E_{out} = \frac{\alpha}{2}\sqrt{P_0}e^{i\omega_0 t} \bullet \left[e^{-i(m_1 \cos\omega_1 t + m_2 \cos\omega_2 t)} + e^{-i(m_1 \cos\omega_1 t + m_2 \cos\omega_2 t + \theta)}\right] \quad (2)$$

In Eq. (2), α is the insert loss, and m_p is defined as

$$m_p = \pi \cdot V_p / V_\pi \qquad p = 1,2 \quad (3)$$

where V_π is the half-wave voltage of the modulator, and θ is defined as

$$\theta = \pi \cdot V_{DC} / V_\pi \quad (4)$$

where V_{DC} is the DC bias voltage.

Then applying the Bessel function expansion to Eq. (2), without considering high-order component we can obtain

$$P_{out} = \frac{1}{2}\alpha^2 P_0 e^{i\omega_0 t}[1 + \cos\theta J_0^2(2m)] \quad (5)$$

where $m_1 = m_2 = m$ is assumed.

3.2. Theory for PM

The output of the phase modulator is that:

$$E_{out} = \sqrt{P_0}e^{i\omega_0 t}e^{i(m_1 \cos\omega_1 t + m_2 \cos\omega_2 t)} \quad (6)$$

Then applying the Bessel function expansion to Eq. (6), we can obtain

$$E_{out} = \sqrt{P_0}e^{i\omega_0 t} \bullet \sum_{n=-\infty}^{\infty}\sum_{l=-\infty}^{\infty} J_n(m_1) J_l(m_2) e^{i(n\omega_1 t + l\omega_2 t + n\pi/2 + l\pi/2)} \quad (7)$$

It is showed from Eq. (7) that, PM generates an infinite number of sidebands.

After the filter, without considering high-order components we obtain the filtered optical field as:

$$E_{out} = \sqrt{P_0}e^{j\omega_0 t} a_0 J_0^2(m) \quad (8)$$

where a_0 is the optical carrier band coefficient and $m_1 = m_2 = m$ is assumed.

3.3. Comparison between MZM and PM

By comparing Eq. (5) and Eq. (8), we could find that the DC bias impresses the output signal effectively.

Then, if high-order component is considered, Eq. (8) is rewritten as:

$$\begin{aligned} E_{out} = \sqrt{P_0}e^{j\omega_0 t}\Big\{ & a_0 J_0^2(m) + a_2 J_1^2(m) e^{i(\omega_1 t + \omega_2 t + \pi)} \\ & - a_0 J_1^2(m)\left(e^{i(\omega_1 t - \omega_2 t)} + e^{i(\omega_2 t - \omega_1 t)}\right) \\ & + a_1 J_1(m) J_0(m)\left(e^{i(\omega_1 t + \pi/2)} + e^{i(\omega_2 t + \pi/2)}\right) \\ & - a_1 J_2(m) J_1(m)\left(e^{i(2\omega_1 t + \pi/2 - \omega_2 t)} + e^{i(2\omega_2 t + \pi/2 - \omega_1 t)}\right) \\ & + a_2 J_2(m) J_0(m)\left(e^{i(2\omega_1 t + \pi)} + e^{i(2\omega_2 t + \pi)}\right)\Big\} \end{aligned} \quad (9)$$

Then, it can expressed as

$$\begin{aligned} I \propto\ & I_0 + I_1(\sin\omega_1 t + \sin\omega_1 t) \\ & + I_3\left[\sin(2\omega_1 - \omega_2)t + \sin(2\omega_2 - \omega_1)t\right] \end{aligned} \quad (10)$$

where $I_3 = 0$ should be established to suppress high-order IMD.

Thus, $a_2 = 3a_0$ should be established.

Journal of Physics: Conference Series **679** (2016) 012008 doi:10.1088/1742-6596/679/1/012008

4. Conclusion

In this paper , Ka and Ku bands links via PM for shore-ship communications are modeled. With this construction the system can be realized easily, and high-order inter-modulation distortion (IMD) and the insert loss can be suppressed effectively.

Acknowledgment

Thanks for the support of Joint Laboratory of Ocean-based Fight Vehicle Measurement and Control.

References

[1] Wei Zhaolin, Hu Shuling 2013 Development of Research on Phase Modulation to Intensity Modulation Conversion *Laser & Opto-electronics Progress* **50** 1103
[2] Peng Li, Lianshan Yan 2013 Improvement of linearity in phase-modulated analog photonic link *Opt. Lett.* **38** 2391
[3] ZHU ZiHang, ZHAO ShangHong 2013 Modelling and optimization of optically preamplified inter-satellite microwave photonics links *SciChina* **43** 217
[4] Li Xuan, Zhao ShangHong 2013 Analysis of Inter-satellite microwave photonic of two-frequency-band signals *Laser & Opto-electronics Progress* **50** 506
[5] Zhu Zihang, Zhao Shanghong 2013 Optimization of optically preamplified inter-satellite micro-wave photonics links with two radio-freguency signals input *Acta Optica Sinica* **33** 217
[6] LI Xuan, ZHAO ShangHong 2013 Performance analysis of inter-satellite microwave photonic link under multiple carrier input *Opto-Electronic Engineering* **40** 86
[7] Wang Weinan, Li Yanhe 2007 Modelling of fiber radio link using Mach-Zehnder modulator *Acta Optica Sinica* **27** 159

Analytic Performance of Monopulse Spread Spectrum Tracking System in Multiple-target Scenario

Bo Cong, Yuanxin Qu and Yongliang Zhang

China Satellite Maritime Tracking and Control Department, China

E-mail: 1151625288@qq.com

Abstract. With the application of spread spectrum technique in satellite tracking, the accuracy of CDMA-based multiple-target tracking needs to be analyzed. In this paper,we present the analytic form of the multiple-target tracking performance through mathematical deduction, together with simulation results.

1. Introduction

By adopting spread spectrum technique, the ground tracking station can track multiple targets in one single beam[1]. Only by intuitive, we can know that in monopulse system, the tracking accuracy will be worse as the number of targets increases. However, there is no analytic form of the momopulse multiple-target tracking accuracy in the literature. When tracking multiple targets with low elevation angle, the multipath effect will make the problem more complicated.

2. Problem Modeling

We assume that the antenna beam points precisely at one target and there are N other targets with individual PN code and the same frequency. It should be noted that the tracking system is not radar, but a kind of satellite communication system adopting monopulse tracking technique, borrowed from the radar community.

For the sake of simplicity, we assume the PN codes are perfectly orthogonal and the power transmitted by the $N+1$ targets are the same.

The received downlink signals and noise can be expressed as

$$S_i(t) = AP_i(n)\cos(\omega t + \phi_i) \tag{1}$$

$$n(t) = n_i(t)\sin(\omega t + \phi_n) + n_q(t)\cos(\omega t + \phi_n) \tag{2}$$

In which

$$\sigma^2_{n(t)} = \sigma^2_{n_i(t)} = \sigma^2_{n_q(t)} \tag{3}$$

3. Angle Error Caused by Multiple Targets and Noise

We assume that among the N other targets, the ith one deviates from the beam axis with an angle θ_i and the difference slope is K, then the difference signal produced by the ith target is

$$S_i(t) = K\theta_i AP_i(n)\cos(\omega t + \phi_i) \tag{4}$$

The spread spectrum gain is denoted as W, then, after de-spread, filtering and normalization, the interference voltage induced by the ith target is $K\theta_i\sin\phi_i/\sqrt{W}$ .The interference voltage induced by the noise is $n(t)/A$

The total interference voltage is

$$\frac{1}{A}n(t)+K\sum_{i=1}^{N}\frac{\theta_i\sin\phi_i}{\sqrt{W}} \tag{5}$$

The angle error caused by the interference voltage is

$$\Delta\theta=\frac{1}{KA}n(t)+\sum_{i=1}^{N}\frac{\theta_i\sin\phi_i}{\sqrt{W}} \tag{6}$$

If the N+1 targets' relative locations and the phases of their signal are fixed, then $n_i(t)$ is Guassian random process with the average is zero and ϕ_i is uniformly distributed in $[-\pi,\pi]$, other variables are constant.

$$E(\Delta\theta)=\sum_{i=1}^{N}\frac{\theta_i\sin\phi_i}{\sqrt{W}} \tag{7}$$

$$\sigma_{\Delta\theta}=\frac{\sigma_n}{KA} \tag{8}$$

If the N+1 targets' relative locations are fixed but the phases of their signal are random, then

$$E(\Delta\theta)=0 \tag{9}$$

$$\sigma_\theta=\sqrt{\frac{\theta_{3dB}^2}{K_m^2SNR}+\sum_{i=1}^{N}\frac{\theta_i^2}{2W}} \tag{10}$$

If any target's relative location and carrier phase are random, then

$$E(\Delta\theta)=0 \tag{11}$$

$$\sigma_\theta=\theta_{3dB}\sqrt{\frac{1}{K_m^2SNR}+\frac{N}{6W}} \tag{12}$$

4. Angle Error Caused by Multipath Effect

When the elevation is low, the tracking accuracy will be deteriorated by multipath effect[1-2]· Denote the elevation of the antenna is *E*, the height of the antenna is *h*, then the phase difference between the line-of-sight signal and the reflected signal is

$$\phi=\frac{2\pi}{\lambda}2h\sin E+\phi \tag{13}$$

We denote the reflective index as ρ .The angle error and its variance can be obtained as

$$\Delta\theta=\frac{\rho\sqrt{G_\Delta}}{K}\left(\sin\theta_0+\sum_{i=1}^{N}\frac{\sin\theta_i}{\sqrt{W}}\right) \tag{14}$$

$$\sigma_\theta=\frac{\rho\theta_{3dB}}{2K_m}\sqrt{\frac{G_\Delta}{G_\Sigma}}\sqrt{1+\frac{N}{W}} \tag{15}$$

References

[1] Yefu Zhao *et al* 2013 *Radio Tracking System*(Beijing, Defense Industry Press) p 429
[2] Jiaxing Liu 2011 *Spacecraft TT&C and Information Transmission Technology*(Beijing , Defense Industry Press) p 386

Journal of Physics: Conference Series **679** (2016) 012010 doi:10.1088/1742-6596/679/1/012010

Parameter Estimation and Data Management System of Sea Clutter

Bo Cong, Qingguang Duan and Yuanxin Qu

China Satellite Maritime Tracking and Control Department, China

E-mail: 1151625288@qq.com

Abstract. In this paper, a parameter estimation and data management system of sea clutter is described, which can acquire the data of sea clutter, implement parameter estimation and realize real-time communications.

1. Introduction

When detecting a target over the sea surface by radar, the echoes from the sea surface, i.e., the sea clutter will also be received by the receiver and the radar's detection performance will be deteriorated. To ease the effects of sea clutter, characteristics that differentiate it from the useful signal must be found and analyzed[1].

Many different distribution models have been proposed and applied in order to simulate the sea clutter by computer. Since the characteristics of sea clutter are very complicated and depend on many factor such as polarization, grazing angle, wind, temperature and location, etc., one still need to acquire the real sea clutter data and estimate its parameters and find the most suitable distribution model to apply. To manage and exchange those data effectively, database and real time communication is indispensable.

2. System Design

In the design scheme, a list of sea clutter's property is established, which include date, environmental factors, operation frequency, pulse width, polarization, grazing angle, etc. After that, a parameter list is established which includes the PDF(probability density function), the fitting curve, spectral characteristics, etc.. The diagram of the system is shown in figure 1.

3. Parameter Estimation of Sea Clutter

In this system, important parameters of sea clutter data are estimated, the probability density distribution curve and spectral curve are fitted.

3.1. The Probability Density Distribution of Sea Clutter's Amplitude

Firstly, the peak point is found and the data of N repetition period in the same range unit and azimuth unit are examined. The procedure is as following.

(1)Find the maximal and minimum value of the N data point and implement 2048-point quantization of the difference between them.

(2)Calculate the probability of the point falling into one of the 2048 intervals and the estimated curve of probability density distribution is obtained.

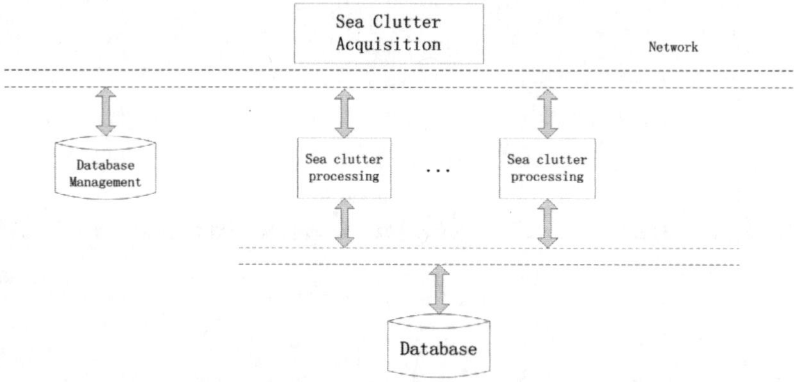

Figure 1. The diagram of the parameter estimation and data management system.

3.2. Curve Fitting of the Probability Density Distribution

Several distribution models have been proposed to describe the sea clutter's probability density function, such as Rayleigh distribution, Log-normal distribution,Weibull distribution and K-distribution[2]. Fitting these distribution curves to our estimated curve and calculate the average and variance of the error, we can find which distribution is the most closed.

3.3. Estimation of Spectral Characteristics

Data of N repetition period are exploited to estimate the spectrum of sea clutter, i.e., N-point FFT is implemented. To reduce the leakage in sidelobe, the data is windowed by Hamming function, which is

$$w(n) = 0.54 - 0.46\cos(\frac{2\pi n}{N-1}), 0 \le n \le N-1 \tag{1}$$

3.4. Spectral Curve Fitting

Spectral power density function of sea clutter is usually assumed to be Guassian and expressed as:

$$s(f) = s_0 \exp(-(af / f_{3dB})^2) \tag{2}$$

In which, a is constant and equal to 1.6665, s_0 is the power density at zero frequency. $f_{3dB} = \sqrt{2} \cdot a \cdot \delta_v$ and δ_v is the average velocity of wave, λ is the operation wavelength. The normalized power density is

$$s(f) = \exp(-(af / f_{3dB})^2) \tag{3}$$

4. IPX-based Real-time Communication

In this system, real-time transfer of large data rate will be involved and a longer preparation can be tolerated,so IPX（Internet Packet eXchange） protocol is applied.

The communication software developed for this system has several prominent features: real-time, large capacity and distributed control and extensible.

References

[1] M.I. Skolnik 2001 *Introduction to Radar System(Third Edition)*(New York, McGraw-Hill Companies Inc) p 544

[2] Simon Watts 2012 Modeling and Simulation of Coherent Sea Clutter *IEEE Trans. on Aerospace and Electronic Systems* **48** 1176

A Comparative Study of Ranging Techniques in Deep Space Applications

Bo Cong, Nanping Mao and Xiaohong Tao

China Satellite Maritime Tracking and Control Department, Jiangyin, China

E-mail:1151625288@qq.com

Abstract. In this paper, main ranging techniques in deep space applications are elaborated and analyzed. Their ranging accuracy, acquisition time and complexity are compared. It is shown that the code-tone ranging method has obvious advantage over others in deep space tracking. Start your abstract here…

1. Introduction

In deep space exploration, ranging determines the distance between the explorer and the earth and thus plays a crucial role.

Three techniques usually adopted in the ground station include sequential ranging, pseudo noise ranging, code-tone ranging. Two operation modes, two-way ranging and three-way ranging can be selected to apply in different scenarios. The diagram of the ranging system is shown in figure 1.

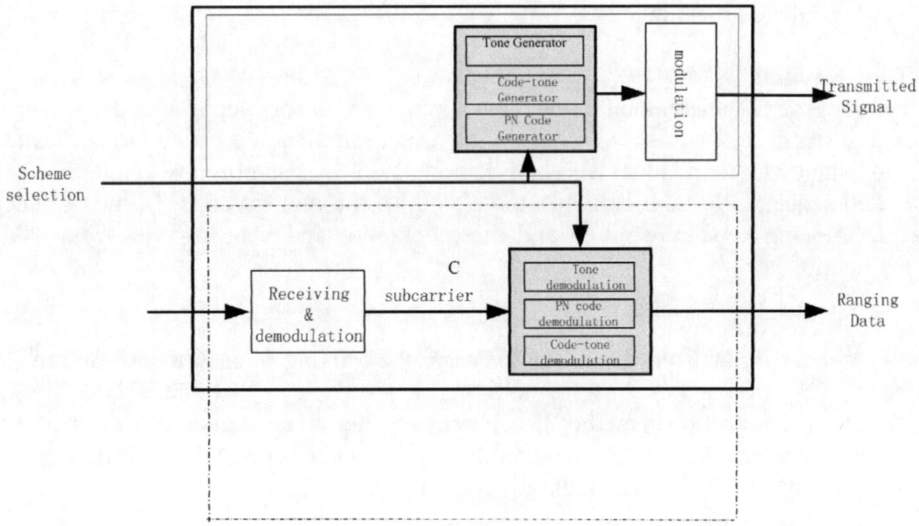

Figure 1. The diagram of ranging system.

In deep space applications, very long distance introduces very large time delay and some protection measures must be utilized to avoid matching error. Besides, in the three-way operation mode, the calibration becomes a great challenge.

2. Sequential Tone Ranging

Due to very large time delay and very weak signal, the processing procedure adopted in deep space application differs much from other conventional ones [1-2]. The time sequence of the transmitted and received tones is shown in figure 2.

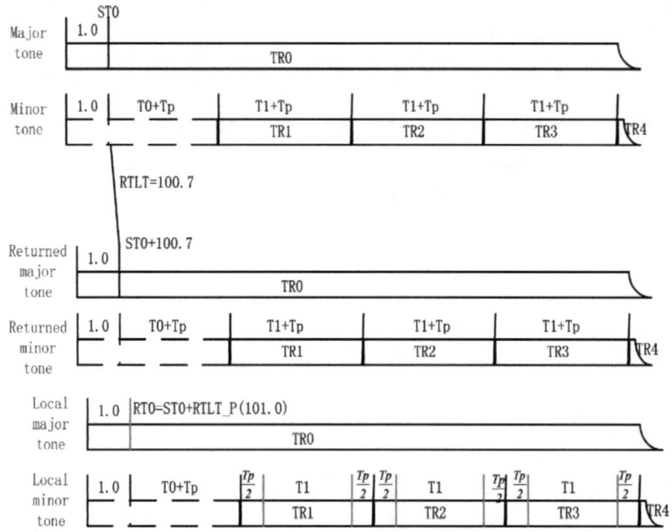

Figure 2. The time sequence of transmitted tones and returned tones.

The frequency of the major tone can be chosen between 100kHz and 500kHz. In the scheme, the major tone is being kept transmitted but the minor tones are sent sequentially. In Fig.2, ST0 is the start time of tone transmitting, RT0 is the start time of tone receiving. RTLT is the two-way time delay, and RTLT_P is the predicted time delay.

3. Code-Tone Ranging

In code-tone ranging [2], there is only one single frequency of tone and the codes are modulated onto the tone. Three steps, i.e., acquisition, disambiguity, measurement are needed to complete the process. In acquisition stage, the tone is received and demodulated. In disambiguity stage, disambiguity codes are transmitted sequentially with certain intervals, up to the highest code. In measurement stage, the highest code is being kept transmitted and the modulation index is lowered from 45° to 22.5° to improve the ranging accuracy.

4. PN Regenerative Ranging

When the spacecraft is far from the ground station, the ranging is easy to fail due to low signal-to-noise ratio.

Compared with conventional methods, coherent regenerative ranging can get better performance since the transponder tracks the ranging signal by phase-locked loop and transmits a pure signal down [2]. The scheme of PN regenerative ranging is shown in figure 3.

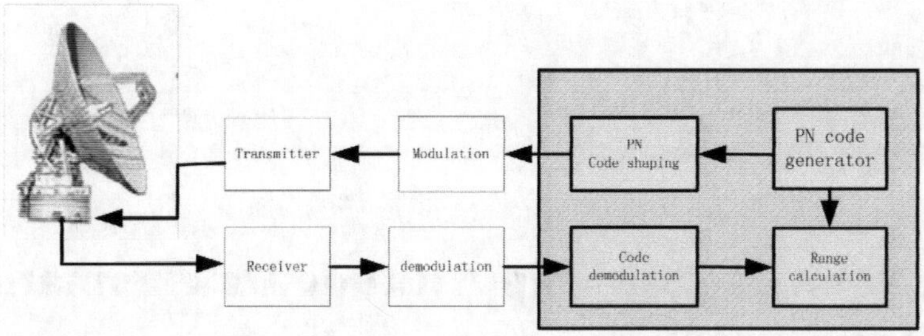

Figure 3. The scheme of PN regenerative ranging.

5. Comparisons of Ranging Techniques

With the same bandwidth and power, tone ranging has a better accuracy than PN ranging. PN ranging has obvious advantages in disambiguity, anti-interference, but longer acquisition time with weak signal. Code-tone ranging combines their advantages but the anti-interference ability is worse than PN ranging, otherwise, most interference can be avoided by choosing tone frequency carefully.

References

[1] Yefu Zhao *et al* 2013 *Radio Tracking System*(Beijing，Defense Industry Press) p 405
[2] Jiaxing Liu 2011 *Spacecraft TT&C and Information Transmission Technology*(Beijing , Defense Industry Press) p 667

Journal of Physics: Conference Series **679** (2016) 012012 doi:10.1088/1742-6596/679/1/012012

Research on Precise Ship Attitude Determination Based on Dual Star Sensors

Hong-wei Jiao, Liang Pan and Tong-shuang Zhang

China Satellite Maritime Tracking and Control Department, Jiangyin, China

E-mail: jhw23@aliyun.com

Abstract. The method of ship attitude determination based on dual star sensors is researched in this paper. Experiment results show that, the method of ship attitude determination based on dual star sensors has high precision of attitude determination of 5 arcsecond, it can be a complementarity of the precise ship attitude determination.

1. Introduction

As high-precision attitude measurement devices, star sensors are widely used in satellite or other spacecraft attitude determination system[1]. But due to the influence of atmospheric refraction and weather condition, the use of star sensor on ground is not deeply researched.

In order to solve the above-mentioned problem, a method of high-precision ship attitude determination based on dual star sensors is researched in this paper. Firstly, the relation between the measuring precision of attitude and the structure of two star sensors is analyzed, and the optimal structure of two star sensors is obtained. Secondly, a new atmospheric refraction correction method which based on reference quantity correction is introduced, which is simple in star apparent zenith distance calculation and dose not require the information of carrier attitude. Finally, the method of ship attitude determination based on dual star sensors is proposed, which fuses the two fields of view of dual star sensors into one and can improve the accuracy of ship attitude determination.

2. Structure of dual star sensors

Dual star sensors have advantages compared with conventional single star sensor, and becomes an important development trend with a better accuracy and reliability. In order to realize the best measurement results, structure optimization for dual star sensor is the first problem in the design and application. The structure optimization for dual star sensor was researched. The results show that dual star sensors have better accuracy with the increase of the angles between optical axes of FOVs, and realize the highest precision when the optical axes are orthogonal, as shown in figure 1.

3. New atmospheric refraction correction method

The atmospheric refraction correction is the most important problem when using star sensor on ground[2]. In order to simplify the atmospheric refraction correction process and improve the independent attitude measurement capability of star sensor, a new atmospheric refraction correction method which based on reference quantity correction is proposed. The proposed method is simple in star apparent zenith distance calculation and dose not require the information of carrier attitude, the refraction correction does not need to solve the dual nonlinear equations, it only needs simply addition.Simulation and experimental results show that the results of attitude estimation using the reference quantity refraction correction method and the observed quantity refraction correction method is consistent with the results, the relative error between them is within 0.1 arcsecond. For star sensor attitude measuring application, the proposed refraction correction method does not need external posture information, is simple and no loss of attitude measuring precision.

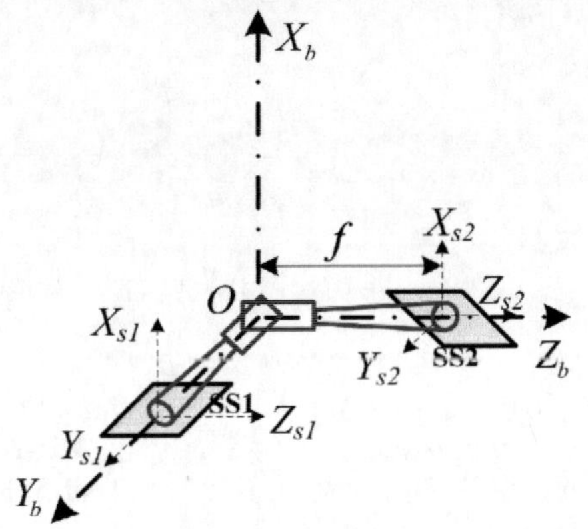

Figure 1. Structure of dual orthogonal star sensors.

4. Measurement model of dual star sensors

The attitude measurement model of star sensor is shown as formula(1):

$$w=Av \tag{1}$$

As for two star sensors, the measurement model is not variational, but the observed quantities in one star sensor coordinate should be transformed into the other star sensor coordinate. The details is shown as follows.

5. Experiment results

Experiment results show that, the method of ship attitude determination based on dual star sensors has high precision of attitude determination of 5 arcsecond, as shown in figure 2, it can be a complementarity of the precise ship attitude determination.

6. Conclusion

The method of ship attitude determination based on dual star sensors is proposed, which fuses the two fields of view of dual star sensors into one and can improve the accuracy of ship attitude determination. Experimental results show that, the method of ship attitude determination based on dual star sensors has high precision of attitude determination of 5 arcsecond, it can be a complementarity of the precise ship attitude determination.

Journal of Physics: Conference Series **679** (2016) 012012 doi:10.1088/1742-6596/679/1/012012

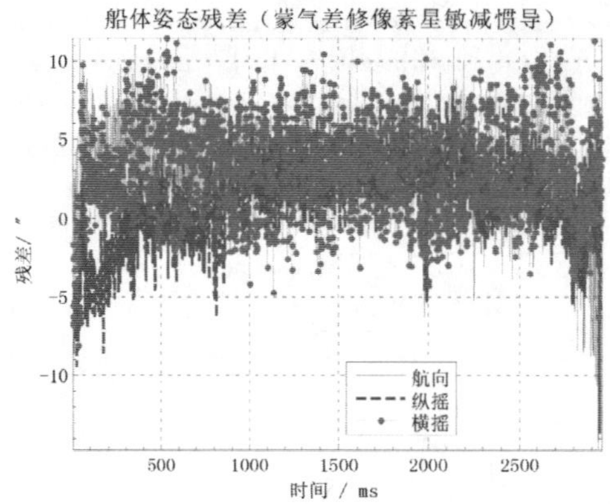

Figure 2. Ship attitude error of dual star sensors.

References

[1] Ju G, Junkins J L. 2003 Overview of star tracker technology and its trends in research and development *Advances in the Astronautical Sciences* **115** 461

[2] Wang Haiyong, Lin Haoyu and Zhou Wenrui 2011 Technology of Atmospheric Refraction Compensation in Starlight Observation *Acta Optica Sinica,* **31** 1002.

Research on Effect of Radar Fixing Error in Navigation on Delay of Plane Flow

Yi Zhang[1], Leilei Chang[1], Jiping Cao[1] and, Xiaodong Ling[2], Wei Wang[3]

1 High-Tech Institute of Xi'an, Xi'an, Shaanxi 710025, P. R. China
2China Satellite Maritime Tracking and Control Laboratory, Wuxi 214400, P. R. China
3Military Representative in Wuxi of the Military Representative Bureau in Nanjing, Wuxi 214400, P. R. China

e-mail address

Abstract. Fixing error of radar introduces some deviation into measuring value from the true. Regular interval between consecutive planes must be extended to accommodate air traffic control regulation. Mathematical relationship between fixing error and buffer time was deduced. Then, the variation of plane flow delay was researched in different condition of fixing error by means of simulation and practical calculation. It is shown that the effect of fixing error on plane flow delay is apparent compared with the determined situation. Also, the amount of delay is not only related to fixing error but also to average arrival rate. So, the fixing error factor should be considered when scheduling and delay analyzing is implemented. At the same time, it is indicated that decreasing the radar detecting error is necessary to reduce the plane flow delay.

1. Introduction

The main purpose of air traffic flow management(ATFM) includes: ensure the aircraft safety, supply accurate information for aircraft, optimize the air traffic flow, decrease the delay, increase the utilization of airport and airspace. The terminal is the critical area in ATFM because the traffic jam often happens in this area. For safety, an interval must be kept in two continuous landing aircraft. The size of the interval is a main factor for delay. There are lots of methods for sequencing in terminal area, such as FCFS algorithm , the time advance algorithm , the restraint position exchange algorithm, fuzzy recognition algorithm[1], genetic algorithm[2-3], sliding time window algorithm[4] . However, these algorithms have a common feature, that is the deviations between measured value and the true value of aircraft position is not considered. They believe that the actual position of the aircraft is the same as the measured position. The effect of fixing error of radar on aircraft flow delays is ignored.

Positioning errors may lead to violations of minimum safe spacing between adjacent aircraft which provided by ATC. In practice, in order to control the violation probability within a certain range, the controller will insert a buffer distance between the adjacent aircrafts. The insertion of a buffer distance is equivalent to an increase of the minimum safe spacing between adjacent aircraft. It is necessary to study the fixing error of radar for aircraft safety spacing and aircraft delays in navigation.

Interval in this paper mainly refers to longitudinally interval.

2. The fixing error and buffer interval

2.1. Fixing error and factors

Wake is the main factor affecting the aircraft interval. To avoid the wake of previous aircraft, the landing aircrafts in terminal area required to maintain a safe interval. Under the condition of radar control, radar detects the respective positions of the two adjacent aircrafts, and then calculates the interval based on the detected positions. When the fixing error of radar is ignored, the actual interval between the adjacent aircraft is the same the actual interval. At this point, the aircraft can fly in accordance with the minimum safe separation. But in practice, since radar is limited by its resolution, there will be error between the detected position and the true position, i.e., positioning error, and the two points as shown in Figure 1. Since the two points are close, it may be beyond its ability to distinguish them. The radar will mistakenly believe that they are the same point and display one point on the screen. Another aspect of this issue, the radar will treat the point within a certain range as the same point, it is the fixing error. Because of the fixing error, spacing between aircraft detected by radar will have a certain bias. The detected interval may be greater than the actual interval may also be less than the actual spacing. In this case, if the flight is scheduled in accordance with interval detected by the radar, there will be danger that the later aircraft is involved by previous aircraft's wake turbulence.

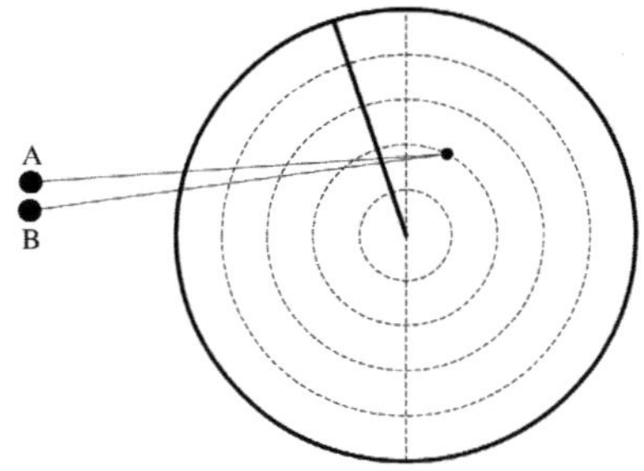

Figure1 Fixing error of radar

In practice, there are many factors that can have an impact on radar fixing error, including[5-6]:

(1) Equipment factors: the position indicating error, including airborne navigation equipment error, instrument error and display error, and the time error in air traffic control system.

(2) Meteorological factors: weather conditions where the airplane is exposed, mainly refers to wind.

(3) the human factor: controllers and pilots, including the reaction time of controllers and pilots, the communication delay, aircraft position estimation error, manipulation error and mental state.

2.2. The insertion of buffer interval

Studies have shown that the actual position obeys the normal distribution[7], and the centre is the detected position, as shown in Figure 2.

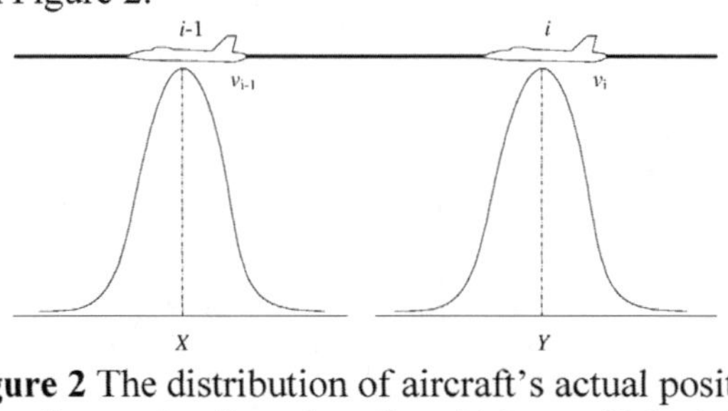

Figure 2 The distribution of aircraft's actual position

Suppose that there are two adjacent landing aircrafts which are flight i-1 and flight i. The detected positions are X and Y respectively. The position of previous fight i-1 has a certain bias from the actual one because of the fixing error of radar. According to the distribution, X is a normally distributed random variable. It consists of two parts. One part is the detected position, which is a certain value x. Another part is a random value due to the fixing error. That is

$$X = x + \tilde{X} \quad (1)$$

The position of later flight can be expressed as follow

$$Y = y + \tilde{Y} \quad (2)$$

The minimal safe separation distance of two aircraft by the rule of ATC is $S_{i-1,i}$ .So

$$S_{i-1,i} = y - x \quad (3)$$

Suppose that the standard deviation of $\tilde{X}$ and $\tilde{Y}$ is $\sigma_{\tilde{X}}$ and $\sigma_{\tilde{Y}}$ respectively. That is $\tilde{X} \sim N(0,\sigma_{\tilde{X}}^2)$, $\tilde{Y} \sim N(0,\sigma_{\tilde{Y}}^2)$

In order to control the probability of violation of the minimum safe separation within a certain range[8-9], the controller will insert a buffer distance in interval of the adjacent aircraft because of the fixing error. The insertion of buffer distance is shown in Figure 3.

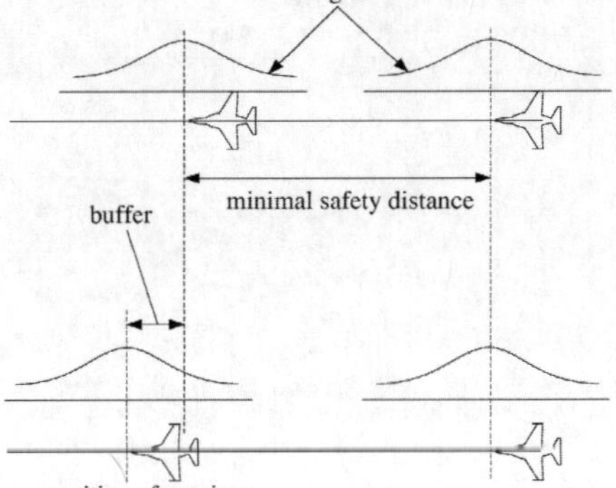

Figure 3 The insertion of buffer distance

The relation of the variables can be expressed as follow

$$p_v = P\{Y - X + DBF_{i-1,i} \le S_{i-1,i}\} = P\{(y+\tilde{Y})-(x+\tilde{X})+DBF_{i-1,i} \le S_{i-1,i}\} \tag{4}$$

The formula (5) can be obtained form the (3) and (4)

$$p_v = P\{\tilde{Y} - \tilde{X} \le -DBF_{i-1,i}\} \tag{5}$$

Form the knowledge of the probability theory, $\tilde{Y} - \tilde{X} \sim N(0,\sigma_{\tilde{X}}^2 + \sigma_{\tilde{Y}}^2)$, so the formula (5) is

$$p_v = P\{\tilde{Y} - \tilde{X} \le -BF_{i-1,i}\} = \int_{-\infty}^{-DBF_{i-1,i}} \frac{1}{\sqrt{2\pi(\sigma_{\tilde{X}}^2 + \sigma_{\tilde{Y}}^2)}} e^{-\frac{t^2}{2\left(\sigma_{\tilde{X}}^2+\sigma_{\tilde{Y}}^2\right)}} dt = \Phi\left(\frac{-DBF_{i-1,i}}{\sqrt{\sigma_X^2+\sigma_Y^2}}\right) \tag{6}$$

Among the formula (6)

$$\Phi(x) = \frac{1}{2\pi}\int_{-\infty}^{x} e^{-\frac{t^2}{2}} dt \tag{7}$$

When the aircrafts lands in the same airport, they use the navigation equipment and obey the same landing procedures. It follows $\sigma_{\tilde{X}} = \sigma_{\tilde{Y}} = \sigma$. So the formula (7) is

$$p_v = \Phi\left(\frac{-DBF_{i-1,i}}{\sqrt{2}\sigma}\right) = 1 - \Phi\left(\frac{DBF_{i-1,i}}{\sqrt{2}\sigma}\right) \tag{8}$$

Then

$$DBF_{i-1,i} = \sqrt{2}\sigma \cdot \Phi^{-1}(1-p_v) \tag{9}$$

Suppose that the velocity of the flight is V_{IAS} , the distance error between detected position and actual position can be converted to time deviation t_e by the velocity V_{IAS} . That is

$$DBF_{i-1,i} = \sqrt{2} \cdot V_{IAS} \cdot t_e \cdot \Phi^{-1}(1-p_v) \tag{10}$$

Form the formula (10), there are three factors which effect the size of buffer distance.

(1) The velocity of the flight

(2) The fixing error of radar

(3) The probability of violation of minimal safety distance.

Suppose that $V_{IAS} = 300kn$, the value of required buffer distance can be obtained when the t_e and p_v are different values, which is shown in Figure 4.

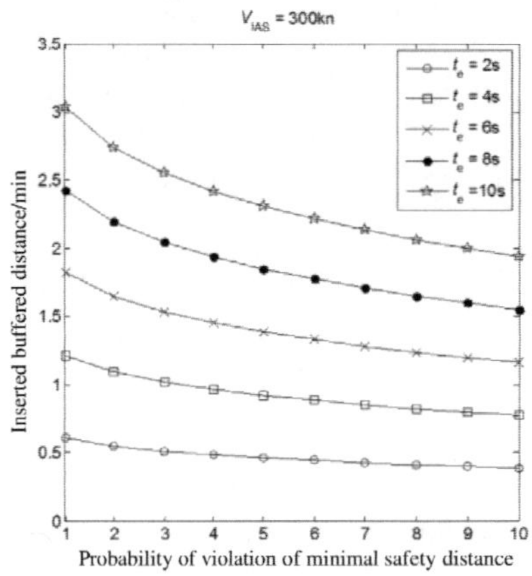

Figure 4 The relation between buffer distance and probability of violation of minimal safety distance

3. The delay effect from the fixing error

3.1. Instance selection and simulation

According to queuing theory, landing aircraft flow in the airport terminal area is Poisson flow. By the characteristics of the Poisson flow, the interval of continuous landing aircraft flow obeys the negative exponential distribution, the average inter-arrival time is the reciprocal of the average arrival rate, which can produce terminal area arrival flow of aircraft is expected to arrive time (Estimated Time of Arrival, ETA).

As the flight of aircraft generates wake, the previous aircraft will affect the later one. International Civil Aviation Organization (ICAO) set a safety interval between different types of aircraft in windless condition. And the corresponding time of the interval is shown in Tab. 1[10].

The controller will give each aircraft a Scheduled Time of Arrival(STA) based on each aircraft's ETA by certain rules and an algorithm, to minimize the waiting time of every aircraft. In the process of implementing the calculation and scheduling delays, the most considered factor is adjacent safe separation between aircraft.

Table 1 The minimal interval time of different aircraft(s)

		later		
		heavy	large	small
previous	Heavy	113	135	170
	Large	89	89	110
	small	83	83	94

The insertion of buffer distance representing an increase of safe separation between aircraft. When the arrival rate is low, the controller will not consider the insertion of buffer distance. When the arrival rate is high, the average interval is short, the controller have to consider the insertion of buffer distance. And the impact on aircraft's waiting time increases. In order to study the effect of buffer distance insertion on the waiting time, the different arrival rate is set.

Because the size of the buffer controller added to all aircraft are the same, in order to facilitate the calculation, STA airplane sequence is calculatedusing FCFS collation. In simulation, the average arrival rate is set as 10 to 100/hr, the range of ETA time of is 1 hour, the value of the fixing error takes 2-10 seconds. Simulation takes 100 times. According to the ATC, the probability of violation does not exceed 5%.

For the flow containing M aircrafts, the effect of fixing error on the whole flow can be obtained by the next formula

$$W_{ext} = \frac{\sum_{i=1}^{M}\left(STA_i^{BF} - STA_i\right)}{\sum_{i=1}^{M}\left(STA_i - ETA_i\right)} \tag{11}$$

In formula (11), STA_i contains no error, that is the STA of aircraft I when $t_e = 0$, and STA_i^{BF} is the STA contains error.

3.2. Delay analysis

By MATLAB Simulation, the delay increment induced by fixing error of radar is shown in Figure 5. As can be seen from Figure 5.

(1) The waiting time increases as the fixing error increases. Meanwhile, the extent of delay is related to average arrival rate too.

(2) When the fixing error is fixed, the waiting time increases as the average arrival rate increases. When the average arrival rate is 30-40/h, this effect is the most obvious; when the average arrival rate is 6/h, this effect is tend to stable, at this time, the waiting time of the flow increase about 1.9 percentage points when value of the fixing error increase 2.

(3) When the arrival rate is fixed, the increment of waiting time of the flow is proportional to the size of fixing error.

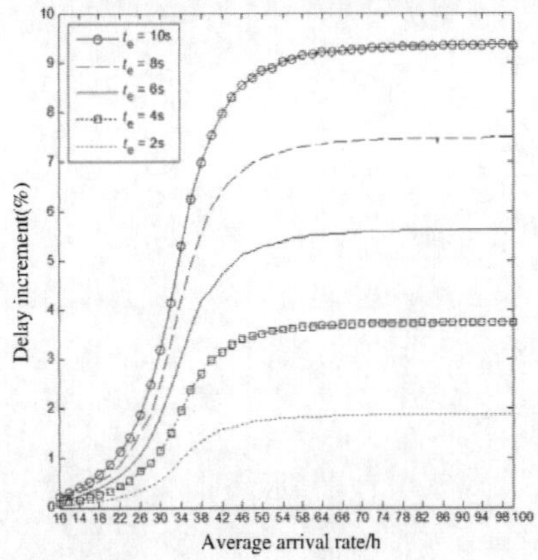

Figure 5 The effect of fixing error on delay

4. Conclusions

The results show that: with respect to the deterministic case, effect of fixing error of radar on the delay of aircraft is apparent. Also, the delay has a greater relationship with the flow density. When the fixing error is fixed, the waiting time increases as the average arrival rate increases. When the arrival rate is fixed, the increment of waiting time of the flow is proportional to the size of fixing error. It shows that only the determined case is considered is not enough, the fixing error of must be considered.

References

[1] X. K. Xu, Z. C. Li, "Application of fuzzy pattern recognition method to final approach

sequencing", Journal of Civil Aviation University of China, 22(02), 1-5, 2004.

[2] Vic Ciesielski, Paul Scerri. Real Time Genetic Scheduling of Aircraft Landing Times[C]. IEEE International Conference on Evolutionary Computation. Anchorage, USA, 1998: 360-364.

[3] Salvatore Capri, Matteo Ignaccolo. Genetic algorithms for solving the aircraft-sequencing problem: the introduction of departures into the dynamic model[J]. Journal of Air Transport Management, 2004, 10(5): 345-351.

[4] F. Ding, E. M. He, P. L. Wu, "Sequencing algorithm for waiting queue of aircraft in ATM", 19(3), 456-460, 2001.

[5] Z. N. Zhang, X. H. Li, L. L. Wang, "Computational method of longitudinal safety separation based on flight following theory", Journal of Traffic and Transportation Engineering, 8(3), 73-76, 2008.

[6] Y. Zhang, N. Chi, J. Hou, "Model of Radar Locating Error Caused by Atmosphere Refraction", Electronics Optics & Control, 16(7), 65-67, 2009.

[7] K. Tysen Mueller, Richard Bortins, David R. Schleicher, et al. Effect of Uncertainty on En Route Descent Advisor(EDA) Predictions[C]. AIAA 4th Aviation Technology, Integration and Operation Forum, Chicago, Illinois, 2004: 1-18.

[8] C. Z. Zhu, N. Cheng, Q. Li, "The four-dimensional trajectory prediction in terminal airspace", Journal of System Simulation, 22(Suppl. 1), 163-190, 2010.

[9] X. He, Y. X. Han, Z. J. Lai, "Capacity estimation model of airport runway", Aeronautical Computing Technology, 40(01), 42-45, 2011.

[10] Heming Chen, Yiyuan J. Zhao. Dynamic Real-time Scheduling of Terminal TrafficAIAA Guidance, Navigation and Control Conference, Portland, Oregon, USA, 1-15, 2011.

8th International Conference on Advanced Infocomm Technology (ICAIT 2015) IOP Publishing
Journal of Physics: Conference Series **679** (2016) 012014 doi:10.1088/1742-6596/679/1/012014

Research on Temperature Modeling of Strapdown Inertial avigation System

Xiao uan uang, Li ian Zhao[1], RuXiang Xu, and eng ang
1 China Satellite Maritime Tracking and Control Department, Jiangyin, Jiangsu, 214431, China

E-mail: eemctt@163.com

Abstract. Strapdown inertial navigation system with laser gyro has been deployed in space tracking ship and compared with the conventional platform inertial navigation system, it has substantial advantage in performance, accuracy and stabilization. Environmental and internal temperature affects the gyro, accelerator, electrical circuits and mechanical structure significantly but the existing temperature compensation model is not accurate enough especially when there is a big temperature change.

1. Introduction

The application field of Strapdown Inertial Navigation System （SINS）, used for navigation and stabilization, is now changed from airplane to naval ships, tactical missile, terrestrial positioning and orientation systems[1]. With the mature of technology, the application of ship-based RLG SINS has been developed very fast and becomes more and more important nowadays[2].

The influence of the temperature in inertial navigation system has two aspects: temperature change on inertial device and temperature field.In mechanically dithered SINS, while used in actual environment, the temperature variation of surroundings and heating in operative system itself will bring about changes in temperature of laser gyroscope and accelerometer themselves as well as relevant electric circles, which will finally cause variation of zero offset and scale factor of laser gyroscope and accelerometer.The calibration value is different, so that the system needs more time to finish initial alignment[3]. Laser gyro and quartz flexible accelerometer are the core of the system, and are also the temperature sensitive device, which is one of the most important factors to determine the accuracy, reliability and stability of the system.

In practical applications, the temperature model is not so precise when the temperature field changes radically; it cannot completely remove the measurement error. Considering that, we should analyze the measurement model and calculation model, conclude the characteristics of some inertial element parameter, and find out an optimization method for temperature model.

2. Analysis of Temperature haracteristic

2.1. Temperature characteristics of laser gyro

In a certain range of temperature and accuracy, it can be considered that the laser gyro is not affected by temperature, but the effect of temperature must be considered when the temperature is high or the accuracy requirement is high. Temperature in SINS will influence the data accuracy and perform in the following aspects: temperature field, gas fluent field, physical characteristics and geometrical features.

Journal of Physics: Conference Series **679** (2016) 012014 doi:10.1088/1742-6596/679/1/012014

The relationship between the temperature and the output of a certain type of laser gyro is investigated. The results are shown in Figure 1. It is shown that the influence of temperature on the test accuracy is not negligible.

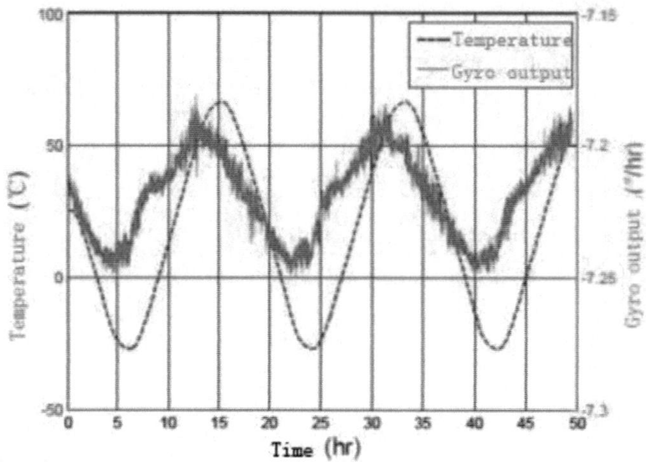

Figure 1. The output of a laser gyro and temperature change curve chart.

2.2. *Accelerometer temperature characteristic*

The accelerometer is the key device of the strapdown inertial measurement device based on Newton's theory of inertia, which has the characteristics of analog input, high performance and high accuracy. It can be used to measure the carrier's line acceleration, and then through the two points to get the trajectory of the carrier (the carrier's speed and distance), combined with gyroscope information, access to the carrier's full information[4].

During the working process of the accelerometer, there are two main reasons resulting in changes of temperature: one is self heating; one is the environmental temperature change.

The relationship between the output and the temperature change of a certain type of accelerometer is investigated. The results are shown in Figure 2.Temperature variation has a huge affection in accelerometer ' s output, its temperature characteristic also influence the initial alignment and course accuracy of SINS greatly.

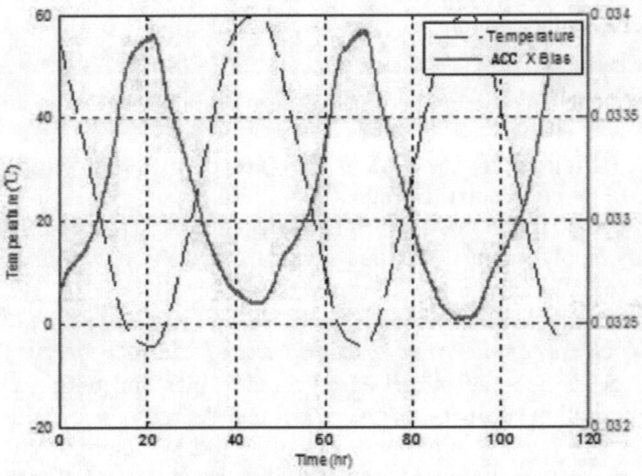

Figure 2. The output and temperature variation curve of a certain type of accelerometer.

3. Temperature model

3. 1. Establishment of zero temperature model

Through the analysis of temperature characteristics, the influence of thermal environment on the accuracy of the system is mainly derived from the temperature error caused by the inertial sensor and its propagation in the system, which can make the parameters of the laser gyro and accelerometer significantly changed, and the accuracy of the output data is significantly affected. According to the experimental analysis of the thermal environment, the scaling factor of laser gyro and accelerometer is smaller, and the main factor is the zero deviation of the gyroscope and accelerometer. It is also a hot spot of the current research to establish the temperature model of laser gyro zero bias, and the method of the software to compensate the economic and practical. In this paper, the improvement of the temperature model of the laser gyro is studied, and the improvement of the temperature model of the accelerometer is proposed.

The temperature of zero bias is mainly reflected by the temperature change, temperature change rate, temperature gradient and so on.Zero bias temperature error compensation model can be expressed as:

$$Bias = B_0 + \varphi(T) + v(T) + \delta(\Delta T) \quad (1)$$

B0: Constant value part of gyro's zero offset

$\varphi(T)$: Influence to gyro's zero offset due to tempera-ture variation

$v(T)$: Influence to gyro's zero offset due to temperature change rate

$\delta(\Delta T)$: Influence to gyro's zero offset due to temperature gradient

When environmental temperature changes slowly and the temperature gradient is stable, the effect of temperature variation on the zero deviation of the gyroscope is considered. The static temperature model of zero bias is established by polynomial fitting method. The formula (1) can be simplified as:

$$Bias = a_0 + a_1 T + a_2 T^2 + \ldots + a_m T^m \quad (2)$$

Journal of Physics: Conference Series **679** (2016) 012014 doi:10.1088/1742-6596/679/1/012014

According to formula (2) design temperature test, the zero deviation value at different temperatures was obtained, and the static temperature model was established:

1. In the temperature range [T_{min}, T_{max}],every temperature ΔT select a temperature point, measure the gyro bias output, and then calculate the temperature of gyro output as the average value of the temperature of the gyro bias.

2 Solve the model coefficients a_0, a_1, a_2... In the case of different temperature points, the model coefficients are determined by polynomial fitting.

3 Determine the number of polynomial m, achievement through regression coefficient of significant test. The actual number of polynomials used are two, three, four polynomial, and can be determined according to the accuracy of compensation.

3.2. *Improvement of temperature compensation model*

In the use of the zero bias temperature model based on zero bias and temperature, the model has some limitations, and the compensation precision of the model can decrease or even diverge with the change of environment temperature. The survey ship equipment of strapdown inertial navigation system is of high accuracy, influence of temperature proposed higher requirements, especially the wide temperature range high accuracy requirements, with the expansion of the scope of application, application environment complex, need to study is of higher temperature compensation model.

Temperature variation and temperature change rate are both considered in temperature model(3); also have improved the static model and added first order term of temperature change rate.

$$\varepsilon = a_0 + a_1 T + a_2 T^2 + a_3 T^3 + a_4 \frac{dT}{dt} \tag{3}$$

The forecast model(4) contains cross term and temperature change rate term, it has built a functional relationship of zero offset with the following factors: temperature, high order term of temperature change rate and its cross term[6].

$$\begin{aligned}\varepsilon = {} & a_0 + a_1 T + a_2 T^2 + a_3 T^3 + a_4 (\frac{dT}{dt}) + a_5 (\frac{dT}{dt})^2 \\ & + a_6 (\frac{dT}{dt})^3 + a_7 T (\frac{dT}{dt}) + a_8 T^2 (\frac{dT}{dt}) + a_9 T (\frac{dT}{dt})^2\end{aligned} \tag{4}$$

Therefore, we can take further consideration to build a temperature model (5) by using present data and historical moment temperature sequence.

$$\varepsilon_t = a_0 + \sum_{i=1}^{m} a_i \cdot T_{t-(i-1)} \tag{5}$$

ε_t、T_t: gyroscope zero offset and measure value of temperature on the N sampling

$T_{t-(i-1)}$: measure value of temperature on the [t-(i-1)] sampling

a_0、a_i: the coefficient of temperature compensation model

m: Total number used in this model, including present and historical temperature points $1 \le i \le m$

In order to verify the effect of the temperature models (5) as mentioned above, temperature experiment has been designed.

The result of compensation in model(1) indicates a relatively large residual; this model shows a low accuracy in temperature compensation in condition of fast temperature change rate. To a certain extent, Model (3) and (4) contain temperature change rate term, which enhanced their compensation ability in a fast temperature change rate. Both of them can well compensate the phase difference between gyro's output and temperature, much better than model (1). Model(5) is a time series model and has the same accuracy as model(3) when the temperature change rate changes fast. Model(3), model(4) and model(5) have the same compensation accuracy which is superior to model(1); improvement of temperature model in this passage is proved to be effect.

4. Temperature e periment

To verify the effect of four kinds of temperature model,a temperature test is designed. The system is placed in the temperature box, temperature change is set as shown in figure 3. The temperature change rate is 13 ℃/h. The temperature range is 30 ℃~60 ℃. Keep temperature for 2 hours at the temperature boundary. Obtaine the temperature model coefficient.

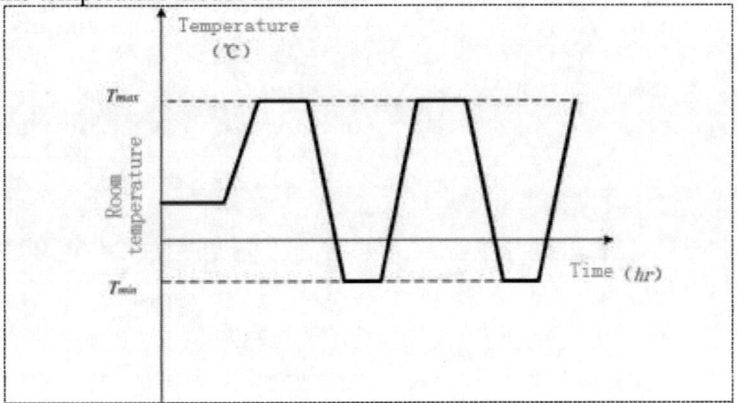

Figure 3. High and low temperature experiment.

Modeling results of the model (2):

$$\varepsilon = -7.1510 + 3.0739\times10^{-4}T + 3.5414\times10^{-6}T^2 - 5.5958\times10^{-8}T^3 \quad (6)$$

Modeling results of the model (3):

$$\varepsilon = -7.1518 + 2.7846\times10^{-4}T + 2.6327\times10^{-6}T^2 - 3.6996\times10^{-8}T^3 + 1.78\frac{dT}{dt} \quad (7)$$

Modeling results of the model (4):

$$\varepsilon = -7.1521 + 2.6981\times10^{-4}T + 3.4340\times10^{-6}T^2 - 4.3623\times10^{-8}T^3 + 2.0341(\frac{dT}{dt}) + 78.3937(\frac{dT}{dt})^2 - 0.0252T(\frac{dT}{dt}) + 1.7346\times10^{-4}T^2(\frac{dT}{dt}) \quad (8)$$

Considering the facts of data calculation and compensation accuracy, the model temperature point number m=25 is selected. Modeling results of the model (5):

$$\varepsilon_t = a_0 + \sum_{i=1}^{m} a_i \bullet T_{t-(i-1)} \quad (9)$$

a0=-7.1494 a1=-0.0481 a2=-0.0307 a3=0.033 a4=-0.0475 a5=0.0409 a6=-0.0259
a7=0.029 a8=-0.0082 a9=-0.0099 a10=0.0101 a11=-0.0197 a12=0.0195 a13=-0.0064
a14=0.0046 a15=-0.0231 a16=0.0269 a17=-0.0025 a18=-0.0165 a19=0.0157 a20=0.0008
a21=-0.0216 a22=0.0215 a23=-0.0258 a24=0.0389 a25=0.0446

The output of the gyro is compensated by type (6), type (7), type (8) and (9) model. The results are shown in Figure 4.

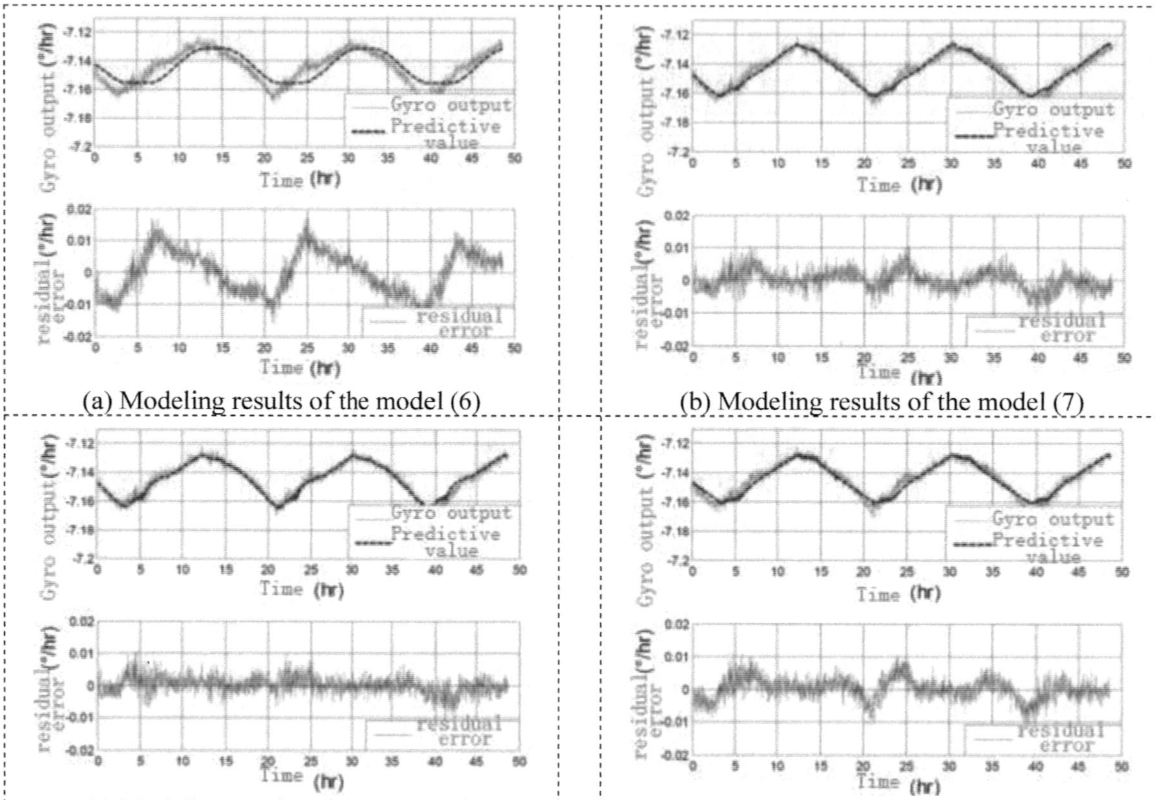

(a) Modeling results of the model (6)
(b) Modeling results of the model (7)
(c) Modeling results of the model (8)
(d) Modeling results of the model (9)

Figure 4. Modeling results of the models.

It can be seen that in the process of the experiment, the temperature change rate is great(13℃/hr), and the compensation results of model (6) have a large residual error, and the model has a low compensation accuracy when the temperature difference is large.

The temperature change rate are added to model (7) and model (8). In a certain extent, the compensation ability of the model is improved. The two models are better than the model (6).

Model (9) is a time series model, and the compensation precision is similar to the model (7) when the temperature change rate is great.

Model (7), (8), (9) are better than the model (6), this indicates the model proposed in the 3.2 section is effective. The specific values are shown in Table 1.

Table 1. Statistical analysis of model compensation before and after.

	Model(6)	Model(7)	Model(8)	Model(9)
Before compensation	0.0115	0.0115	0.0115	0.0115
After compensation	0.0067	0.0031	0.0027	0.0034

5. onclusions

This paper analyses the temperature characteristics of CLG as well as accelerometer, proposes the improvement method for present temperature model, and verifies the compensation effect through experiments. By means of comparing, the temperature model added with time series and temperature change rate performs better than before, it shows excellent environmental suitability, also available for complex temperature conditions such as wide temperature range and fast temperature variation rate.

References

[1] B.B Xie redln and P.M Kuku Lief 1992 *Laser Gyro and its application*(Russian: Aviation Industry Press)

[2] Wei Guo 2013 *Research on some key technologies of mechanically dithered ring laser gyro SINS system*(doctoral dissertation of NUDT)

[3] Wang Yu 2005 *Preliminary investigation on mechanically dithered ring laser gyro SINS system* (Changsha: doctoral dissertation of NUDT)

[4] Wang Renlin 1982 *Accelerometer* (National Defence Industry Press)

[5] Yan Junxuan, Zhang Ke 2011 *J. Research on temperature compensation of ring laser gyroscope. Computer measurement&Control* **1** 6

[6] Zhang Pengfei 2006 *Research on mechanically dithered ring laser gyro SINS system and its temperature compensation method* (Changsha: doctoral dissertation of NUDT)

[7] Yu Xudong, Wang Yu, Wei Guo, Zhang Pengfei and Long Xingwu 2011 *J. Novel temperature modeling and compensation method for bias of ring laser gyroscope based on least-squares support vector machine. Chinese Optics Letters* 051201

[8] Wei Guo, Li Geng, Wu Yan and Long Xingwu 2011 *J. Application of Least Squares-Support Vector Machine in system-level temperature compensation of ring laser gyroscope. Measurement* **44** 1898

[9] Jiahe X, Yongyuan Q and Rui L 2008 *Modeling temperature data of RLG's scale factor using LS-SVM//Knowledge Acquisition and Modeling Workshop*(IEEE International Symposium on. IEEE)

[10] Junchao C and Jiancheng F 2012 *Comparison of compensation methods on RLG Temperature error and their application in POS//Instrumentation and Control Technology (ISICT)* (8th IEEE International Symposium on. IEEE)

Accuracy Evaluation of Tracking Equipment Based on Star-Station- Difference Technique

Li ian Zhao, Xiao uan uang, Liang Pan, and RuXiang Xu
China Satellite Maritime Tracking and Control Department, Jiangyin, Jiangsu, 214431, China

E-mail: eemctt@163.com

Abstract. An approach based on the star station difference technology is proposed for accuracy evaluation of tracking and controlling shipboard equipments in this paper. The proposed method has the advantages of simple equipment, convenient measurement, and low requirements on environmental conditions.

1. Introduction

The accuracy evaluation of tracking equipment is a important work. To ensure the calculating accuracy and dependability, the ground stations are established in the near region to provide the GPS difference messages. In this way, region restriction, high cost, low efficiency are the prominent problems.

This paper proposes a method for accuracy evaluation of tracking equipment based on star-station-difference technique. The proposed method has the advantages of simple, convenient, precise, automatic, and low requirements on environmental conditions [1] .

2. The star-station-difference techni ue

By the star station difference technology, the GPS difference correction signals broadcast by satellites instead of the stations on the ground. It makes the receivers can receive the difference correction signals and high precision positioning is provided on the open sea. In the Last century ninety's, foreign countries began to study the development of Star Station difference GPS. There are many system such as : CDGPS, OmniSTAR system, SeaSTAR system, Skyfix system , StarFire[2,3] system and Veripos system. StarFire[2,3] system, Veripos system and OmniSTAR system are commonly used in China.

The StarFire system is one of the star station difference systems. The whole system consists of five parts: the reference station, data processing center, data immit station, INMARSAT and user station. Global reference station network is composed of dual-frequency GPS receiver which receive signals from GPS satellites. Data from reference station are sent to the processing center. After being processed, the differential correction data is generated. The differential correction data transmit data communications link to the immit station and upload to the INMARSAT in the world. At the same time, user station GPS receiver which having two actually receiving part receive the correction signals. The GPS receiver track all visible satellites and then obtain the measured value of the GPS satellites, the L-band communication receiver correct the data by L-band satellites signals. When the correct data are used in GPS measurement, the real-time high-precise positioning is done.

3. Accuracy testing

Journal of Physics: Conference Series **679** (2016) 012015 doi:10.1088/1742-6596/679/1/012015

3. 1. Precision identification

Precision identification[4] is used to test and evaluate the dynamic performance and accuracy of the aerospace measurement and control equipment, and to analyze the variation of its error, and to test whether the technical indicators meet the design requirements[5]. It is the premise to obtain accurate target data from the measurement and control equipment, which is great significance.

3. 2. Existing problems

GPS precision identification system is the only way to evaluate the accuracy of TT&C equipment. The system need RTK system by land differential measurement station, eliminate the error by using differential station, Airborne GPS, ship borne GPS. Differential positioning technology requirements that the distance between substation and dynamic measure point is short as possible. It is difficult to choose a GPS station for every test of evaluation.It is urgent to find a new technical means to reduce the cost, optimize the system work way and improve the working efficiency.

3. 3. Star Station difference application

Star Station differential system in the application of Survey ship exploration began in 2008. The previous carrying test show that, star station difference technology can effectively improve the position accuracy , and further improve inertial navigation position calibration accuracy, timing accuracy and tracking fixed rail precision. It is of great application value. In order to further deepen the application of the star station differential technique in the survey ship, it has been demonstrated that using star station differential technique in precision identification of the measurement equipment is feasible.

In May 2014, the StarFire system was used in the test of precision identification as a backup method of the GPS precision identification system. Test results show:

a.The dynamic index (nominal value) of Star Station differential equipment meets the requirements of ship navigation and flight dynamics;

b.The data precision of the Star Station differential equipment is quite the same as the accuracy of the existing precision identification system;

c.Star Station differential equipment is feasible in the measurement of the ship's accuracy.

4. Accuracy valuation Scheme

4. 1. Airborne subsystem

The airborne subsystem is composed of GPS antenna, navigation display computers, GPS receivers and data broadcasting station with antenna. The airborne equipments position is shown in Figure 1, two antenna are installed on the plane to receive the GPS difference correction signals. The airborne subsystem measure the real-time plane ' s position, velocity and attitude which broadcasted by the broadcasting station; the navigation display computers show the flight path according the received data.

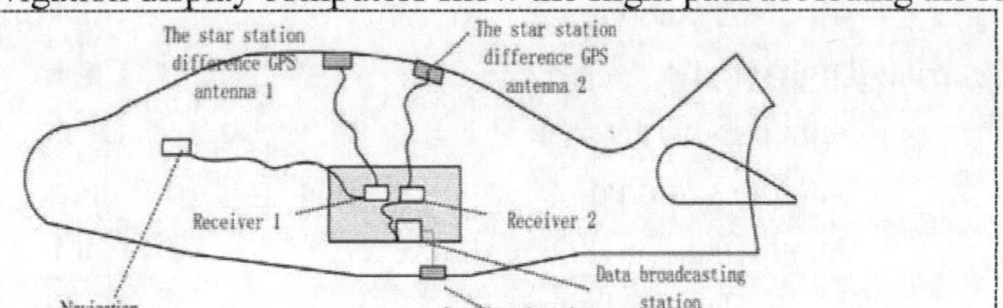

Figure 1. Schematic diagram of airborne equipment installation.

4. 2. Ship carrying subsystem

The shipboard subsystem is composed of GPS antenna, navigation display computers, GPS receivers and data broadcasting station, also with an antenna[3]. The shipboard equipments position is shown in Figure2.

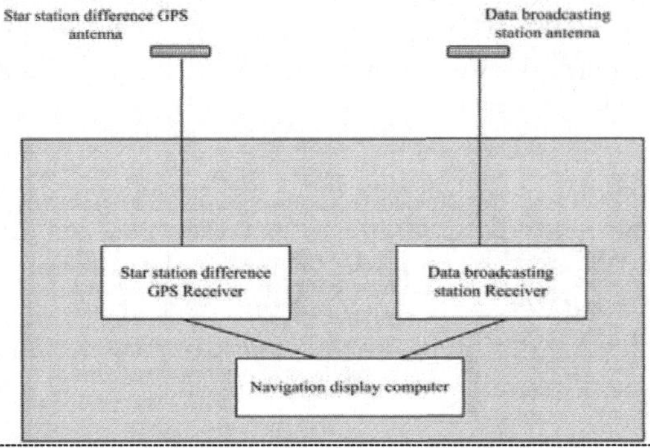

Figure 2. Schematic diagram of shipboard equipment installation.

Compared with the conventional GPS precision evaluation system, the star station difference technology has the following advantages: high real-time pointing precision, simple equipment, convenient measurement, , low requirements on environmental conditions, low economic cost and easy scheme design.

4.3. Data processing subsystem

The data processing subsystem is composed of a high configuration computer. Its function is calculating the position and attitude data of the airborne subsystem and the ship carrying subsystem, and providing high accuracy data for comparison. It can be installed on the land or on the ship.

4.4. System composition

Table 1 includes the equipments of accuracy identification system.

Table 1. The equipments of accuracy identification system.

	equipment	position
Airborne subsystem	Digital radio	Frame
	Digital radio antenna	Lower part of the fuselage
	Satellite station differential antenna	Upper fuselage
	Satellite station differential receiver	Frame
	Air display computer	Cockpit
Ship carrying subsystem	Digital radio	Cabin
	Digital radio antenna	Deck
	Satellite station differential antenna	Deck
	Satellite station differential receiver	Cabin
	Air display computer	Cabin
Data processing subsystem	Data processing computer	Cabin
	Data processing software	Computer

Through the analysis and comparison, it can be seen that the difference between the Star-Station-Difference system and the old identification system.The Star-Station-Difference system has the following advantages:

a. Real time positioning accuracy is high. The high accuracy position data can effectively reduce the influence of position error, so as to improve the accuracy of measurement.

b.The data can be used directly for comparison, reducing the amount of post data processing workload and improving Measurement accuracy.

c. high integration, high reliability, stand-alone operation, simple installation.

d.Reference station is not necessary.

5. Conclusion

This paper proposes a scheme of accuracy evaluation of tracking equipments. By the star station difference technology, the GPS difference correction signals broadcast by satellites instead of the stations on the ground. It makes the receivers can receive the difference correction signals and high precision positioning is provided on the open sea. It makes the dynamic precision evaluation of shipboard equipments feasible and easy.

References

[1] Wei Zhou, Jinming He and Guoyuan Ma 2012 *J. accuracy evaluation method research of tracking and controlling equipment based PPP. Mapping aviso* **5** 5

[2] Lingang Feng and Yonggui Zhao 2006 *J.Star-based Differential GPS System—StarFire. Bulletin of Surveying and Mapping.***11** 6

[3] Yanjun Zhang(translation) 2007 *J. StarFire global high precision wide area differential GPS system. EGP* **1** 145

[4] Haizhong Xie, Shouxing zhang and Xurong Dong 1999 *J. Accuracy evaluation method research of spaceflight tracking and controlling equipments. command technology institute transaction* **1** 30

[5] Wenda Jiang 2002 *Space measurement ship*(National Defence Industry Press)

The modulation and demodulation module of a high resolution MOEMS accelerometer

Xufen Jiao, Jian Bai*, Qianbo Lu, Shuqi Lou
State Key Laboratory of Modern Optical Instrumentation, Zhejiang University, Hangzhou 310027, China

E-mail:bai@zju.edu.cn

Abstract. A MOEMS accelerometer with high precision based on grating interferometer is demonstrated in this paper. In order to increase the signal-to-noise ratio (SNR) and accuracy, a specific modulator and an orthogonal phase-lock demodulator are proposed. Phase modulation is introduced to this accelerometer by applying a sinusoidal signal to a piezoelectric translator (PZT) amounted to the accelerometer. Phase demodulation module consists of a circuit design and a digital design. In the circuit design, the modulated light intensity signal is converted to a voltage signal and processed. In the digital part, the demodulator is mainly composed of a Band Pass Filter, two Phase-Sensitive Detectors, a phase shifter, and two Low Pass Filters based on virtual instrument. Simulation results indicate that this approach can decrease the noise greatly, and the SNR of this demodulator is 50dB and the relative error is less than 4%.

1. Introduction

Accelerometer plays an important role in inertial navigation systems[1-2], biomedical area, military fields and our daily life, such as detecting the respiratory rate[3-6], measuring the Earth's surface and the distribution of the gravity field of the earth[7]. The accelerometers based on traditional methods suffer from low sensitivity, bad reliability, great complexity of signal processing, etc. However, optical accelerometers are able to achieve high resolution and sensitivity compared to the traditional ones. As acceleration is related to the displacement definitely, acceleration can be calculated accurately by measuring the displacement with high precision. A displacement sensor based on a grating interferometer with nanometer-scale displacement resolution has been demonstrated in [8]. Therefore, the optical accelerometer based on grating interferometer is able to achieve high resolution and sensitivity compared to the traditional ones. Furthermore, the MOEMS accelerometer which combines MEMS technology to optical accelerometer is of great significant due to the advantages of small size and light weight.

To further improve the SNR and suppress the noise, phase modulation and demodulation techniques are introduced in fiber-optic communication system[9-10] In this paper, phase modulation is introduced to this accelerometer by applying a sinusoidal signal to a piezoelectric translator (PZT) amounted to the grating, which can increase the signal-to-noise ratio (SNR) and accuracy dramatically by demodulation. Meanwhile, the frequency of the signal is 500 Hz which is away from the main frequency ranges of noise signal. To demodulate the modulated signal, an orthogonal phase-lock amplifier technique based on virtual instrument is applied. The reference signal is a sinusoidal wave with 500Hz frequency and the orthogonal phase-lock demodulator is mainly composed of a Band Pass Filter, two Phase-Sensitive Detectors, a phase shifter, and two Low Pass Filters.

2. Principle

In this section we discuss the principle of the MOEMS accelerometer based on grating interferometer, then demonstrate how the sensor works with the modulation and demodulation module.

2.1. Sensing principle

As shown in figure 1, the MOEMS accelerometer is based on a grating interferometer. The incident coherent laser beam is diffracted on the surface of an amplitude grating with duty ratio of 0.5. A part of light is directly reflected and serves as the diffracted beam 1; the other part reaches the mirror bonded to the mass, then is reflected back to the grating and becomes the diffracted beam 2. Both Acceleration and displacement have a linear relationship, when acceleration is applied to the mass, the displacement changes correspondingly. Then acceleration of the mass can be measured by detecting the change of the brightness of the interference fringes. Since the intensity becomes weaker when the number of interference order increases and the 0th order is hard to measure due to its position, the selected order is $\pm 1st$. The relationship between the intensity of the first interference orders and the acceleration can be described as:

$$I_{\pm 1} = \frac{2I_{in}}{\pi^2}(1-\cos 2kd) = \frac{2I_{in}}{\pi^2}[1-\cos\frac{4\pi}{\lambda}(d_0+\Gamma a)] \tag{1}$$

where a is the acceleration, $I_{\pm 1}$, λ, d_0, Γ are the intensity, the wavelength of the incident coherent laser beam, the initial distance between the grating and proof mass, the scaling factor, respectively.

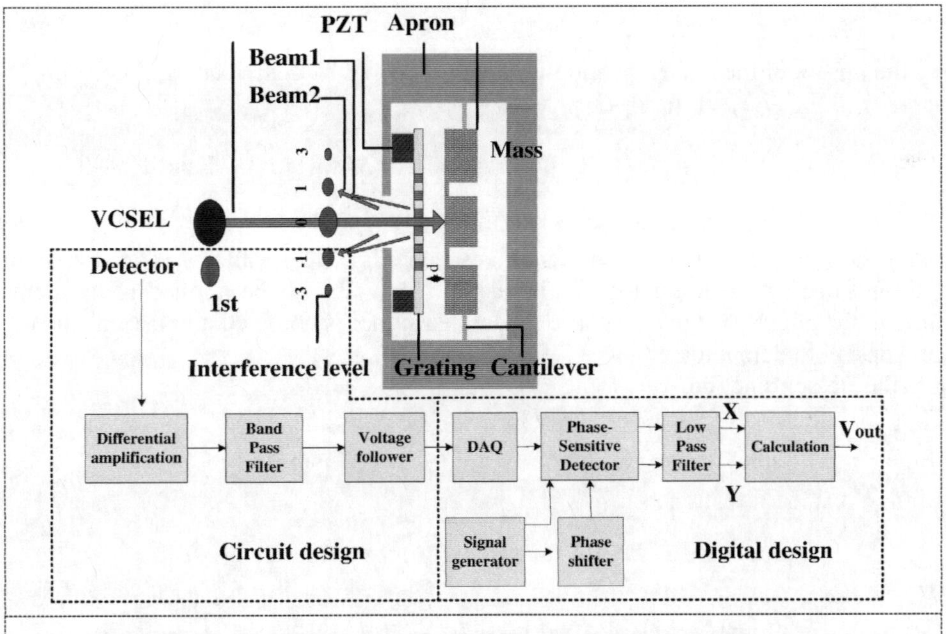

Figure 1. The schematic diagram of the MOEMS accelerometer

2.2. Phase modulation principle

For slow frequency signal or DC signal, the weak signal will be buried, as random drift and 1/f noise brought by ambient environment will be amplified by DC amplifier[11].To eliminate the impact and decrease the signal distortion, phase modulation is applied. The basic principle of phase modulation introduced in the grating interferometer can be described as follows: output a sinusoidal signal to a power-driven module to drive a PZT, then the length of the PZT is modulated by the signal.

Meanwhile, the distance between the grating and mirror is modulated by the sinusoidal signal and the optical path of diffracted beam 2 is changed. Since the phase shift between beam 1 and beam 2, the modulated intensity of the first interference orders received by the PD could be described as:

$$I_{\pm 1} = \frac{2I_{in}}{\pi^2}(1-\cos 2kd) = \frac{2I_{in}}{\pi^2}\{1-\cos[\frac{4\pi}{\lambda}(d_0+\Gamma a)+M\sin\omega_0 t]\} \tag{2}$$

where M is the modulation depth and ω_0 is the modulation frequency. The modulated phase can be expressed as $M\sin\omega_0 t$.

2.3. Phase demodulation principle

Phase demodulation is used to extract the carrier signal of displacement which is related to the acceleration. Since the modulated intensity of the first interference orders is very weak and varies with time periodically. Thus, correlative detection technology is introduced in detection of feeble signal to improve the signal-noise ratio. The demodulation module consists of a signal processing circuit design and a digital design based on virtual instrument.

In the part of circuit design, a photodetector converts the light intensity to a voltage signal. Because the output electrical signals from the photodetector are pretty small, the modulated signals should be amplified by precise differential amplifier. The Band Pass Filter is applied to reduce the impact of noise signal. The power supply design is an adjustable voltage source which uses a LC- filter circuit and can provide variable types of voltages. The useful modulated signal could be described as:

$$V_1(t) = A\times I_{in}\{1-\cos[\frac{4\pi}{\lambda}(d_0+\Gamma a)+M\sin\omega_0 t]\} \tag{3}$$

where A is the circuit amplification factor.

Considering the impact of the noise, the modulated signal could be described as:

$$\begin{aligned} V(t) &= V_1(t)+n(t) \\ &= A\times I_{in}\{1-\cos[\frac{4\pi}{\lambda}(d_0+\Gamma a)+M\sin\omega_0 t]\}+C\sin(\omega_n t+\theta_n) \end{aligned} \tag{4}$$

where $n(t)$, $V_1(t)$, ω_n , C , θ_n are the noise signal, the useful signal, the frequency of useful signal, the amplitude and the initial phase of noise signal, respectively. Meanwhile, a reference signal whose frequency is the same as the frequency of the useful signal should be applied in the demodulation system. Through the circuit design, the phase modulated signal is obtained and used as the input signal for the digital design to demodulate.

According to the Bessel function,

$$\begin{aligned} \cos(M\sin\omega_0 t) &= J_0(M)+2J_2(M)\cos 2\omega_0 t+\ldots \\ \sin(M\sin\omega_0 t) &= 2J_1(M)\sin\omega_0 t+2J_3(M)\sin 3\omega_0 t+\ldots \end{aligned},$$

Then equation (4) can be rewritten as:

$$V(t) = A\times I_{in}\left\{\begin{aligned} &1-\cos\frac{4\pi}{\lambda}(d_0+\Gamma a)\left[J_0(M)+2J_2(M)\cos 2\omega t+\ldots\right] \\ &+\sin\frac{4\pi}{\lambda}(d_0+\Gamma a)\left[2J_1(M)\sin\omega t+2J_3(M)\sin 3\omega t+\ldots\right] \end{aligned}\right\}+C\sin(\omega_n t+\theta_n) \tag{5}$$

In the digital design, phase-lock demodulator is mainly composed of a Band Pass Filter, two Phase-Sensitive Detectors (PSD), a phase shifter as well as two Low Pass Filters. One reference signal $V_2(t)$ can be described as:

$$V_2(t) = B\sin(\omega_0 t+\theta) \tag{6}$$

where B is the amplitude of reference signal.

The phase sensitive detection (PSD) process is shown in figure 2. The dotted line describes the power spectral density of 1/f noise and white noise which focus on low frequency. After phase modulation, the frequency spectrum of the useful signal removes to the frequency around 500Hz where the noise signal is quite weak. Because of the PSD, the signal whose frequency is the same as the frequency of the reference signal will be changed to frequency-doubled signal and DC signal which can be described as:

$$V_{PSDout} = R_{V_1V2} + R_{V_2n} = V_1(t)V_2(t) + n(t)V_2(t) \tag{7}$$

The $R_{V_1V_2}$ can be described as:

$$R_{V_1V_2} = ABI_{in}\left\{\begin{array}{l} \sin(\omega_0 t+\theta) - \cos[\frac{4\pi}{\lambda}(d_0+\Gamma a)]\{J_0(M)\sin(\omega_0 t+\theta) \\ +J_2(M)[\sin(3\omega_0 t+\theta)-\sin(\omega_0 t\text{-}\theta)] \\ +J_4(M)[\sin(5\omega_0 t+\theta)-\sin(3\omega_0 t\text{-}\theta)]+...\} \\ +\sin[\frac{4\pi}{\lambda}(d_0+\Gamma a)]\{J_1(M)[\cos(2\omega_0 t+\theta)+\cos\theta] \\ +J_3(M)[\cos(4\omega_0 t+\theta)+\cos(2\omega_0 t-\theta)] \\ +J_5(M)[\cos(6\omega_0 t+\theta)+\cos(4\omega_0 t-\theta)]+...\} \end{array}\right\} \tag{8}$$

$R_{V_1V_n}$ can be described as:

$$R_{V_2n} = -C[\cos(2\omega_0 t+\theta_n)+\cos\theta_n] \tag{9}$$

After LPF, frequency-doubled signal is filtered out. Then the demodulated signal becomes DC signal:

$$V_{out1} = -ABI_{in}\sin[\frac{4\pi}{\lambda}(d_0+\Gamma a)]J_1(M)\cos\theta \tag{10}$$

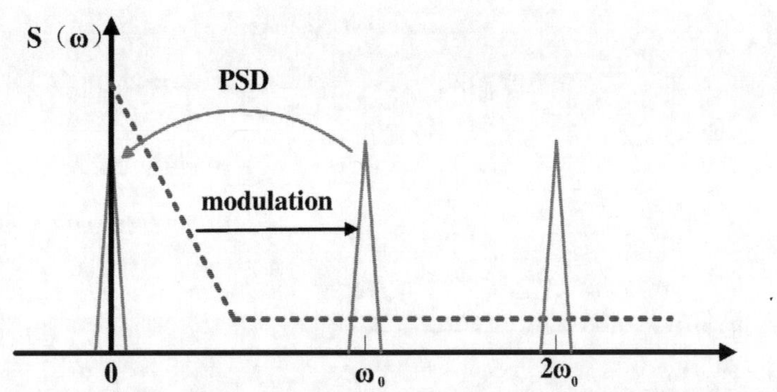

Figure 2. The phase sensitive detection process

When the phase shifter is applied to the reference channel, the orthogonal reference signal $V_3(t)$ has the form:

$$V_3(t) = B\cos(\omega_0 t+\theta) \tag{11}$$

Then the output signals after LPF can be expressed as X and Y :

$$\begin{aligned} X &= -ABI_{in}\sin[\frac{4\pi}{\lambda}(d_0+\Gamma a)]J_1(M)\cos\theta \\ Y &= -ABI_{in}\sin[\frac{4\pi}{\lambda}(d_0+\Gamma a)]J_1(M)\cos(\theta+90^0) \end{aligned} \tag{12}$$

Finally, the amplitude of the output signal V_{out} can be expressed as:

$$V_{out} = \left(X^2+Y^2\right)^{\frac{1}{2}} = \left|ABI_{in}\sin[\frac{4\pi}{\lambda}(d_0+\Gamma a)]J_1(M)\right| \tag{13}$$

Journal of Physics: Conference Series **679** (2016) 012016 doi:10.1088/1742-6596/679/1/012016

where V_{out} is the phase demodulated signal related to acceleration.

Therefore, acceleration could be calculated by equation (13) through the measurement of the demodulated signal.

3. Simulation results and discussions

In this paper, modulation and demodulation module are designed and simulated to increase the performance of the high resolution MOEMS accelerometer.

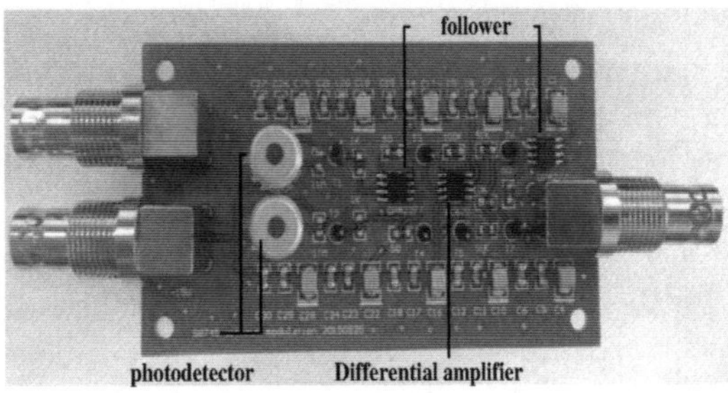

Figure 3. Signal generator based on LabVIEW

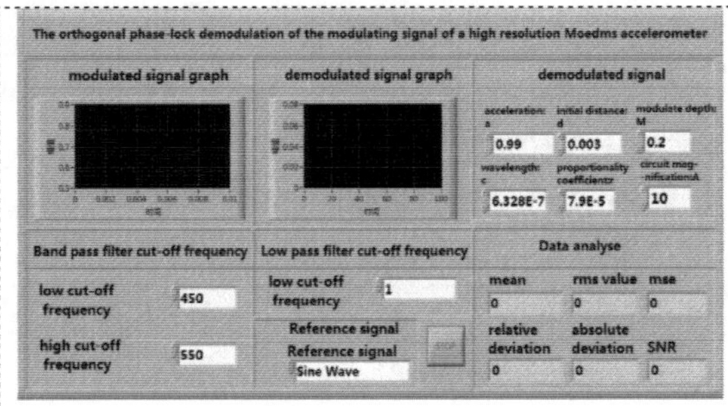

Figure 4. The front panel of the demodulation module

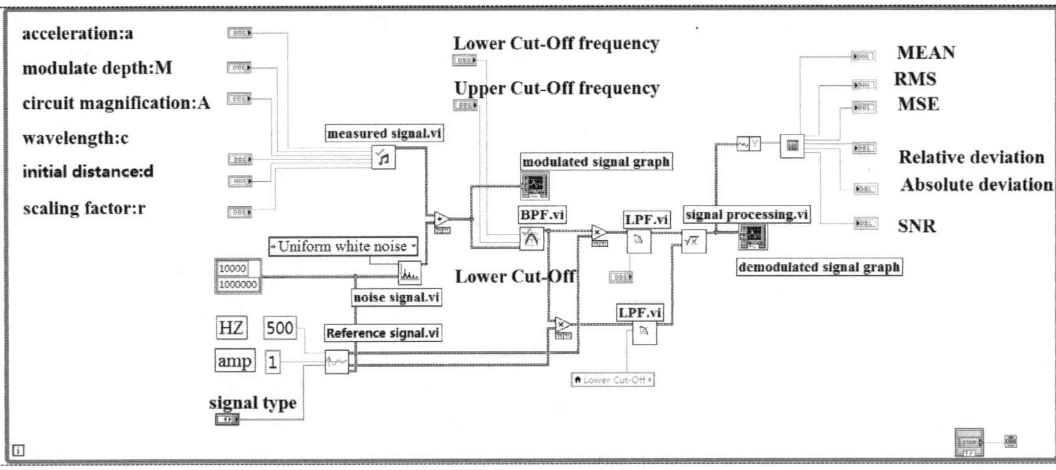

Figure 5. The block diagram of the demodulation module based on LabVIEW

In the circuit design, PZT is driven by a power-driven module. The intensity of the first order is received by S8745-01 which is a photodetector (PD) with a preamplifier. The other PD detects the intensity of the ambient light. Then the two detectors convert light intensity signals to electrical signals. After a differential amplifier which contains AD620, the influence of the ambient light can be deceased. The modulated signal output from follower which is composed of an OPA277, is acquired by a DAQ (NI USB-6212). The circuit design of signal processing is shown in figure 3.

The front panel and block diagram of the digital demodulation module are shown in figure 4 and figure 5.The whole simulation module contains the following parts: a signal generator, a reference channel, a Band Pass Filter, two PSD and Low Pass Filters. Real-time signal will be shown in the waveform graph of the front panel. The Lower and upper cut-off frequencies of the BPF are 450Hz and 550Hz. The cut-off frequency of the LPF is 1Hz. Better magnitude-frequency characteristics obtained when the type of BPF and LPF is Butterworth.

Journal of Physics: Conference Series **679** (2016) 012016 doi:10.1088/1742-6596/679/1/012016

As is shown in figure 6, the signal generator based on LabVIEW generates the modulated signal which is not a standard sinusoidal signal as shown in figure 7. In order to obtain the real signal and eliminate the distortion, the sampling frequency is set to 10KHz, which is much higher than the frequency of the modulated signal.

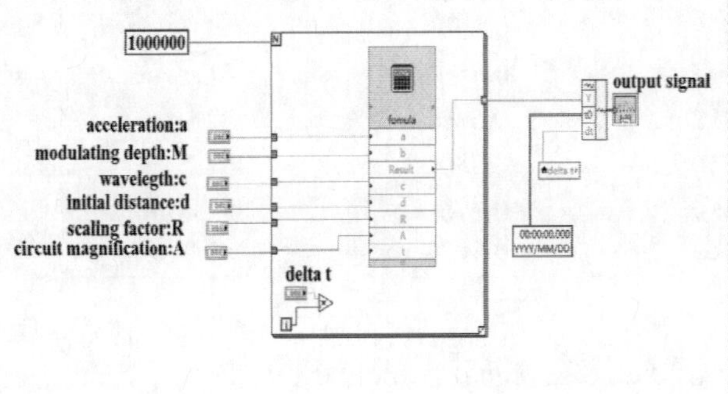

Figure 6. Signal generator based on LabVIEW

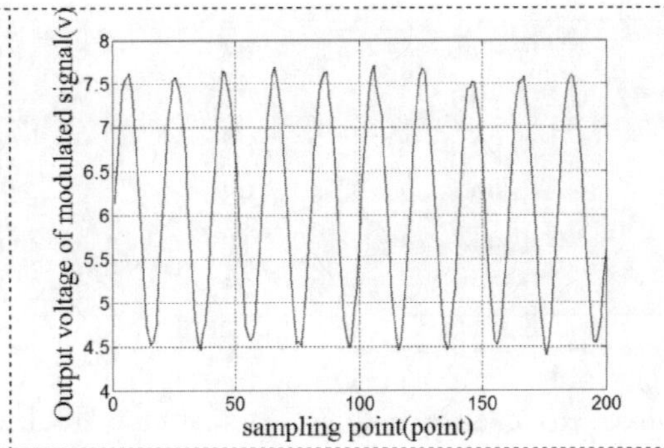

Figure 7. Modulated signal simulated by LabVIEW

The reference signal channel is divided into two parts. Both parts consist of a reference signal generator, but one of them contains a phase shifter. In the front panel, reference signal could be selected as sinusoidal signal or square signal. The frequency and amplitude of both two orthogonal reference signals are 500Hz and 1V respectively. The phase difference between the two reference signals is 90 degree. The demodulated signal is measured by the orthogonal component without phase adjustment between the modulated signal and reference signal. In this case, the precise demodulated signal is produced. The block diagram of reference signal channel is shown in figure 8.

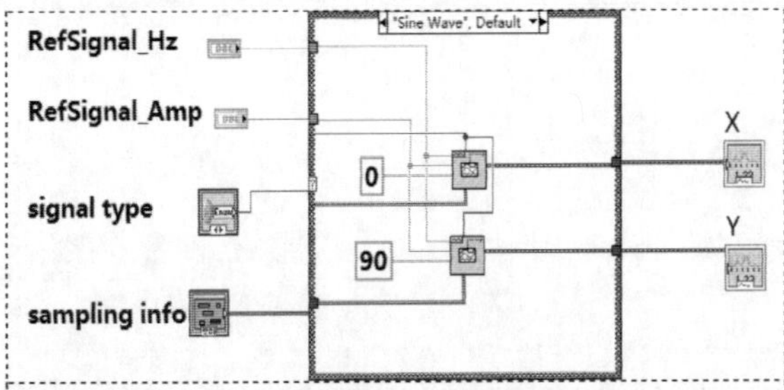

Figure 8. The block diagram of reference signal channel.

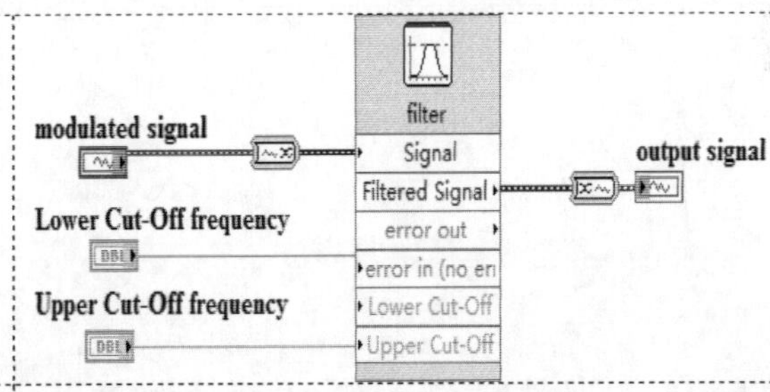

Figure 9. The block diagram of the Band Pass Filter

In terms of the Band Pass Filter and the Low Pass Filter, the common hardware active filter which consists of operational amplifiers and R, C components, is easy to implement but the parameter modulation is difficult to conduct[12]. For digital filter, there is no need to consider the impact caused by surrounding environment and stray capacitance. Also, it can provide different filtering types, orders and topologies such as Butterworth, Chebyshev, Bessel. The cut-off frequency of the BPF and LPF could also be changed in terms of different input signals. The block diagram of Band Pass Filter is shown in figure 9.

In the part of PSD design, a digital multiplier is used instead of analog multiplier chip. After the Low Pass Filters, DC signals are obtained. SNR and relative deviation are displayed in the front panel.

A series of output voltages is obtained when the input acceleration is changed. Demodulated signal is buried in the noise, whose amplitude is 0.1 V, the RMS (Root-Mean-Square), SNR (Signal-to-noise ratio), max relative deviation and absolute deviation are presented in Table1.

Table1. The results after phase demodulation

Acceleration (g)	RMS (V)	Max relative deviation (%)	Absolute deviation (mm)	SNR (dB)
0.990	7.043	0.365	0.260	54.3
0.991	7.013	0.647	0.450	51.6
0.992	7.071	0.952	0.670	52.3
0.993	6.980	0.550	0.380	51.7
0.994	7.103	0.389	0.276	52.4
0.995	6.955	0.463	0.322	53.8

From table 1, we can see the max relative deviation of the output signal is below 1% and the signal-noise ratio reaches up to 50 dB when the amplitude of the noise signal is about 0.1V.

Figure 10 shows the relationship between the acceleration and output voltage after LPF when the acceleration is 0.992g. DC signal is reserved after LPF. Figure 11 represents the acceleration measurement result which is the absolute value of a sine wave signal. In this figure, there is little difference between the two curves and the output voltage is well matched to equation (13). Through the simulation of demodulation, it is clear that the measured value is close to the theoretical output value.

The simulation shows the max relative deviation is below 1% and the SNR of this demodulator is about 50dB. This design is quite fit for the phase modulated signal whose frequency is 500Hz.

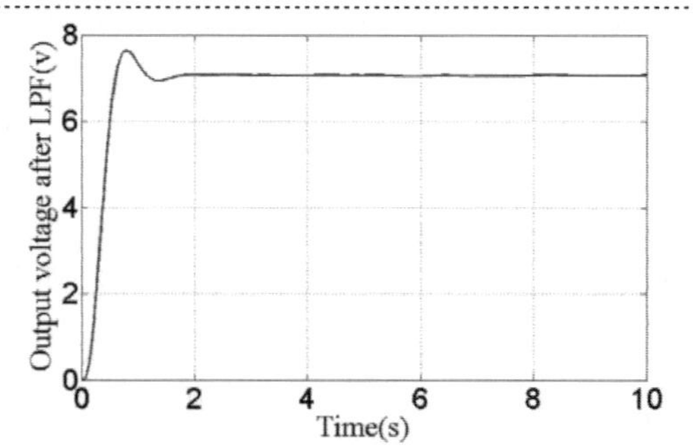

Figure 10. The DC signal obtained after LPF when the acceleration is 0.992g

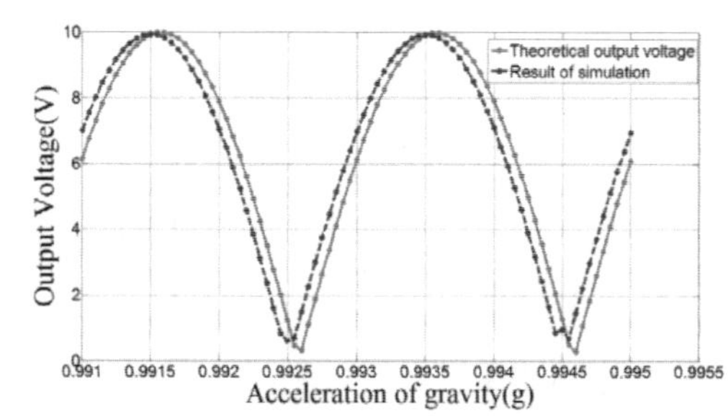

Figure 11. Acceleration measurement result

4. Conclusions

In this paper, we have simulated a modulator and an orthogonal phase-lock demodulator to demodulate the modulated signal of a high resolution MOEMS accelerometer which is based on phase-sensitive detection technique. The theoretical analysis and corresponding LabVIEW simulation are presented in this paper. The simulation shows the max relative deviation of the acceleration is below 1% and the SNR of this demodulator is about 50 dB. This modulation and demodulation module is quite fit for this specific MOEMS accelerometer.

5. References

[1] Tan C W and Park S. 2005 Design of accelerometer-based inertial navigation systems *IEEE. T. Instrum. Meas.* **50** 2520-30

[2] Hopkins R, Miola J, Setterlund R, Dow B and Sawyer W 2005 The Silicon Oscillating Accelerometer: A High-Performance MEMS Accelerometer for Precision Navigation and Strategic Guidance Applications *Proceedings of the 61st Annual Meeting of The Institute of Navigation (Cambridge, MA, 27-29 June 2005)* pp 1043-52

[3] Phan D H, Bonnet S, Guillemaud R, Castelli E and Pham Thi N Y 2008 Estimation of respiratory waveform and heart rate using an accelerometer *Biomedical Imaging: From Nano to Macro, 2008. ISBI 2008. 5th IEEE International Symposium on (Paris, 14-17 May, 2008)* pp 4916-19

[4] Dianne S, Ward, Kelly R, Evenson, Amber, Vaughn, Anne Brown, Rodgers, Richard P and Troiano 2005 Accelerometer use in physical activity: best practices and research recommendations *Med. Sci. Sport Exer.* **37** 582-588

[5] Salarian A, Russmann H., Vingerhoets F J, Burkhard P R and Aminian K 2007 Ambulatory Monitoring of Physical Activities in Patients With Parkinson's Disease *IEEE. Trans. Biomed Eng* **54** 2296-99

[6] Kau L J and Chen C S 2015 A Smart Phone-Based Pocket Fall Accident Detection, Positioning, and Rescue System *IEEE. J. Biomed. Health.* **19** 44-56

[7] Shuangshuang Z, Juan Z, Changlun H, Jian B and Guoguang Y, 2012 Optical accelerometer based on grating interferometer with phase modulation technique *Appl. Opt.* **51** 7005-10

[8] Shuangshuang Z, Changlun H, Jian B, Guoguang Y and Feng T 2012 *Appl. Opt.* **50** 1413-16

[9] Winzer P J and Essiambre R J, 2006 Winzer P J, Essiambre R. Advanced Optical Modulation Formats *Proceeding of the IEEE* **94** 952-985

[10] Lin Q, Chen L and Wu X 2010 A high-resolution fiber optic accelerometer based on intracavity phase-generated carrier (PGC) modulation *Meas. Sci. Technol.* **22** 1-6

[11] Jinzhan G 2002 *Detection of Weak Signals* (Beijing: Tsinghua University Press) p 171

[12] Liu Y L and Zhang R 2012 AD630 Lock-in Amplifier Circuit for Weak Signal *Adv. Mater. Res.* **482-484** 975-980

Mid-infrared fiber optical parametric oscillator using a three-hole suspended-core chalcogenide fiber

Hangyu Bai, Xiong Yang, Yizhen Wei, Wei Pan and Shiming Gao*

Centre for Optical and Electromagnetic Research, State Key Laboratory of Modern Optical Instrumentation, Zhejiang University, Hangzhou 310058, P. R. China

gaosm@zju.edu.cn

Abstract. A three-hole As_2S_5 suspended-core fiber with a zero-dispersion wavelength of 2 μm is proposed to construct a mid-infrared (MIR) fiber optical parametric oscillator (FOPO). The numerical model of the MIR FOPO is established and a tuning range from 2 to 5 μm for the idler light is predicted.

1. Introduction

Mid-infrared (MIR) lasers are of great interest due to their various applications in spectroscopy, medicine, and military. However, the output wavelength range of a conventional MIR laser is limited and it is hard to find a material to cover a wide wavelength region in the MIR band. Contrastingly, fiber optical parametric oscillator (FOPO) using a pump to generate a pair of signal and idler through four-wave mixing in a cavity has been proved to be a promising way to generate broadband MIR lasing. For MIR FOPO, chalcogenide fibers are a kind of promising nonlinear materials due to its low intrinsic loss and high nonlinearity in MIR band [1].

The generation of the pair of signal and idler is controlled by the phase-matching condition. Only when the signal and idler wavelengths together with the pump wavelength satisfy the phase-matching condition, the signal and idler can be efficiently created. In order to meet the phase-matching condition, the pump wavelength should be located near the zero-dispersion wavelength (ZDW) of the used fiber. However, the ZDW of bulk chalcogenide glass is around 5 μm or even longer and it is a key problem to shift the ZDW to 2 μm or even shorter. The suspended core structure is a feasible way to realize wide band ZDW shift [2]. Recently, thulium-doped fiber lasers (TDFLs) provide high-quality lasing from 1.8 to 2.1 μm with high average power and high slope efficiency [3], which becomes a promising choice to serve as the pump of a chalcogenide FOPO. In this paper, we proposed a MIR FOPO using a three-hole As_2S_5 suspended-core fiber (SCF) with a core diameter of 2.6 μm as the gain medium. A tuning range of 2-5 μm can be achieved by tuning the TDFL from 1.9 to 2 μm.

2. Simulation Model

Figure 1(a) shows the dispersion curve for the bulk chalcogenide glass and the SCF whose cross-section structure is shown in the inset of Figure 1(a). The dispersion curve is calculated using the material refractive index of As_2S_5 in [4]. Tuning the core diameter while the width of the bridge is fixed as 0.6 μm, one can find that the ZDW of the SCF will moves to shorter wavelength as the core diameter reduces. When the core diameter is set to 2.6 μm, a ZDW equal to 2 μm can be obtained. Such a SCF is suitable for a MIR FOPO pumped by a TDFL. Considering a degenerate four-wave mixing (DFWM) into the effective phase-matching equation, one can get the signal and idler wavelengths those satisfy the effective phase-matching condition when the pump power is 1 W, as shown in Figure 1(b). The idler wavelength can vary from 2 to 5 μm if a TDFL with a tuning range of 1.9-2 μm is used as the pump.

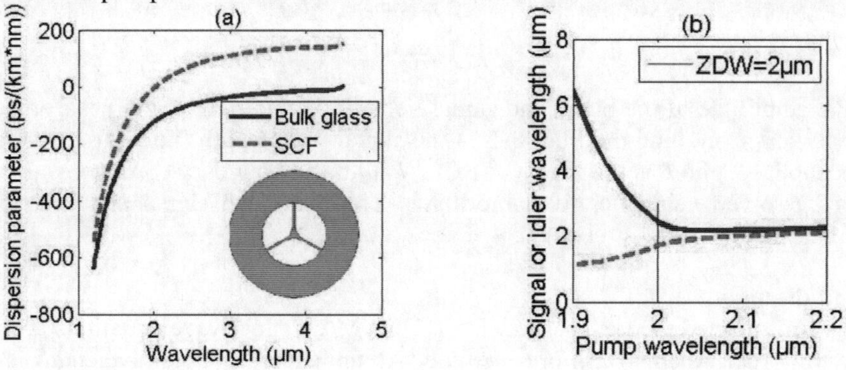

Fig. 1. (a) Dispersion performance and the designed SCF structure; (b) Phase-matched signal and idler wavelength as a function of the pump wavelength when pump power is 1 W.

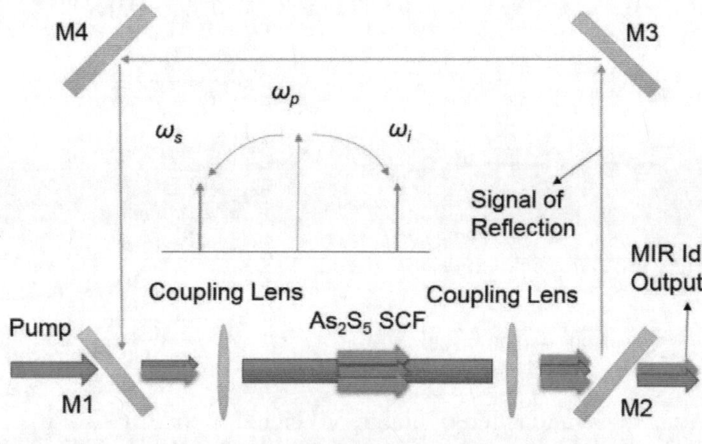

Fig. 2. FOPO structure.

The FOPO structure used for analysis is illustrated in Figure 2, a single resonant cavity is designed, which ensure the signal light resonant in the cavity and keep other lights pass the gain medium only once. The resonant cavity consists of four mirrors: M1, M2, M3 and M4. All of these mirrors are of high reflectivity at the signal wavelength. M1 is antireflective at the pump wavelength and M2 is antireflective at the idler wavelength. Using the coupled-wave equations for DFWM [5] and taking two-photon absorption (TPA) effect into consideration, the nonlinear parameter is expressed as

$$\gamma_j = \frac{n_2 \omega_j}{cA_{effj}} + i\frac{\beta_{TPAj}}{2A_{effj}} (j = p, s, i) \tag{1}$$

where n_2 is the nonlinear refractive index coefficient, c is the speed of light, $\gamma_{p,s,i}$, $\omega_{p,s,i}$, $\beta_{TPAp,s,i}$, $A_{effp,s,i}$ is the nonlinear parameter, the angular frequency, the TPA coefficient and the effective mode area for the pump, signal, and idler lights, respectively. A_{eff} is the effective mode area defined as:

$$A_{eff} = \frac{\left[\iint |F(x,y)|^2 \, dxdy\right]^2}{\iint |F(x,y)|^4 \, dxdy}. \tag{2}$$

where $F(x, y)$ is the normalized mode distribution.

Assuming that the pump is provided by a nanosecond pulse laser, it can be considered as a quasi-CW pump and the time derivative in the coupled-wave equation can be ignored. The cavity length can be tuned by moving the two mirrors M3 and M4 to ensure the synchronization between the pump and the reflected signal light. Therefore, the phase of the signal at the beginning of each round trip is reasonable to be set as 0. For such a FOPO configuration, the boundary conditions for the amplitude of each interacted wave is expressed as

$$\begin{aligned} A_s(z=0,t) &= \sqrt{1-\alpha_{linear}}\,|A_s(z=L,t)| \\ A_p(z=0,t) &= \sqrt{P_p} \end{aligned} \quad (3)$$

Here, $A_{p,s}$ is the amplitude of the pump and signal light, P_p is the input pump peak power, α_{linear} is the additional cavity loss excluding the fiber loss at the signal wavelength (i.e., introduced by the mirrors and the fiber coupling), and L is the length of SCF. With these boundary conditions, the coupled-wave equations can be solved using iterative algorithms. The generated signal and idler powers can be numerically obtained.

3. Results and discussion

Using the above-mentioned theoretical model, the DFWM coupled-wave equations are solved using a home-built fourth-order Adams-Bashforth method. Assuming the pump wavelength is 1.95 μm, and the phase-matched signal and idler wavelengths are calculated to be 1.318 and 3.749 μm, respectively. Taking this case as an example, the performance of the MIR FOPO is analyzed. The numerical parameters used in the simulation are as follows: $n_2 = 0.54 \times 10^{-17}$, $A_{effs} = 4.308$ μm^2, $A_{effp} = 4.646$ μm^2, $A_{effi} = 6.113$ μm^2, $\beta_{TPAs} = 0.102$ cm/GW, $\beta_{TPAp} = 0.0072$ cm/GW, $\beta_{TPAi} = 0.00005$ cm/GW, $\alpha_s = \alpha_p = \alpha_i =$ 1 dB/m, $\alpha_{linear} = 50\%$, and $\Delta\beta = -2\gamma_p P_p$.

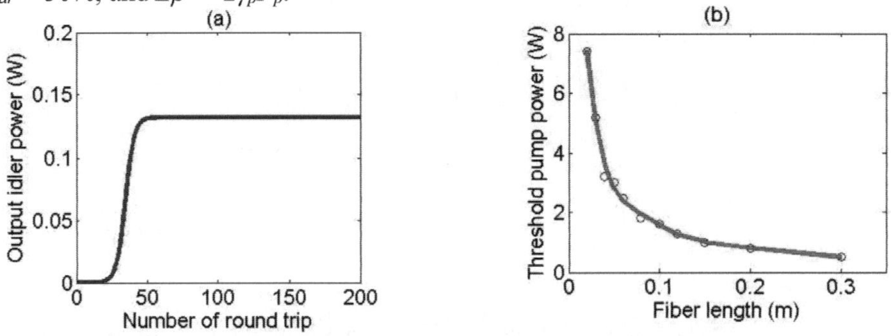

Fig. 3. (a) Output idler power as a function of number of signal round trip for Pp = 1 W and L = 0.2 m; (b) Threshold pump power as a function of fiber length.

To verify the stability of the model, the number of cycles for oscillation is simulated, as shown in Figure 3(a). It is shown that the signal eventually reaches stable oscillation with the MIR output of 0.131 W for $P_p = 1$ W and $L = 0.2$ m. Figure 3(b) shows the pump threshold of the FOPO as a function of the fiber length. Increasing the fiber length from 0.02 to 0.3 m, the threshold pump power rapidly decreases from 7.4 to 0.5 W. It means that the requirement to the pump power will be reduced if a longer SCF is used.

In conclusion, a MIR FOPO is suggested utilizing a three-hole As_2S_5 SCF with a core diameter of 2.6 μm and pumping by a nanosecond TDFL with output wavelength around 1.95 μm. The idler wavelength can cover the wavelength band of 2-5 μm if the pump wavelength is tuned from 1.8 to 2 μm and the pump threshold can be as low as 0.5 W.

Acknowledgment

This work was supported by the Zhejiang Provincial Natural Science Foundation of China (Grant No. LY14F050006), the China Postdoctoral Science Foundation (Grant No. 2014M551730), and the Fundamental Research Funds of Zhejiang University.

References

[1] I. Savelii, O. Mouawad, J. Fatome, B. Kibler, F. Désévédavy, G. Gadret, J-C Jules, P-Y Bony, H. Kawashima, W. Gao, T. Kohoutek, T. Suzuki, Y. Ohishi and F. Smektala 2012 *Opt. Express* **20** 27083.

[2] L. Dong, B. K. Thomas and L. Fu 2008 *Opt. Express* **16** 16423.

[3] T. S. Mccomb, R. A. Sims, C. C. C. Willis, P. Kadwani, V. Sudesh, L. Shah and M. Richardson 2010 *App. Opt.* **49** 6236.

[4] T. Cheng, R. Usaki, Z. Duan, W. Gao, D. Deng, M. Liao, Y. Kanou, M. Matsumoto, T. Misumi, T. Suzuki and Y. Ohishi 2014 *Opt. Express* **22** 3740.

[5] H. Cheng, Z. Luo, C. Ye, Y. Huang, C. Liu and Z. Cai 2013 *App. Opt.* **52** 525.

Journal of Physics: Conference Series **679** (2016) 012018 doi:10.1088/1742-6596/679/1/012018

The Concept of Photonics-Based Virtual Ground Tracking Station

Y Liu, B Cong, Y M Nie, J He and X Q Wang

China Satellite Maritime Tracking and Control Department, Jiangyin, China ,myflyingheart1976@hotmail.com

Abstract. In this paper, we propose the concept of virtual ground tracking station for space missions. Based on microwave photonics, the virtual tracking station can realize spatial diversity, antenna arraying, dynamic resource allocation and distributed signal processing. Compared with conventional design, the flexibility, efficiency and performance can be significantly improved.

1. Introduction

For long time application of ground tracking station in our country is the single station mode. That is a firm mode using multiband radio space TT&C equipment such as S C X Ka. All single station equipments must have Uplink and Downlink channel, terminal unit of signal and data. It needs not only plenty of money human resources and maintenance cost but also more modification investments to meet demand of growing techniques.

Even under such background of high investment, the whole coverage rate by ground tracking station is just 20%～30%, far behind rate of near 90% by space tracking station. So a new plan of big center small station has been put forward ten year ago. Its essential meaning is to move terminal unit of signal and data from station to center and station is just served as transparent channel of microwave[1]. Such layout not only makes the new mode of unmanned remote control settled, but also realizes unified processing information by same center without intermediate links in order to enhance the effectiveness highly.

This paper develops a new approach to the plan of big center small station that using photonics technique to construct a total new ground tracking mode named virtual TT&C which breaks the boundary between stations and centers. By the reflexive configuration between not the virtual station but the virtual point and the physical center, the new approach will make ground tracking improved toward higher level of distributed collaborativeTT&C.

2. General Design Considerations of Virtual TT&C

Please follow these instructions as carefully as possible so all articles within a conference have the same style to the title page. This paragraph follows a section title so it should not be indented.

2.1. Basic Concept Of Virtual TT&C

In this scheme, tracking equipments can be dynamically allocated, combined, networked and operate in a distributed/cooperative fashion. The basis for achieving these special characteristics is constructing the intensive optical networks between base points and centers. For example, 200～300 base points which are non-substantieve fiber optical node in the whole local area of our country is being built and 5 existing space tracking control center which serve as substantieve entity will be kept to play a role of switching center in the whole network.

Journal of Physics: Conference Series **679** (2016) 012018 doi:10.1088/1742-6596/679/1/012018

In the new design, existing ground tracking stations are connected with optical fibres， and all connected with the control centre. Every antenna modulates its received signal on the optical fiber[2] and transferred to the control centre where signal processing and data processing are implemented. In addition, control center must have plenty of front-end vehicular equipments which keep only a small quantity of radio frequency(RF) unit such as Antenna Servo Feed System, Low Noise Amplifier Power Amplifier. According to the principal of Radio Over Fiber system microwave signal is transimtted by optical fiber under the direct modulation of optical signal carrier by photo-electric and electric-photo conversion[2]. The concept of ground station is virtualized by such a function reconfiguration so as to name virtual TT&C.The diagram of virtual tracking system is shown in Figure.1.

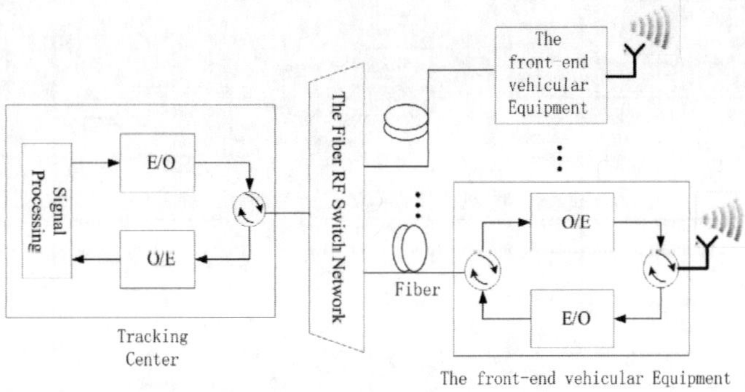

Figure 1. Principle of inveracious TT&C system

2.2. Technical Advantage

There are two technical advantage of virtual TT&C. One is optimal allocation of physical factors in ground tracking resources by dynamic allocating position, quantity and specific type of front-end vehicular equipments based on its own features of aircraft, just like flight characteristics, load feature, frequency band and modulation features, etc. The second advantage is downsizing staff massively so it makes human factors in ground tracking resources optimate too.

2.3. Design Considerations

Downlink channel of front-end vehicular equipments are composed of the antenna, the source feed network, Low Noise Amplifier and the optical transmitter. Solid integrated connection is appropriate to the source feed network and Low Noise Amplifier to reduce insertion loss. There are three key links which is the optical transmitter, the fiber slip-ring and the fiber RF switch network on the whole system design.

For the optical transmitter which refers to electric-photo conversion unit(E/O), although common intensity modulated direct detection（IM/DD）is simple, but the influence of chirp effects on the RF signal phase is severe for F-P cavity has more longitudianl-mode laser, the wider spectral bandwidth, and the larger injection current. So IM/DD is unfit for long-distance transmission which exeeds 500km from base points and centers. On the other hand, virtual TT&C must alter from point-to-point into optical interconnection(EPON or GPON) which needs plenty of optical cross-connection in the intermediate switching node. It will changes bandwidth great, therefore orienting the future external modulation is suitable. Single longitudinal mode laser（SLM）plus Mach-Zehnder mode filter(MZM) might be considered. Planar waveguide integrated electro absorption modulated laser (EML) is recommended for the reason that EML has become commercialized, higher Side-Mode Suppression Ratio by engraving the Prague grating and the electric absorption waveguide on the same substrate. In addition, this structure can be no need for MZM which benefits for installation on the antenna front-end as a result of its smaller sized.

As the E/O covernsion unit installed on the antenna front-end and together rotating with the antenna, simultaneously the location of the fiber optic connector terminal is fixed, there needs a fiber slip-ring to connect the antenna motion and static part. But The fiber coupling efficiency is greatly affected by the degree of alignment of two fiber end faces, so there is a very high requirement for the machining accuracy of the antenna structure. At present the more mature solution is using a self focusing lens to convert the conical light into the parallel light to increase the coupling efficiency. There are documents presented using Finel lens to produce multiple reflection optical path between the axis and off-axis to improve the coupling efficiency further[3].

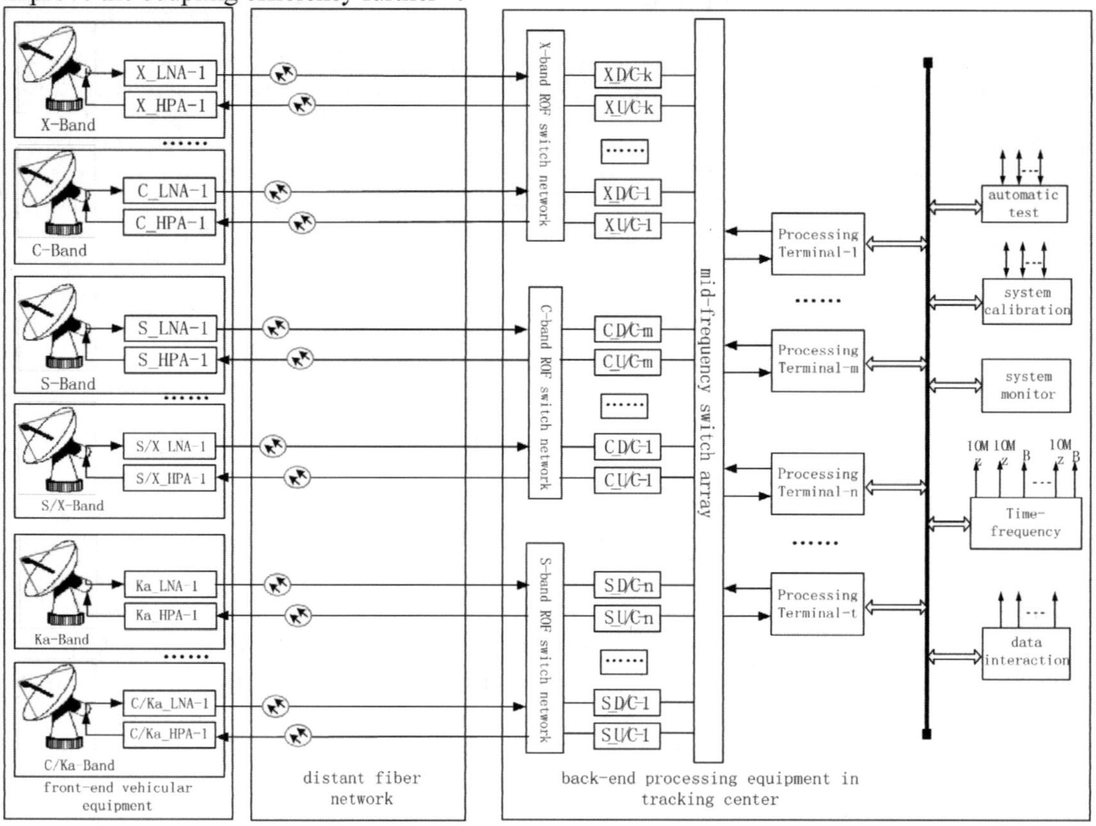

Figure 2. The typical composition frame of RF resources reconstruction system

As before said, the reflexive configuration by virtual TT&C is mainly reflected in the dynamic configuration and invocation between the front-end vehicular equipments and the back-end processing unit in the tracking center. This configuration and invocation is realized through the fiber RF switch network. As shown in Figure.2, the fiber RF switch network Directly connect the other end of the optical fiber, This means that its switching function can only be done in the optical domain. It is not complicated achieving it in the technology as long as the star optical coupler or a plurality of ordinary 2 x 2 directional coupler, a plurality of light switches can take a variety of complex optical switching array. The degree of commercialization of the above coupler and optical switch is very high. Need to focus on is the insertion loss, channel isolation and dispersion characteristics. As the ordinary double cone star coupler is limited by the light splitting ratio, its insertion loss and channel isolation is restricted each other that is not conducive to the realization of the scale of 20 x 20 optical switching network. So the array waveguide gratingstar coupler (AWG) which is the new integrated optical results may considered. As its input, output waveguide normal direction is directly to the other side of the phase center, the power of the free space coupling region can be well distributed to each branch, especially to the edge branch. So under the same insertion loss, the channel isolation degree can be improved by 10 dB.

The dispersion characteristics of the coupler will give a large phase and delay variation to the optical signal, so the timing precision of microwave signal will be influenced. It has little effect on the microwave signal of C and S band but it can not be ignored of Ka, X band. At present, the effective method is to use the Ka or X frequency conversion to C or S band to transmit. This, of course, is at the expense of increasing the complexity of the device.

3. Conclusion

Virtual TT&C proposed in this paper orienting in the new direction of "big center small station", breakthrough the current fixed station mode of construction bottleneck in accordance with the specific characteristics of the aircraft to dynamically set and control resources. On the overall design method, the microwave photonic technology is put forward, which is directly on the upper and lower link of the RF terminal to complete the microwave signal of the electro optic and photoelectric conversion. In detail design process it put forward three following measures.
(1) Single longitudinal mode modulation of microwave signal is achieved by using EML laser.
(2) Fiber slip rings is achieved by using the self focusing lens or lens Finel multire flection technique.
(3) AWG star coupler is used to realize optical signal switching array of 20 * 20 or above, and the effect of switching element dispersion on phase and delay is overcome by using the Ka or X frequency band to C or S band.

Reference

[1] Jiaxing L 2008 On Developing China's Ka-band Tracking Network *J. Telecommunication Engineering.* **48** 90
[2] Rong Y 2011 Optical Fiber Communications *Mechanical Engineering Press.*
[3] Dagong J , Hongxia Z and Wencai J 2004 Double channel rotating connector with passive optical structure *J. Optoelectronic Engineering.* **52** 17.

Journal of Physics: Conference Series **679** (2016) 012019 doi:10.1088/1742-6596/679/1/012019

Analysis on high-altitude earth Orbit Satellite Determination

J He, Y W Hou and L Yang

China Satellite Maritime Tracking and Control Department, Jiangyin, China ,icejialinhe@sina.com

Abstract. The difference is introduced between approx circular apogee orbit and approx circular perigee one by error transmitting at first. Then the characteristic of secant compensation is analysed when radar tracking object with high elevation. And two kinds of orbit force be pressed to, their perturbation influence and their earth-core angles are explained. And then the series of emulation results are shown including error data emulated with Monte Carlo method, the influence of the velocity increment from the ejecting force of spring while satellite-rocket separating and their perturbation influence and the length of influence of the data arc. Then decision analysis of Wald method and Bayesian statistics rule and the results from the two rule are introduced . So the suitable orbit determination decision is put forward from the decision method. Finally the result is tested reasonable and feasible via the real data. In the end it is useful to reference to make orbit decision in short injection of circular orbit far from the earth for calculating concurrently precise and timely.

1. Introduction

There are different results in same arc length of different kinds of orbit . Even to the same orbit there are different results in same arc length at different space position. So it is important to orbit that how long or how many points should be use. Also it is urgent to know that what precision should be. And a set of analysis method are explained in this paper.

2. Analysis of tracking characteristic

2.1. Analysis of error transmitting

It is obvious that different tracking equipments have different characteristic when tracking aircraft of space at different point. For example, when satellites are launched into schedule orbit at altitude about from 200 ~ 300 kilometers to 36000 kilometers the tracking distance changes from 300 ~ 3000 kilometers to 37000 ~ 40000 kilometers. And the location error gets obviously bigger, especially in REA single measuring frame the angle error alter great with larger rang.

$$\begin{bmatrix}\sigma_{x_s}^2\\ \sigma_{y_s}^2\\ \sigma_{z_s}^2\end{bmatrix}=\begin{bmatrix}\cos^2 E\cos^2 A\sigma_R^2+R^2\sin^2 E\cos^2 A\sigma_E^2+R^2\cos^2 E\sin^2 A\sigma_A^2\\ \sin^2 E\sigma_R^2+R^2\cos^2 E\sigma_E^2\\ \cos^2 E\sin^2 A\sigma_R^2+R^2\sin^2 E\sin^2 A\sigma_E^2+R^2\cos^2 E\cos^2 A\sigma_A^2\end{bmatrix}$$

$\sigma_R,\sigma_E,\sigma_A$ means error of measuring element,

$\sigma_{X_s},\sigma_{Y_s},\sigma_{Z_s}$ means location error of aircraft .

2.2. Analysis of tracking with high elevation

Journal of Physics: Conference Series **679** (2016) 012019 doi:10.1088/1742-6596/679/1/012019

When Radar tracking objective secant compensation[1] are added in azimuth spur track. When elevation is up to 70 degree ,the Radar 's capability comes down in a sort of way, such as the largening dynamic lag error , the stronger influence of ship sway, the bigger random error with blowing up yawp and much more area of no sensitivity in electrocircuit exist. When radar running in azimuth direction its balance adjustment becomes worse.

The influence of secant compensation not only aggravate the tracking balance but also make the remainder of error exaggerated when the radar is with high elevation. Even when elevation is up to 70 degree, the remainder of error exaggerated to 2.9 times refer to table 1.

Table 1. Value of SecE

E	45°	60°	70°	75°
SecE	1.414	2.00	2.924	5.759

3. Analysis of orbital characteristic

3.1. Analysis of error transmitting

A s a rule, the force element of the space craft is analysed firstly. Commonly, gravitation is the main force to all earth satellites. And other orbit perturbation force are used to get more accurate results of satellite of orbit determination, orbit prediction and orbit maneuver for researchers.

For example, the atmospheric drag is the most important force of orbit perturbation for low earth orbit (LEO) comparing to the force of the solar light pressure which is almost ignored .But for geosynchronous-orbit satellite (GEO) the influence of the force of the solar light pressure is so obviously mighty that it is never to be dispensable. At the same time the force of the atmospheric drag is infirm enough to be neglectable .

3.2. Analysis of the perturbation influence of the force

Secondly, figures are listed for explaining the difference of the orbital variety between approx circular apogee orbit and approx circular perigee one. The semi-major axis of orbit of approx circular apogee orbit varies less than that of approx circular perigee orbit. View fig.1, the change of semi-major axis of orbit of approx circular apogee orbit is about 4 kilometres while that of approx circular perigee orbit is up to 10 kilometres within 24 hours. It is shown in fig.2 that the alteration of orbit elements is very small within 80 seconds .

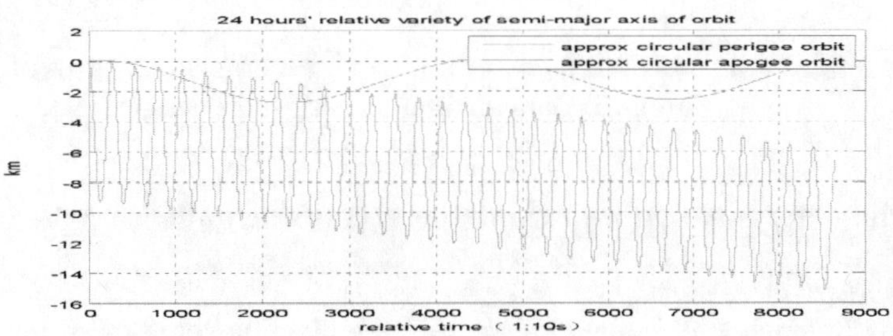

Figure 1. Variation of orbit determination within 24 hours

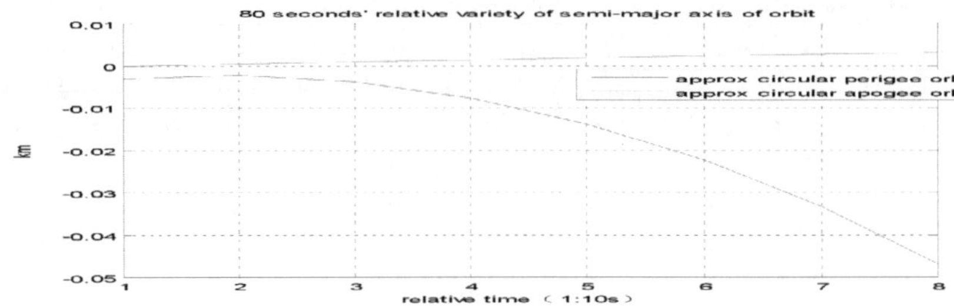

Figure 2. Variation of orbit determination within 80 second

3.3. Analysis of measuring data arc

The data with different earth-core angle is with difference influence on the orbit determination. Figure.3 is shown that there is big difference between approx circular apogee orbit and approx circular perigee one with data arc from 60 seconds to 600seconds.And the earth-core angle of approx circular perigee orbit is far away langer than that of approx circular apogee one in same hours.

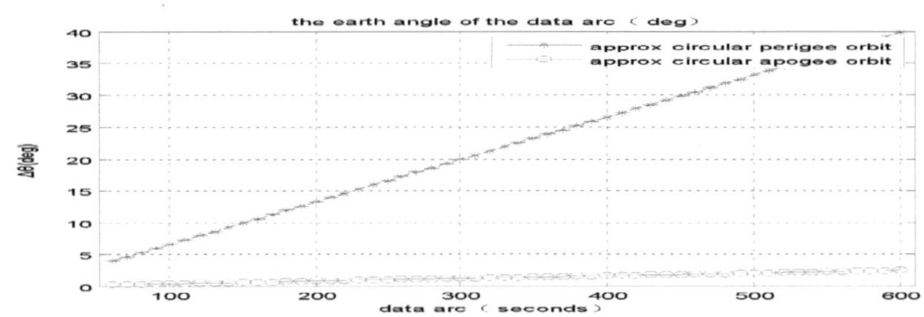

Figure 3. Data arc and its earth-core angle

4. Analysis of orbit determination with emulational tracking data

4.1. Analysis of velocity increment

If the satellite gets 0.5m/s velocity increment from the ejecting force of spring while satellite-rocket separating. It can change the orbit of the satellite. So it must be modified that when the results of orbit are calculated[2]. At the same time we can infer that the attitude data influences the value of orbit determination refers to the table 2 and the table 3.

Table 2. Attitude data in the correction of velocity increment

order	Elevation (°)	Yaw (°)	Scroll (°)
1	151.6	1.8	91.1
2	150.5	14.9	83.8
3	149.3	20.8	80.3
4	148.4	24.3	78.0

Table 3. Orbit determination value from velocity increment

order	Δa(km)	Δe（e-4）	Δi(°)	ΔΩ(°)	Δλ(°)
1	11.3619	-2.7	-0.00527	0.000967	0.000852
2	12.8367	-3.0	-0.00342	0.000628	0.000108
3	13.2882	-3.2	-0.00251	0.00046	-5.1E-05
4	13.4954	-3.2	-0.00193	0.000355	-9.3E-05

4.2. Analysis of measuring data added error

Fifty groups of 80s observation data added errors are emulated from the standard orbit. Errors are created with the method of Monte Carlo. Then data are used to determine the results of the orbit. Finally the difference between the standard orbit and the results from the emulational data are listed in the table 4.

Table 4. Deviation of different emulational data

	Δ a(km)	Δ e（10-04）	Δ i(°)	Δ Ω(°)	Δ λ(°)
Mean	0.6	-0.180	0.015722	-0.006259	-0.03034
variance	28.6	6.789	0.030131	0.005635	0.006559

4.3. Analysis of measuring data arc

Data arc of observation from 120s to 900s with errors are emulated. And every data arc includes fifty groups refer to 4.2. Then the trend of data is shown in figure 4 and figure 5.

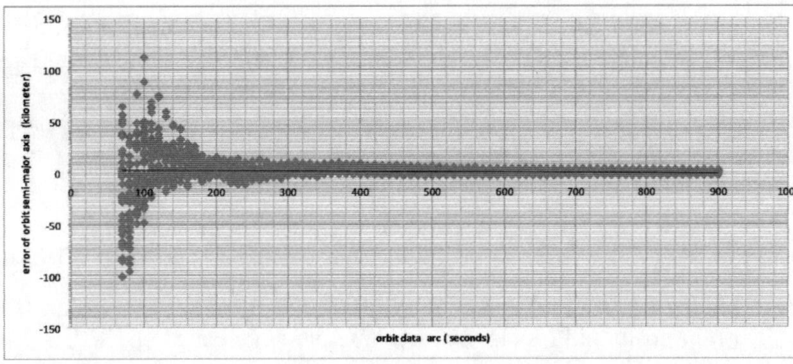

Figure 4. Difference of orbit determination with different data arc

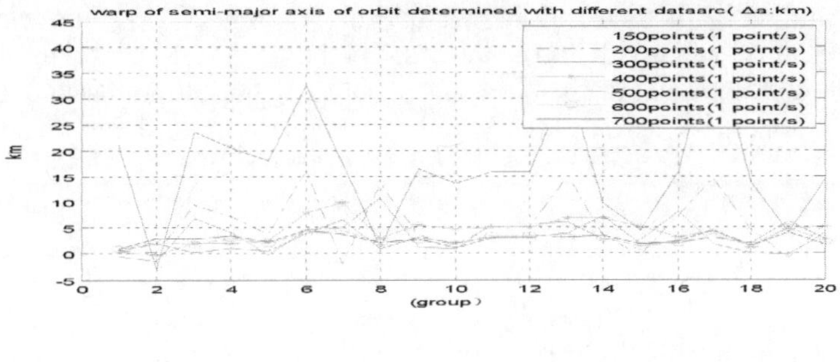

Figure 5. Difference of orbit determination with different data arc

Journal of Physics: Conference Series **679** (2016) 012019 doi:10.1088/1742-6596/679/1/012019

5. Analysis of decision [3]

5.1. Decision

Decision means the passing of judgment on an issue under consideration before it occurs. Decision can be classified into definite model and probability model.

5.2. Wald method

Wald method is a rule of selecting the good from the bad. It is also called Max-min rule .Under the rule the decision-maker will pick out the proceeding with the maximal income from the least. Premising that there are m kinds of spare plans and n sorts of nature the value of profit and loss is F_{ij} . And $F_{\min}(a_i)$ stands for the least profit with the plan of i and the nature of a_i .

$$F_{\min}(a_i) = \min_{\theta_j \in \Theta} F_{ij} = \min[\, F_{i1}, F_{i2}, \cdots, F_{in}]$$

If F^* is the profit of optimization plan it can be

$$F^* = \max_{a_i \in A} F_{\min}(a_i)$$

$$= \max_{a_i \in A} [F_{\min}(a_1), F_{\min}(a_2), \cdots, F_{\min}(a_m)]$$

The plan corresponding F^* is the best choice according as the Wald rule .
If the function of profit and loss means the loss, the model is

$$F_{\max}(a_i) = \max_{\theta_j \in \Theta} F_{ij} = \max[F_{i1}, F_{i2}, \cdots, F_{in}] , \quad F^* = \min_{a_i \in A} F_{\max}(a_1) = \min[F_{\max}(a_1), F_{\max}(a_2), \cdots, F_{\max}(a_m)] .$$

5.3. Bayes method

Bayes rules mean every nature will be assign certain probability to show its generant possibility. The Value of expectation is standard to all the possible results. A nd its weight is evaluated according to its probability.

First ,the expected profit and loss is computed under every action, called $E(a_i)(i = 1, 2, \cdots, m)$, so

$$E(a_i) = \sum_{j=1}^{n} F_{ij} P(\theta_j)(j = 1, 2, \cdots, n)$$

As to the Wald method F_{ij} is the value of profit and loss .And $P(\theta_j)$ is the probability of the nature of θ_j . So there are m kinds of profit and loss if there are m kinds of preparing plan for selecting. So the plan with the max or min profit and loss is the recommended one based on the Bayes rule.

$$E^* = \max_{a_i \in A} E(a_i) , \quad E^* = \min_{a_i \in A} E(a_i) .$$

5.4. Decision examples

There are eight kinds of spare plan for selecting with nine orbit error zone in one plan. 'a1'is the result of orbit with 120 points, 'a2' is the result of orbit with 150 points,' a3' is the result of orbit with 200 points, 'a4' is the result of orbit with 300 points, 'a5' is the result of orbit with 350 points, 'a6' is the result of orbit with 400 points , 'a7' is the result of orbit with 500 points, 'a8' is the result of orbit with 600 points. The zone from zero to 45 degree is divided into nine, shown in table 5.

Table 5. Orbit determination arc desicion

a_i	Nature (θj)								
	5	10	15	20	25	30	35	40	45
1	0.25	0.1	0	0.05	0.25	0.15	0.1	0	0.1
2	0.2	0.05	0.15	0.3	0.15	0	0.1	0	0.05
3	0.5	0.3	0.1	0.1	0	0	0	0	0
4	0.65	0.3	0.05	0	0	0	0	0	0
5	0.45	0.55	0	0	0	0	0	0	0
6	0.65	0.35	0	0	0	0	0	0	0
7	0.85	0.15	0	0	0	0	0	0	0
8	1.00	0	0	0	0	0	0	0	0

Wald rule ,

$$F^* = \min[F_{\max}(a_1), F_{\max}(a_2), \cdots, F_{\max}(a_m)]$$
$$= \min_{a_i \in A} F_{\max}(a_8) = 5(km)$$

Bayes rule ,

$$E^* = \min_{a_i \in A} E(a_i) = 5(km)$$

Based on the rule of Wald or Bayes, the plan 'a8' is the recommended one.

But if we want to get the result with both suitable precision and the shortest data arc another plan will be the best. So we suppose the warp of 10 kilometers is the suitable precision , the effect is 0.8 when the time longer than 300 seconds because that will effect at a discount and the time less than 300 seconds is perfectible 1. Results of profit and loss is shown in table 6.From the table 6 the plan 'a5' is the most suitable one .

Table 6. Results of profit and loss

ai	1	2	3	4	5	6	7	8
$F_{\max}(a_i)$	45	45	20	15	10	10	10	5
$E(a_i)$	17.5	19.3	9	7	7.8	6.8	5.8	5
assignment	0	0	0	1	1	1	1	1
effect	1	1	1	0.8	0.8	0.8	0.7	0.7

It is tested that the plan 'a5' is the most suitable one with real data. The error of orbit determination is 5 kilometers or so.

Reference

[1] En-dian H, Pei L and Rui C 2011 The Indetermination Factor Analysis of High Elevation Arc Segment Radar Traching *J. Science Technology and Engineering*. (Bei jing) **11** 1671

[2] Jing H, Jin-biao Z, Yang L and Zheng-xu D 2012 Solution to problems of jumping dots in radar′ s digital guidance *J. Telecommunication Engineering*. (Cheng du) **52** 323

[3] Ying L 2005 *Statistical Analysis of Decision* (Tian Jin:tianjin university) chapter 4 pp 88–95

Journal of Physics: Conference Series **679** (2016) 012020 doi:10.1088/1742-6596/679/1/012020

A Broadcast Channel Assignment Mechanism based on The Broadcast Tree for Multi-radio Multi-channel Wireless Mesh Networks

Yan Xia[1] and Yingzhi Zeng[2]

[1]College of Computer Science and Electronics Engineering, Hunan University, ChangSha, China
[2]Information Center, National University of Defense Technology, ChangSha, China
E-mail: xiayan@hnu.edu.cn

Abstract. This paper proposed a broadcast Channel Assignment Mechanism on base of optimized Broadcast Tree for wireless Mesh network (WMN), which is created by Branch and Bound Method. The simulations show that our algorithm not only reduces the broadcast redundancy but also avoids the potential channel interferences produced by unnecessary relay nodes.

1. Introduction

Broadcast communication is widely used in many applications of wireless Mesh network (WMN), such as control instructions and data release. How to allocate the appropriate channel for the Mesh network nodes in the Broadcast communication is an important issue in the research of the WMN channel allocation technology.

Broadcast is a kind of communication mode which is widely used in network protocols. The original working characteristics of wireless broadcast communication can also provide support for many applications of WMN. If all nodes in WMN use the same channel, the data packets only need to be received by all the other neighbor nodes, so the broadcast protocol design for the single channel WMN only needs to take node as the center. However, the whole network uses a single channel radio is bound to produce channel interference. A certain number of nodes will be unable to properly receive the broadcast message due to channel interference. Re-broadcasting will increase the network communication overhead.

In multi-radio multi-channel WMN, wireless radio can use different channels while mesh nodes can transmit data with multiple wireless radios at the same time. Two nodes in the interference range can also use different channels to broadcast. If all the wireless radios of a mesh node use different channels and does not carry out the channel adjustment for broadcast communication, this mesh node have to broadcast data communication in multiple channels to transmit packets to all of its neighbors. This will undoubtedly weaken the effective use of the broadcast transmission characteristics and the wireless broadcast advantage (WBA) characteristics of the wireless transmission medium in a certain extent.

Due to the limitations on broadcast redundancy, channel interference and impossibility in reaching largest coverage in relay nodes selections in existing approaches based on leveled channel distribution algorithms, we did not follow the common method to level and then seek relay nodes, but provided a new algorithm for optimize Broadcast Tree based on Branch and Bound Method, and then fulfilled the optimization in selection of relay nodes via the optimization and pruning for Broadcast Tree, and thus completed the largest coverage. Then we designed a broadcast Channel Assignment Mechanism with low rebundancy on base of optimized Broadcast Tree. It not only reduces the broadcast redundancy but also avoids the potential channel interferences produced by unnecessary relay nodes.

2. Related work

The research of multi-channel technology from the perspective of multicast and broadcast in WMN is still in the initial stage. There are some references to the broadcast problem in wireless Mesh networks, but most of them are only limited to the single channel, or based on the premise that the channel allocation is known.

Many researchers have studied the low redundancy broadcast mechanism for the single channel network environment. These methods mainly include the probability method [1], the method based on the cover [2], the cluster method [3] and the method based on connected dominating set [4,5,6].

The simplest way of broadcasting is flood. Flood is not required to maintain the topological structure of the network and the related routing calculation. Only the nodes that receive the information are requested to transmit the data to their neighbor nodes by broadcast. Paper [7] points out that the cost of flooding is high, and it will lead to the broadcast storm problem, that is, the serious redundancy, channel competition and conflict. The experimental results show that the channel allocation strategy has a great influence on network performance.

There are some studies on how to design a low redundancy and multi-channel broadcast in the given network channel allocation. P.Kyasanur proposed multi-channel flooding broadcast [8]: when the node first received a broadcast message, the data copies of the broadcast message will be rebroadcast in each channel queue. This kind of multi-channel flooding broadcast will undoubtedly lead to new forms of broadcast storm, the redundancy, conflict and competition of a large amount of broadcast data packets are inevitable.

Qin Shaohua proposed a distributed broadcast tree construction algorithm [9], which is simple and easy to implement. However, this paper mainly studies the construction of the broadcast tree. It does not combine multi-channel assignment with broadcasting. This paper also does not make full use of the advantages of a neighbor node that uses the same channel in wireless communication to receive broadcast messages at the same time.

Guokai Zeng proposed Level (Level-CA) based on the hierarchical channel assignment algorithm [10]. This algorithm is based on the breadth first search strategy, and the Level-CA hierarchy is the most classical method of building the broadcast tree. The drawback of Level-CA is that the channel interference of the same level nodes may be formed, and the available multi-channel resources cannot be fully utilized.

3. A New Broadcast Channel Assignment Mechanism

3.1. Problem Analysis

According to the network environment of multi-radio and multi-channel, the corresponding communication broadcast tree is created in this paper, and the appropriate relay nodes are optimized, and the redundancy broadcast communication is reduced. Based on the optimized communication broadcast tree, the corresponding channel assignment scheme is designed for the multi-radio broadcast tree node and relay node.

The broadcast source node S and the whole network nodes constitute an undirected connected graph G= (V, E, Ch_S).V is a set of nodes, E is a collection of all links, and Ch_S is the available channel numbers of the connected dominating set G.

G= (V, E) is given, if set S satisfies the following two conditions: $S \subseteq V$ and $S \neq \phi$, if $\forall x \in (V - S)$, the S is the dominating set of G,S=DS (G). If node $x \subseteq S$, or one neighbor node of x belongs to the dominant set S, it is known that the node x is dominated by the S cover. If the S=DS (G), the sub graph derived from S is a connected graph. There exists a path between each pair of nodes from S, such as node x and y, then the dominating set S is a connected dominating set (CDS). If the connected dominating set has the least number of nodes, it can be called the minimal connected dominating set MCDS. In this section, the WMN is abstracted into a connected undirected graph, and the optimization problem of multi hop broadcast tree in WMN is summarized as a classic MCDS (Minimum Connected Dominating Set) problem.

Taking into account the establishment of each edge in the graph G contains two nodes on a link should be adjusted to the same communication channel at the same time, the minimum connected dominating set problem is also equivalent to an edge coloring problem. In this paper, we study the combinatorial optimization problem of minimum connected dominating set and channel assignment problem. As we all know, the minimal connected dominating set problem and the edge coloring problem have been proved to be NP complete problem in graph theory.

3.2. Definitions

Through our in-depth analysis, the problem is summarized as follows: find a MCDS, based on this MCDS, the broadcast source node can quickly reach all nodes in the network, and ensure that the entire network of radio and communication overhead is minimal. Therefore, from any one of the broadcast source node s, some neighbor nodes of s, should be chosen to act as the relay nodes. After assigned different channels, those relay nodes continue to broadcast until the original broadcast messages reach all nodes in the network G.

AA(**A**ctive **A**rray): the current working state for the active node set, the array initialization phase is empty, that is $\{\phi\}$.

UA(**U**nActive **A**rray): the current working state for the non-active node set, the array initialization phase is empty, that is $\{\phi\}$.

DSA(**D**ominating **S**et **A**rray): DSA (Set Array Dominating), the active node set (dominating set) with the current working state is broadcast or re-broadcast(relay), in the initialization stage there is only broadcast source node S, {S}.

Related attributes Euv: if the two adjacent Mesh nodes v and u are both in DSA, then the link Luv between v and u should in DS, Euv=1, otherwise Euv=0.

Taking into account the broadcast characteristics of multi-channel allocation, one-hop neighbor node coverage set of nodes V is divided into three kinds, N_1(v)， N_2(v)和 N_3(v).

One-hop neighbor nodes of node v and node v consists of a collection of N[v]， $\mathrm{N[v]} = \mathrm{N(v)} \cup \{\mathrm{v}\}$.

One-hop neighbor nodes group of node v is N_1(v), $N_1(v) = \{u \in N(v) \mid N(u) \backslash N[v] \neq \emptyset\}$.

One-hop neighbor nodes table of node v is N_2(v), $N_2(v) = \{u \in N(v) \backslash N_1(v) \mid N(u) \cap N_1(v) \neq \emptyset\}$.

One-hop leaf nodes table of node v is N_3(v): $N_3(v) = \{N(v) - (N_1(v) \cup N_2(v))\}$.

The following conclusions can be deduced from the definition given above, $\mathrm{N(v)} = N_1(v) + N_2(v) + N_3(v)$.

The roles of these nodes in broadcast communication are different, and the status of the nodes are also different. In the forwarding process of the broadcast message, the nodes in $N_1(v)$ will act as relay nodes, so they should be attached to DSA. The nodes in $N_2(v)$ do not have to be involved in the relay

Journal of Physics: Conference Series **679** (2016) 012020 doi:10.1088/1742-6596/679/1/012020

operation, which can be used as the receiving nodes of the broadcast message which have been broadcasted by member nodes of $N_1(v)$.Acting as the leaf nodes, the nodes in $N_3(v)$ would be the receiving nodes of the broadcast message and also do not have to be involved in the relay operation.

3.3. Method Description

The branch and bound method is proposed by Richard Kapp. It is a widely used algorithm, and its basic idea is to search for all feasible solutions (limited) space of constrained optimization problem. In general, the optimal solution can be obtained by using the branch and bound method.

Any one broadcast source node, first of all, should select the relay nodes from its neighbor nodes N (s).After allocating appropriate communication channels, those relay nodes continue to broadcast, until the broadcast message reaches all nodes in the network G. In this process, two key issues need to be solved: the selection of relay node and channel assignment. We use the branch and bound method to optimize the channel allocation. We design a broadcast tree creation algorithm based on branch and bound method, and design the channel assignment mechanism in the construction process of broadcast tree. Our new algorithm, called BTCAM, is the abbreviation for Branch and Bound Method Broadcast Tree based Channel Assignment Mechanism.

The problem can be described as: **CA-broadcast**=max{cx: x∈S}.The problem is divided into a plurality of branch, and the union of all branches is S. For any i,j=1,...k， when i≠j, $S^i \cap S^j=\{0\}$。

In section 3.2, the neighbor nodes V is divided into three types of roles. According to the different roles of the node, the corresponding branch is carried out. This can simplify the number of branches, especially for dense network topology.

The current problem solution Sv can be divided for $Sv=Sv^1 \cup Sv^2 \cup Sv^3$, which Sv^1 is solution set for $N_1(v) = \{u \in N(v) \mid N(u) \backslash N[v] \neq \emptyset\}$, Sv^2 is solution set for $N_2(v) = \{u \in N(v) \backslash N_1(v) \mid N(u) \cap N_1(v) \neq \emptyset\}$, Sv^3 is solution set for $N_3(v) = \{N(v) - (N_1(v) \cup N_2(v))\}$. The sub problems are satisfied the following relations: $Sv^1 \cap Sv^2=\{0\}$, $Sv^1 \cap Sv^3=\{0\}$, $Sv^2 \cap Sv^3=\{0\}$.

According to the branch and bound method, the pruning strategy is described as follows.

Theorem Pruning Strategy. As a branch divided in the broadcasting process, If Sv^i satisfies any of the following three conditions, it will be pruning.

(1) Not feasible: Sv^i solution is empty.
(2) Optimality: an optimal solution has been found for the current sub problem CA-broadcast(v);
(3) Optimal value comparison: a solution for the sub problem CA-broadcast (v) has been included in some existing solution.

As the basic thought of the new Mechanism, the branch and bound method is used to realize the broadcast tree pruning and optimization, in order to reduce the number of relay nodes and reduce the broadcast redundancy and reduce channel interference in the neighbor broadcast communication links.

3.4. Construction of Broadcast tree and Assignment of channels

The broadcast source node S is the execution node of the corresponding algorithm of BTCAM. After determining the DSA member, the broadcast tree can be created in the next stage. Based on the broadcast tree, the channel assignment of the transmission link can be carried out in the third stage.

The input part of the BTCAM algorithm includes the following:

(1) G={V， E}, V is a set of nodes, E is a collection of all links.
(2) Each node maintains a neighbor information within two hops.
(3) The broadcast source node S.

The output part of the BTCAM algorithm includes the active node set {DSA}.

Due to the space, the algorithm details are omitted. The main steps of BTCAM are described as follows.

- Initialization of active node sets of work state, initial AA={S}；

- Initialization of active node set (dominating set) with the current working state is broadcast or re-broadcast(relay), initial DSA={Ø};
- Initialization of the non-active node set, initial UA={Ø};
- Sets the relay identifier bit for all the nodes to 0, initial Flag=0;
- By the topology discovery and HELLO communication, the adjacent nodes can establish respectively N[v]、N1(v)、N2(v)、N3(v);
- Add a new DSA node x, add the node's neighbor table N (x) to join the AA set;
- According to the situation of neighbor table coverage and pruning strategy in section 3.3, increase the members of new DSA and UA members;
- Set the relay identifier bit of the relay node Flag=1;
- After executing the channel assignment function Assign-BC(Broadcasting node ID, receiving nodes list), the relay node will be assigned its broadcasting channel;
- The node of N2(x) executes the channel assignment function Assign-BC(Broadcasting node ID, receiving nodes list). Add N2(x) into UA set.
- The node of N3(x) executes the channel assignment function Assign-BC(Broadcasting node ID, receiving nodes list).Add N3(x) into UA set.
- All nodes in figure G, either in DSA, or in UA.

4. Simulations

Based on multi-radio multi-channel WMN simulation platform, we build the experimental environment. In the experiment, the NS2 platform is used to set up the wireless Mesh network (900× 900 m^2). The Mac layer protocol used in the simulation is 802.11a, the number of non superimposed channels is 12, and the channel bandwidth is 54Mbps. Network randomly generated multiple independent constant rate data stream (CBR).

Considering the structure of the algorithm, the Level-CA algorithm and BTCAM algorithm are compared from three major parameters. These parameters are network throughput performance under different available channel numbers, Broadcast delay under different node numbers and the proportion of the broadcast nodes to the whole network nodes under different node numbers.

As shown in figure1 to 3, experimental results show that the BTCAM algorithm has better performance than the Level-CA algorithm.

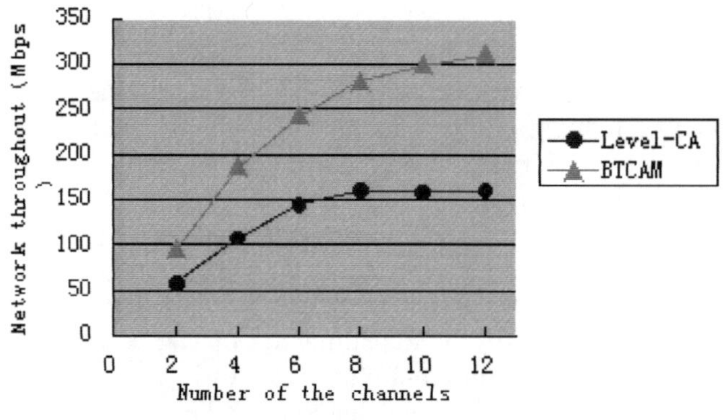

Fig. 1. Comparison of network throughput under different available channels

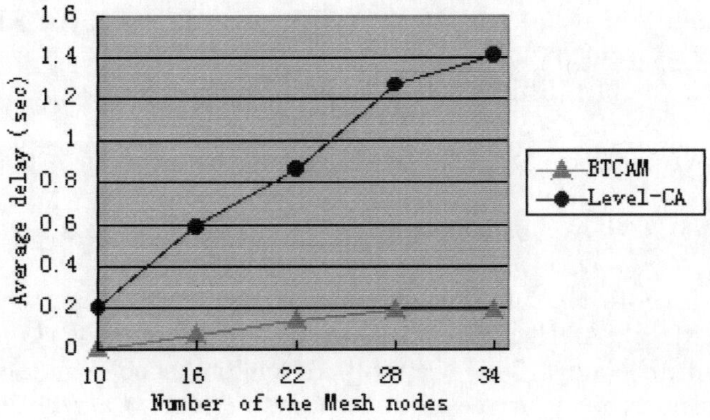

Fig. 2. Comparison of delay under different Numbers of Mesh Nodes

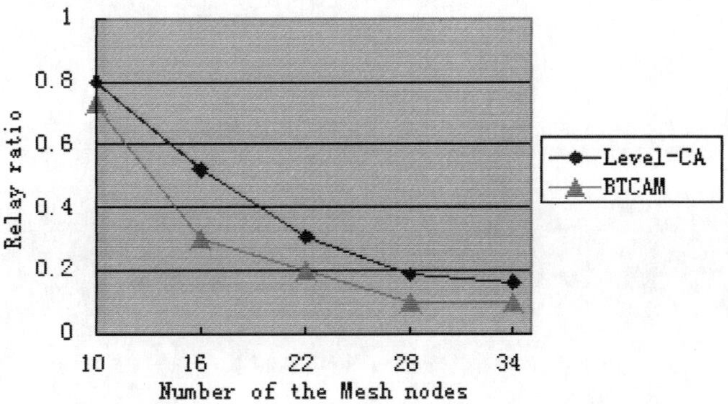

Fig. 3.Transmission ratio contrast under different Numbers of Mesh Nodes

5. Conclusions

Due to the limitations on broadcast redundancy, channel interference and impossibility in reaching largest coverage in relay nodes selections in existing approaches, this paper proposed a new algorithm for optimize Broadcast Tree based on Branch and Bound Method, and then fulfilled the optimization in selection of relay nodes via the optimization and pruning for Broadcast Tree, and thus completed the largest coverage. Based on the optimized Broadcast Tree, a broadcast Channel Assignment Mechanism called BTCAM is designed, which is not only reduces the broadcast redundancy but also avoids the potential channel interferences produced by unnecessary relay nodes. The simulations show that BTCAM can acquire fine network capacity in WMN with different network environment and with different number of available channels.

Acknowledgment. We would like to thank anonymous reviewers for their comments. This work was supported by Young teacher development program of Hunan University and the National Natural Science Foundation of China under Grant No.61170286 and No.61202486.

References

[1] Brad Williams and Tracy Camp 2002 Comparison of broadcasting techniques for mobile ad hoc networks *Proc.Int.Conf.on ACM MobiHoc 2002* pp 194-205
[2] S. Boyd, A. Ghosh, B. Prabhakar and D. Shah 2005 Gossip Algorithms: Design, Analysis and

Applications *Proc.Int.Conf.on IEEE INFOCOM 2005* vol **3** pp 1653-1664

[3] M. Jiang, J. Li and Y. Tay 1999 Cluster Based Routing Protocol (CBRP) *Internet Draft* draft-ietf-manet-cbrp-spec-01.txt

[4] W. Peng and X. L 2001 Efficient broadcast in mobile ad hoc networks using connected dominating sets *Journal of Software* vol **12(4)** pp 529-536

[5] W. Lou and J. Wu 2002 On Reducing Broadcast Redundancy in Ad Hoc Wireless Networks *IEEE Transactions on Mobile Computing* vol **1(2)** pp 111-122

[6] J. Wu and D. Fei 2004 A generic distributed broadcast scheme in ad hoc wireless networks *IEEE Transactions on Computers* vol **53(10)** pp 1343–1354

[7] Sze-yao Ni,Yu-Chee Tseng,Yuh-Shyan Chen and Jang-Ping Sheu 1999 The broadcast Storm Problem in a Mobile Ad Hoc Network *Journal of Wireless Networks* vol **8(23)** pp 151-162

[8] P. Kyasanur and N. Vaidya 2005 Capacity of multi-channel wireless networks: Impact of number of channels and interfaces *Proc.Int.Conf.on ACM MobiHoc 2005* pp 43-57

[9] Shaohua Qin and Dongyan Chen 2012 A Distributed Broadcast Protocol of Wireless Sensor Networks Based on Dynamic Multi-channel MAC Protocol *Proc.Int.Conf.on CWSN 2012* pp 363–370

[10] Zeng Guokai,Wang Bo,Ding Yong,Li Xiao and Matt Mutka 2007 Multicast Algorithms for Multi-channel Wireless Mesh Networks *Proc.IEEE.Int.Conf.on Network Protocols* **21(1)** pp 1-10

Design and implementation of priority and time-window based traffic scheduling and routing-spectrum allocation mechanism in elastic optical networks

Honghuan Wang[1], Fangyuan Xing[1], Hongxi Yin[1]*, Nan Zhao[1] and Bizhan Lian[2]
[1]Lab of Optical Communications and Photonic Technology, School of Information and Communication Engineering, Dalian University of Technology, Dalian, 116034, China
[2]HAEPC Information and Telecommunication Company, Zhengzhou, 450000, China

E-mail: *hxyin@dlut.edu.cn

Abstract. With the explosive growth of network services, the reasonable traffic scheduling and efficient configuration of network resources have an important significance to increase the efficiency of the network. In this paper, an adaptive traffic scheduling policy based on the priority and time window is proposed and the performance of this algorithm is evaluated in terms of scheduling ratio. The routing and spectrum allocation are achieved by using the Floyd shortest path algorithm and establishing a node spectrum resource allocation model based on greedy algorithm, which is proposed by us. The fairness index is introduced to improve the capability of spectrum configuration. The results show that the designed traffic scheduling strategy can be applied to networks with multicast and broadcast functionalities, and makes them get real-time and efficient response. The scheme of node spectrum configuration improves the frequency resource utilization and gives play to the efficiency of the network.

1. Introduction

In recent years, with the rapid development of broadband and high rate services, such as big data, cloud computing, high definition video on demand and Internet of Things, optical networks need to be able to deal with the enormous carrier pressure brought by these large capacity services. Nowadays, the fixed-grid rigid-bandwidth WDM network is widely used to allocate resources. This fixed-grid WDM optical network allocates the same spectrum resources for spectrum requirements of different services at various rate and different bandwidth, which will lead to excessive resource waste when the services need only low-rate and low-bandwidth requirements. Moreover, the waste of resources will show deteriorating trend with the expansion of networks. Therefore, the novel spectrum-sliced elastic optical path (SLICE) network [1, 2] based on programmable devices has been proposed. However, the flexible grid network is also facing new issues, such as traffic scheduling scheme [3] and the effective allocation of network resources, to be addressed. Therefore, an adaptive traffic scheduling policy based on the priority and time window is proposed in this paper. The performance of this algorithm is evaluated in terms of scheduling ratio. Moreover, the routing and spectrum allocation are achieved by using the Floyd shortest path algorithm and establishing a node spectrum resource allocation model based on greedy algorithm. The fairness index (FI) [4] is introduced to improve the capability of spectrum configuration.

2. Adaptive scheduling algorithm

The priority is defined considering bandwidth and duration of a service request, assuming that bandwidths are proportional to data-rates. The model that determines the scheduling order is abstractly expressed as $T(s,d,p,dt,w)$, where s and d are the source node and the destination node, respectively, and p is the traffic priority. dt is the expected starting time of scheduling and w represents time window. Figure 1 denotes the time-window model of scheduling. If the actual scheduled starting moment S_i can move within time window following dt, i.e., $dt+w \geq S_i$, the traffic is scheduled, which stops at E_i. Otherwise, the scheduling fails. When dt is the same, the scheduled starting moment is movable within the individual time window.

Table 1 represents priorities and time-window setting. The traffic with high-priority should be scheduled earlier, while the time efficiency of scheduling should also be considered at the same time. The shorter time window of traffic, i.e., the earlier deadline $dt+w$, can save more idle time for scheduling more traffic. Therefore, scheduling efficiency of each service request consists of priority efficiency and time efficiency. The priority efficiency is expressed as $a_i = p_i$. The time efficiency is expressed as $b_i = e^{(-0.1(dt_i+w-S_i)/(E_i-S_i))}, i=1,2,\cdots$. The overall efficiency of a scheduled traffic is $G_i = a_i + b_i$, which is exploited to measure the performance of the scheduling algorithm. Obviously, this algorithm reflects a better performance with a greater scheduling efficiency. The goal of this algorithm is

$$\sum_{i=1}^{N} G_i = \max$$

$$s.t. \quad S_i \geq dt \quad (1)$$

$$w > 0 \quad (2)$$

$$dt + w \geq S_i \quad (3)$$

where $S_i \geq dt$ represents that the actual starting moment is not earlier than expected starting time of scheduling, $w > 0$ shows that the time window is not less than 0, $dt+w \geq S_i$ means that the actual starting moment is before the deadline of request.

The adaptive scheduling algorithm with time window will have a range and sort criteria to adjust to the actual scheduling order when a large number of service requests arrive synchronously. The method is depicted in **Algorithm 1**. The core of this scheduling policy is the sorting a large number of service requests based on three different criteria, which are the priority rule (PR), the deadline rule (DR), and the priority-deadline rule (PDR). The performance of this algorithm is evaluated in terms of scheduling ratio.

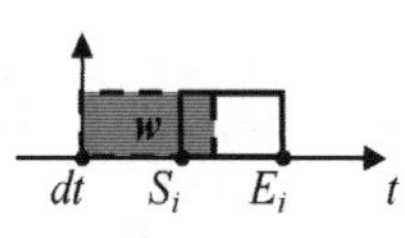

Figure 1. Time-window model of scheduling.

Table 1. Priorities and Time-window Setting.

Requests	Priority	Time Window(ms)
High-bandwidth, Short-duration	4	50
High-bandwidth, Long-duration	3	30
Low-bandwidth, Short-duration	2	20
Low-bandwidth, Long-duration	1	10

Journal of Physics: Conference Series **679** (2016) 012021 doi:10.1088/1742-6596/679/1/012021

Algorithm 1 Adaptive Scheduling Algorithm

Initialize idle time
Read the requests belonging to a scheduling period
Sort the requests by priority or deadline
Take the sorted requests in order
while requests queue is not empty, **do**
  for idle time meet the *dt* and *w* **do**
    (a) A request join executive matrix.
    (b) Waiting for allocation of spectrum resources.
    (c) Update starting point of idle time.
  end for
    (a) Request added to delay matrix.
    (b) Update requests to the next scheduling period
end while
Scheduling finished

3. Routing and node spectrum configuration

Scheduling and resource allocation model in Figure 2 shows that the traffic successfully added to the execution queue will wait for corresponding routing and spectrum allocation according to the source node and destination node.

The shortest path between any two nodes is calculated by using Floyd algorithm [5] in this paper. The spectrum resource distribution of node output links is altered on a regular basis, taking into account the network spectrum consistency and continuity in the perspective of network node spectrum resource reconfiguration. Assume that each node has the capability of center frequency conversion [6]. As shown in Figure 3, the spectrum resource is allocated for the traffic to access the respective links through node i. An appropriate spectrum allocation policy can groom traffic as much as possible, and save spectrum resources substantially, while ensuring the quality of service.

In order to make the output links of node i get equitable distribution of spectrum resources and reduce traffic blocking probability, a node spectrum allocation scheme based on greedy algorithm is proposed. The steps of this algorithm are presented below.

1) Set a certain time period. Calculate the average value of maximum and minimum bandwidth among traffic requests through each link in the set time interval. This average bandwidth B_i is defined as the initial solution of bandwidth for link i.

2) Set an appropriate frequency slot period. Select several links randomly and make them share a same frequency-slot period, and accumulate each B_i of these selected links. If the sum of B_i is less than the frequency slots period, replace each B_i with a larger bandwidth B_i' in the traffic requests queue. If the sum is still lower than the frequency slots period, do the same operation until it is closest to the period.

3) Move central frequencies of traffic to the corresponding spectrum.

4) Use the first-fit spectrum assignment approach to allocate spectrum resources for traffic leading to different links through the node.

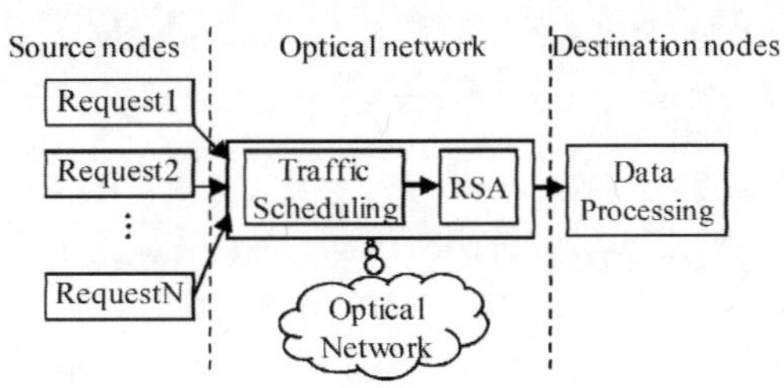

Figure 2. Scheduling and resource allocation model.

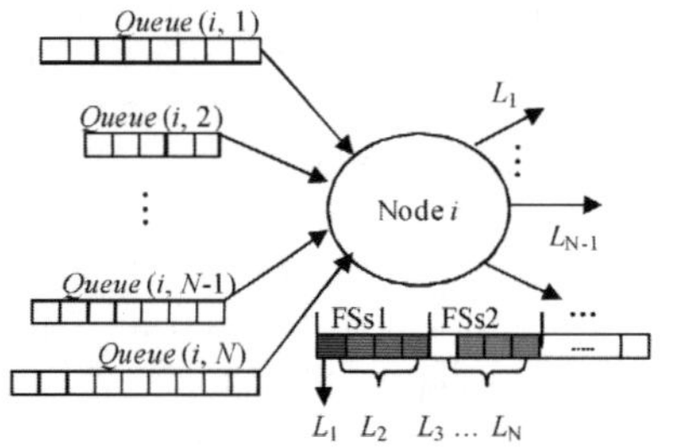

Figure 3. Spectrum allocation model of node *i*.

The effectiveness of the proposed algorithm in terms of better quality of service is modified by assigning broadband for each output links of the node. The fairness index (FI) can be defined as below to measure the equity of spectrum allocation scheme

$$FI = \frac{(\sum_{i=1}^{N}(B_i / A_i))^2}{N\sum_{i=1}^{N}(B_i / A_i)^2}, i = 1, 2, \ldots N \tag{4}$$

where N is the total number of node output links. A_i and B_i are respectively the request spectrum bandwidth and the actual assigned bandwidth for link i. Visibly, FI is a value between 0 and 1. The FI closer to 1 indicates a better fairness between the links and a lower blocking probability. When FI is lower than the threshold, the node spectrum for links is reconfigured.

4. Results and discussions

4.1. Performance of scheduling policy

To verify the performance of the adaptive scheduling algorithm based on priority and time window, the number of priorities is increased by adding the number of bandwidth threshold of request. Six priorities and 50ms scheduling interval are set in the simulation. The scheduling ratio with different priorities is taken as a evaluation index. The result shows scheduling performance comparison curve of three criteria, as shown in Figure 4.

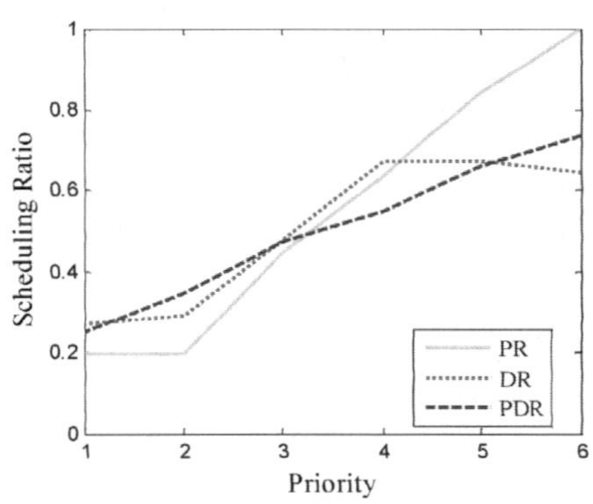

Figure 4. Scheduling ratio vs. priority under three criteria

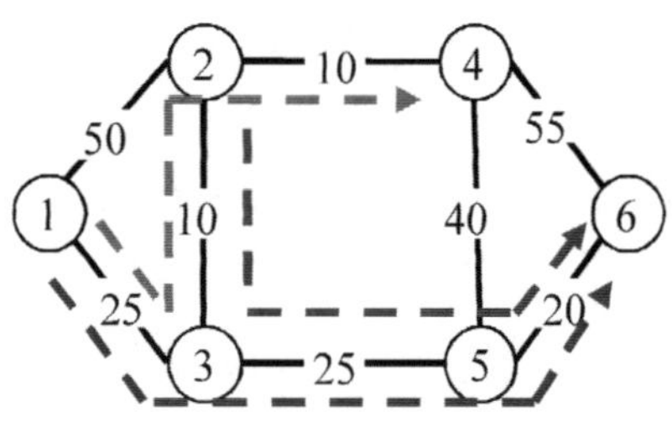

Figure 5. 6-node test network.

Figure 4 shows that scheduling ratio of PR and PDR increases linearly with the growth of priority. The scheduling ratio of DR has the similar trend with the former two criteria when the priority is low, while presenting a downtrend with higher priorities. In this case, scheduling ratio of DR depends primarily on the deadline of traffic requirement. Therefore, different scheduling rules exhibit their own advantages in different requirements. The PDR or DR should be selected to avoid low-priority traffic not to be treated for a long time. The PR presents better scheduling ratio when the priority is high. Therefore, an appropriate sort rule should be chosen to meet the user's multi-media service requests with different granularities. This scheduling scheme is also applicable among multiple networks. The appropriate adjustments for priority or time window of service request can improve the probability of a timely response for the network with real-time requirements.

4.2. Performance of routing and node spectrum allocation

The performance of routing algorithm is verified through a 6-node test network presented in Figure 5. As seen from the weighted graph, Floyd algorithm can accurately calculate the shortest path between any two nodes in the network.

The objective of spectrum allocation is to meet service requirements while reducing the spectral fragmentation. Assume that each node has the capability of center frequency conversion. The traffic passes node i to different destinations through respective output links. Seven output links are set to verify the fairness of node spectrum allocation in the simulation. 50 spectrum slices are set to a frequency slot period (FS) and allocated to partial links randomly. When the spectral occupancy of one FS reaches the upper limit, another FS is assigned to remaining links. Figure 6 (a) shows the status of node spectrum. The white and blue histograms, respectively, represent idle spectrum and request spectrum of links, while the red histogram presents the occupied spectrum within FS. Figure 6 (b) reports the spectrum allocation of output links. Clearly, links 6, 4, 2, 1 share the same FS and links 5, 3 share the second FS, while link 7 is configured in another FS. Result of simulation shows that the node spectrum configuration scheme greatly reduces spectrum fragmentation and improves the utilization of spectrum resources.

The throughput of the node is set to 110 Erlang in the node spectrum allocation simulation based on greedy algorithm. The bandwidth is assigned to each link to satisfy traffic transmission. Fairness Index (FI) varies with variance between the links under node spectrum allocation as shown in Figure 7. When the bandwidth of each link request is similar, the FI is closer to 1, namely, the fairness of allocated spectrum between links is better. If FI is less than a set threshold, reset the combination of links sharing the same FS or shorten FS to reduce the link variance. In this case, less allocated spectrum for one link that affects the quality of service and increases blocking probability is avoided.

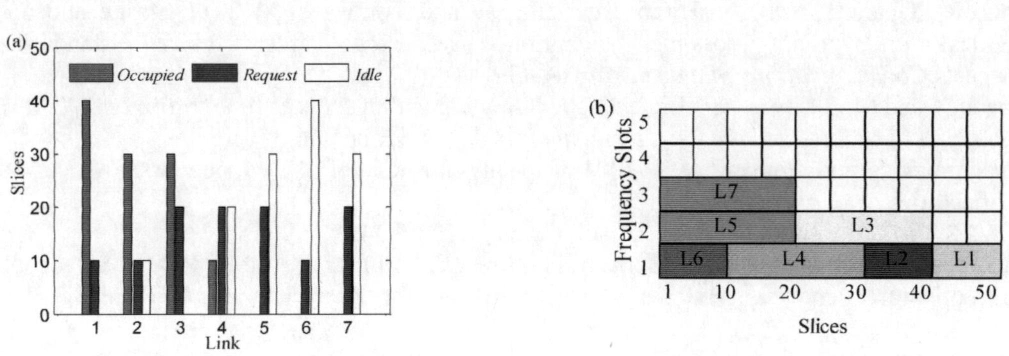

Figure 6. (a) Status of node spectrum (b) Spectrum allocation of output links

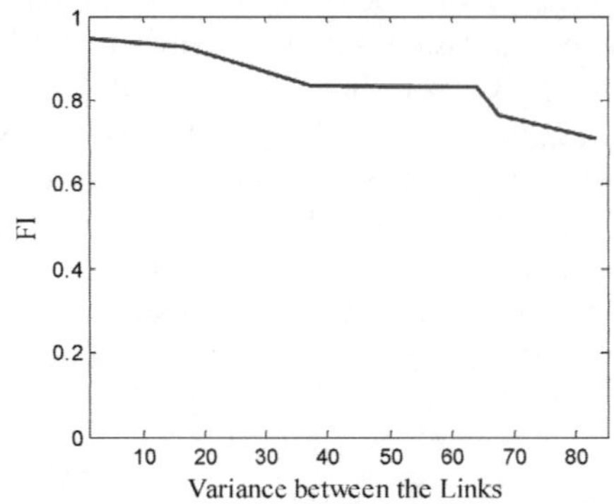

Figure 7. Fairness vs. variance between the links under node spectrum allocation.

5. Conclusion

In this paper, an adaptive traffic scheduling algorithm based on priority and time window is presented. Routing and spectrum allocation for scheduled traffic are achieved with the flexible and reconfigurable network node. The designed traffic scheduling strategy can be applied to the networks with multicast and broadcast, and makes them get real-time and efficient response, which lays the foundation for the scheduling investigation of multimedia services with different granularity in the future network. The scheme of node spectrum configuration improves the frequency resource utilization and gives play to the efficiency of the network, while the fairness of spectrum assignments for services through links leading to different destination nodes, is ensured.

Acknowledgment

This work was supported in part by the National Natural Science Foundation of China (NSFC) under Grants 61071123 and The First HAEPC Science and Technology Project in 2015 under Grant 5217Q014006V.

References

[1] Jinno M, Takara H, Kozicki B, Tsukishima Y, Sone Y and Matsuoka S 2009 Spectrum-efficient and scalable elastic optical path network: architecture, benefits, and enabling technologies *Communications Magazine* **47** 66-73

[2] Jinno M, Ohara T, Sone Y, Hirano A, Ishida O and Tomizawa M 2011 Elastic and adaptive optical networks: possible adoption scenarios and future standardization aspects *Communications Magazine* **49** 164-172

[3] Huang X and Ma M 2005 An efficient scheduling algorithm for real-time traffic on WDM optical networks *The 9th Int. Conf. on Communications Systems* pp 366-370

[4] Thiagarajan S and Somani A K 2001 Capacity fairness of WDM networks with grooming capabilities *Optical Network Magazine* **2** 24-32

[5] Floyd R 1962 Algorithm 97: shortest path *Comm. ACM* **5** 345

[6] Paolucci F, Sambo N, Meloni G, Berrettini G, Fresi F, Poti L and Castoldi P 2014 SDN control of all-optical frequency conversion and defragmentation for super-channels *OFC* pp 1,3, 9-13

Design of Control Plane Architecture Based on Cloud Platform and Experimental Network Demonstration for Multi-domain SDON

Ming Li[1], Hongxi Yin[1]*, Fangyuan Xing[1], Jingchao Wang[2] and Honghuan Wang[1]
[1]Lab of Optical Communications and Photonic Technology, School of Information and Communication Engineering, Dalian University of Technology, Dalian, 116034, Liaoning, China
[2] Institute of China Electronic System Engineering Company, Beijing 100039, China

E-mail: *hxyin@dlut.edu.cn

Abstract. With the features of network virtualization and resource programming, Software Defined Optical Network (SDON) is considered as the future development trend of optical network, provisioning a more flexible, efficient and open network function, supporting intraconnection and interconnection of data centers. Meanwhile cloud platform can provide powerful computing, storage and management capabilities. In this paper, with the coordination of SDON and cloud platform, a multi-domain SDON architecture based on cloud control plane has been proposed, which is composed of data centers with database (DB), path computation element (PCE), SDON controller and orchestrator. In addition, the structure of the multi-domain SDON orchestrator and OpenFlow-enabled optical node are proposed to realize the combination of centralized and distributed effective management and control platform. Finally, the functional verification and demonstration are performed through our optical experiment network.

1. Introduction

Due to the emergence and rapid development of the big data, the cloud computing, the internet of things and video surveillance and other services, the more flexible and efficient optical network interconnection between these remote and geographically distributed data centers are in demand to carry these massive amounts of data, and therefore SDON with centralized management and control arises as the times require [1]. With the dynamic adaptive ability to support high burst traffic, the flexible scalability of network and the elastic adjustment ability of bandwidth, SDON leads the future development direction of optical network [2, 3]. However, the scalability of SDON is limited by the computing and storage capacity of a single SDON controller and the independent path computing element and traffic engineering database are indispensable to assist in the implementation of the control functions, in which case the deployment and operation of the control node become complex. Moreover, the frequent and massive traffic between distributed controllers are increasing, which will result in the burden of network bandwidth.

With converged infrastructure and shared services, cloud computing and cloud storage not only become an important part of information technology, but also are an effective solution to network demands [4]. Through services of cloud, subscribers can access resources on demand from shared infrastructure, such as computing, storage, and networking [5].

The combination of traditional SDON and SDON-supported cloud computing and cloud storage makes full use of powerful computing, management, storage and transaction processing ability in cloud platform, so that the equivalence and coordination of distinct controllers as well as the intraconnection and interconnection of data centers, can be implemented. In addition, the collaborative management, control and path computing ability of SDON are enhanced further, and the interoperability between multi-domain heterogeneous networks is improved, which achieves a high degree of unity of the application, management and control, simplifies the complexity of network facilities and reduces the cost of operation and maintenance. Consequently, a multi-domain SDON optical network architecture based on cloud platform and the centralized and distributed control mechanism are proposed in this paper, the signaling process is detailed, and the functional verification and demonstration are performed in our experimental network.

2. SDON architecture based on cloud control platform

The SDON architecture based on Cloud Control Platform (CCP) is shown in Figure 1, which is the coordination of traditional SDON and cloud platform. The model of SDON based on CCP is divided into three planes, namely, Infrastructure Plane, Cloud Control Plane and Application Plane. The difference with the traditional SDON structure is that the orchestrator and cloud services are added to the cloud control plane, and geographically distributed controllers are gathered in data centers of cloud service. The hierarchical structure of control plane and distributed peer-to-peer architecture of cloud service data centers are reserved to achieve highly integrated distribution and concentration as well as effective management and control. The concentrated orchestrator and controllers in data centers improve the efficiency of data processing and network management and control, while the distributed architecture of the cloud platform data centers also provides a great scalability for the network.

From the vertical perspective, the centralized management and control structure can achieve the local and global optimal control. And from the horizontal point of view, the distributed management and control can overcome the disadvantages of the heavy task, low efficiency, complicated management system and database, which is caused by the centralized management and control. The hierarchical centralized control is realized in the CCP architecture. When a local controller has an enough capability to handle the request effectively, the higher level controller or orchestrator will not interfere. The higher level controller or orchestrator deal well with multi-domain requests only if the management and control task is not borne by a local controller. Meanwhile, hierarchical management systems play different roles to coordinate effectively. In this way, the scale of global management and control database, the time of enquiring and the amount of traffic management information are decreased significantly. These functions can be easily implemented by the efficient computing, management and storage of the cloud platform. This structure is suitable for a large scale, heterogeneous network interconnection, and geographically distributed network.

2.1. Three planes in the CCP architecture

The function of the CCP architecture is realized by three planes. The Infrastructure Plane is a layer in which all types of devices and resources from different vendors are interconnected. The optical network is mainly composed of optical nodes in geographically distributed multi-domains. Optical nodes include OF-switches, optical cross-connectors (OXCs) and optical add/drop multiplexers (OADMs), etc. Each domain, which is divided by different geographical location or other characteristics, is controlled by a single SDON controller. The difference between controllers brings inconvenience for the unified control over multi-domain optical networks. Therefore, the multi-domain SDON orchestrator is introduced into the Cloud Control Plane to solve the problems caused by the difference between controllers.

Journal of Physics: Conference Series **679** (2016) 012022 doi:10.1088/1742-6596/679/1/012022

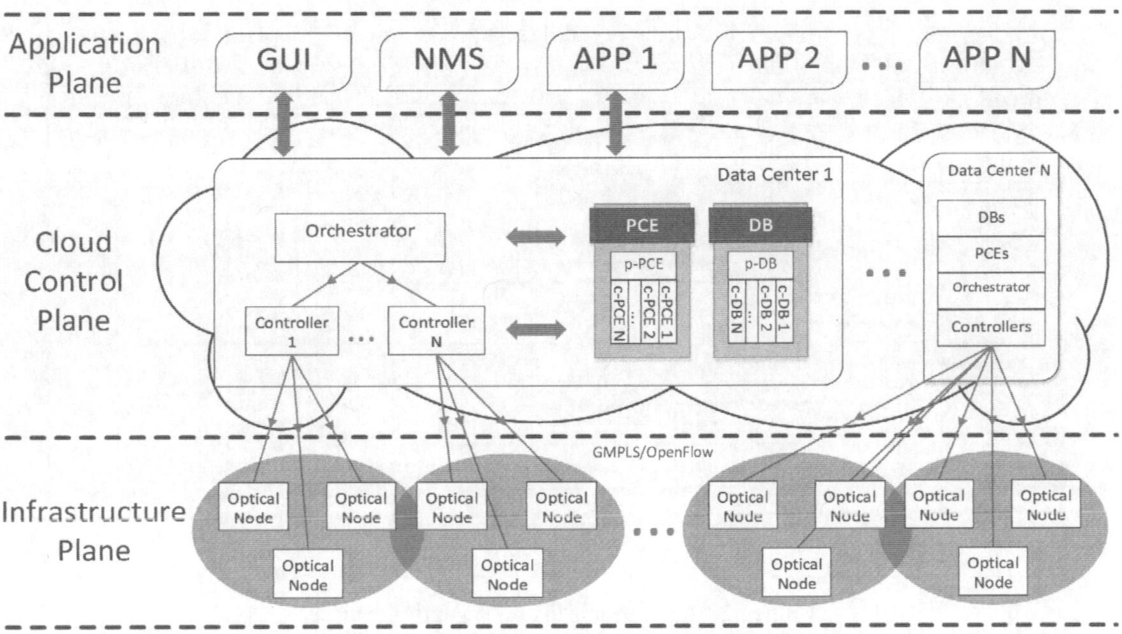

Figure 1. Cloud Control Platform architecture.

In the Cloud Control Plane, the data center is the component element. Date centers are capable of providing basic functions, such as transmission, computing and storage systems. The data center in Cloud Control Plane provides powerful computing capacity and data storage, which solves the capability limit of single control node. PCEs and DBs are corresponding to an individual controller, to handle the requests in respective domains. Meanwhile, parent PCE and parent DB are introduced into the Cloud Control Plane corresponding to the orchestrator, to deal with requests across multi-domain. In a data center of Cloud Control Plane, the controllers from distributed domains are gathered under the control of an orchestrator, which simplifies the complexity of updating and deployment, and reduces the cost of operation and maintenance. In addition, an orchestrator is employed to coordinate different controllers and deal with different protocols, and responsible for the management of global information and resources. The interactive communication between orchestrators from data centers can be implemented by application programming interfaces (APIs). Meanwhile, the information of data center can be exposed to authorized users and other data centers in the form of API.

The Application Plane consists of a plenty of vendors and third-party applications. All applications are supposed to interact with the Cloud Control Plane with the northbound interface (NBI). The graphical user interface (GUI) as an application provides users with a more personalized service experience, while the network management system is also recommended to communicate with the Cloud Control Plane through northbound interfaces, which reflects the openness of the CCP architecture.

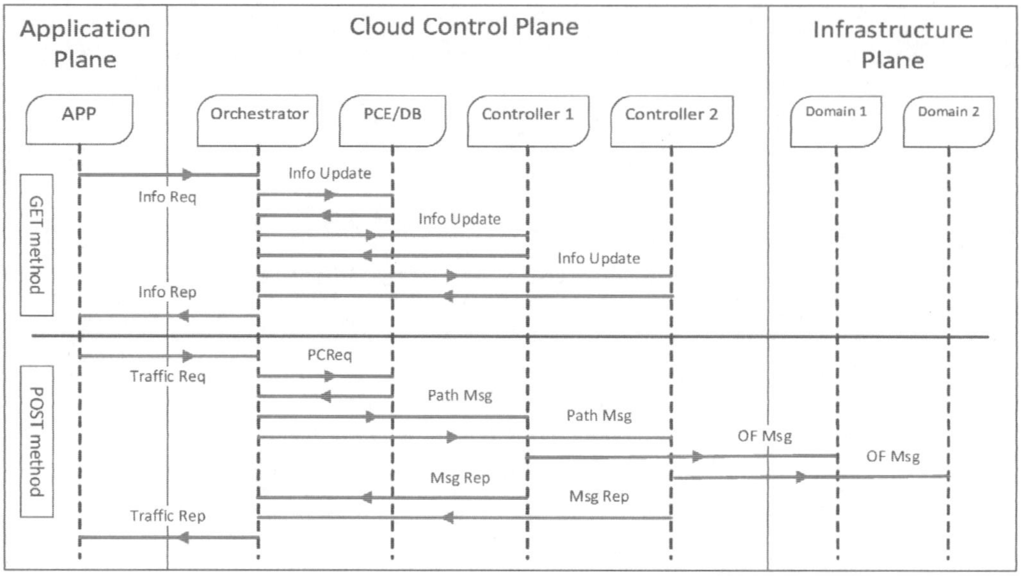

Figure 2. Basic scenarios on signaling process.

2.2. Signaling process in the CCP architecture

Figure 2 illustrates the signaling process of two kinds of operation from the Application Plane, which are the requests sent through GET method and POST method of the REST API, respectively. In Scenario 1, Application Plane sends a GET request to the Cloud Control Plane to obtain information, in which case the Infrastructure Plane is not involved. The process is as follows: the app of client in the Application Plane sends out the request to the Cloud Control Plane through the NBI; the orchestrator in the Cloud Control Plane updates information from the controllers and the PCE/DB; the collected information returns to the app through the NBI. In Scenario 2, the Application Plane sends a POST request through transmitting a request of traffic to the Cloud Control Plane, which carries the information of source and destination nodes for traffic requests, and changes the flow table information of optical nodes in the Infrastructure Plane. The process is as follows: the app sends traffic information to the Cloud Control Plane orchestrator; the orchestrator asks PCE/DB for the path information according to the source and destination of the traffic, then sends requests respectively to the relevant controllers; the controllers change the flow tables of the switch in the Infrastructure Plane with the OpenFlow protocol; after receiving the responds of controllers, the orchestrator replies to the app in the Application Plane.

3. Experimental demonstration of CCP architecture

The function of the proposed architecture has been validated in our SDON experimental network. The overall structure is shown in Figure 3(a). The Application Plane interacts with the Cloud Control Plane through the NBI, while the Infrastructure Plane is controlled by the Cloud Control Plane through the southbound interface (SBI). More detail of the Cloud Control Plane and the Infrastructure Plane is shown in Figure 3(b). OpenStack cloud platform of Juno version is deployed in the IBM Linux server with Intel Xeon E5-2620v2 and 32Gbits RAM as a simplified version of data center. OpenStack platform provides computing nodes, storage nodes and virtual machines (VM), which run as the orchestrator and two SDON controllers (e.g., Floodlight and OpenDaylight). Two controllers are connected to the optical nodes from two respective domains. The structure of optical node is detailed in Figure 3(c). The OpenFlow Agent in the optical node contains Open vSwitch (OVS), which communicate with the controller through OpenFlow protocol. The Protocol Adaptor translates the orders from the OVS into the protocols that can be recognized by the optical devices to response to the instruction from controllers.

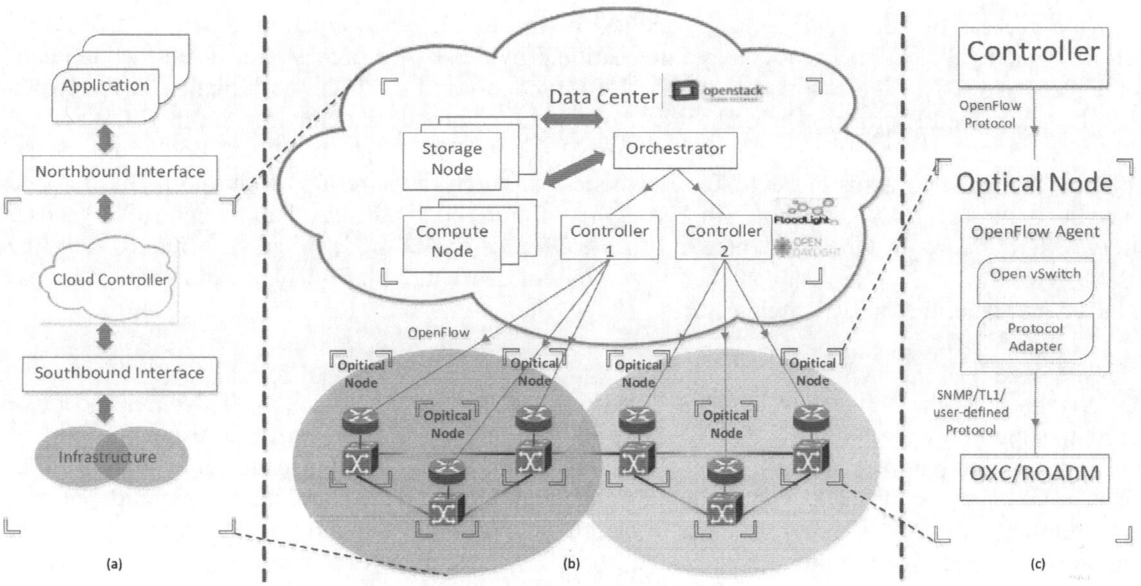

Figure 3. Experimental network scenario

(a) Abstract architecture. (b) Cloud Control Plane and Infrastructure Plane. (c) Optical node structure.

Figure 4. Experiment process. (a) Topology information from the orchestrator in Postman plug-in. (b) HTTP packet in Wireshark. (c) FLOW_MOD packet in Wireshark. (d) Ping process between hosts.

In the process of our experiment, the request of the topology information is sent to the Cloud Control Plane by the client. The topology information provided by the orchestrator is returned to client in the form of JSON. Figure 4(a) presents the HTTP request and the JSON information in the REST API plug-in, named Postman. The orchestrator of Cloud Control Plane enquires the PCE/DB for the solution to the traffic, and then, sends path requests to controllers respectively, and finally returns the replies to the client in terms of controllers. A request captured by Wireshark from the orchestrator to controller 1 through its NBI is shown in Figure 4(b). Figure 4(c) indicates that the controller sends a FLOW_MOD packet in OpenFlow protocol to the OVS of optical node. Finally, in Figure 4(d), the host receives the ping replies from another host in the Infrastructure Plane, which validates the success of the client request in the CCP architecture.

4. Conclusion

SDON represents the future development direction of the optical network, with dynamic adaptive ability to support high burst traffic, flexible scalability of the network and elastic adjustment ability of bandwidth. Cloud platform can provide powerful computing, storage and management capabilities. In this paper, a new type of SDON control plane architecture based on the coordination of cloud platform and traditional SDON is proposed, the structure of the multi-domain SDON orchestrator and OpenFlow-enabled optical node are presented, the efficient management and control in hierarchical centralized structure are realized, and the feasibility and availability of optical network are validated in our experimental network.

Acknowledgment

This work was supported in part by the National Natural Science Foundation of China (NSFC) under Grants 61231011 and The First HAEPC Science and Technology Project in 2015(5217Q014006U).

References

[1] Xia W, Wen Y, Foh C H 2014 A survey on software-defined networking *Communications Surveys & Tutorials* **17** 27-51

[2] Channegowda M, Nejabati R, Simeonidou D 2013 Software-defined optical networks technology and infrastructure: Enabling software-defined optical network operations *Journal of Optical Communications and Networking* **5** A274-A282

[3] Casellas R, Muñoz R, Martínez R 2014 SDN based provisioning orchestration of OpenFlow/GMPLS flexi-grid networks with a stateful hierarchical PCE *Optical Fiber Communication Conference Optical Society of America* Th3I 2

[4] Autenrieth A, Elbers J P, Kaczmarek P 2013 Cloud orchestration with SDN/OpenFlow in carrier transport networks *Transparent Optical Networks* 1-4

[5] Zhao S, Li L, Ling X, Xu C and Yang J Architecture and scheduling scheme design of Tsinghua Cloud based on OpenStack *Journal of Computer Applications* **33** 3335-3338

The design and implementation of TPC encoder and decoder

L J Xiang[1], Z B Wang[2], J B Yuan[1] and L H Zheng[1]
[1] School of Electronic Sciences and Engineering, the National University of Defense Technology, Hunan Chang-sha, 410073, China
[2] National Key Laboratory of Science and Technology on Information System Security, Beijing, 100101, China

Abstract. Based on the analysis and simulation of TPC codec principle and iteration decoding algorithm based on Chase 2, the design and implementation methods of TPC encoder and decoder are presented in this paper. In particular, circuit design and implementation flow of soft-input soft-output modified method in decoder algorithm and iteration decoding are analyzed. Aiming at (64,57,4) TPC, when the iteration times is 3 and the uncertain position number is 3, simulation and implementation results show that at least 6.8dB encoding gain can be obtained under the condition of BER=10^{-6}.

1. Introduction
Turbo Production Code (TPC) [1] comes from the Turbo Convolution Code (TCC) [2]. Block Turbo Code (BTC) was carried out by Pyndiah in 1994, and by using the BTC as sub-code, He designed the Chase decoding algorithm which can achieve SISO iteration[3]. This kind of iteration was applied in TPC[4]. In 1998, a modified Chase decoding algorithm was brought out by Pyndiah. Numerous experiments have proved that, compared with the Turbo code with sub-code of convolution code which generally 6-10 iterations is necessary, TPC with sub-code of blocking code converge faster, and only 4-5 iterations is necessary to approximately achieve the coding gain of the PCCC SISO decoding algorithm. In terms of the decoding, TPC is similar to the TCC. TPC has the performance of higher encoding efficiency and lower decoding complexity and it has lots opportunities in deep space communication, satellite communication and other digital communication systems.

2. Principles of TPC
TPC is a blocking code, generally, two or more blocking codes are coded to two or much bigger dimension coding blocks, and the blocking codes are the sub-code of the TPC, which can be same or difference. Blocking code of Hamming code, Extended Hamming code, BCH code and RS code, etc. are usually adopted as the sub-code.

2.1. Encoding Mechanism of TPC
Suppose two linear block code $C_1(n_1,k_1,d_1)$ and $C_2(n_2,k_2,d_2)$ is used as sub-code, where n is the code length, k is the length of the information, d is the minimum Hamming distance. Two dimensions TPC can be represented by $C_1 \otimes C_2$. The encoding process is as follows.

Firstly, sequentially read the $k_1 \times k_2$ information block as the first k_1 rows and k_2 columns.

Secondly, encode the k_1 row with $C_2(n_2,k_2,d_2)$.

Thirdly, encode the n_2 column with $C_1(n_1,k_1,d_1)$.

Journal of Physics: Conference Series **679** (2016) 012023 doi:10.1088/1742-6596/679/1/012023

The parameter of the TPC is: code length $n = n_1 \times n_2$, information length $k = k_1 \times k_2$, minimum Hamming distance $d = d_1 \times d_2$, and the encoding efficiency $R = R_1 \times R_2$,among which, R_1、R_2 is the encoding efficiency of C_1、C_2 . From the encoding theory, if $t_1 = \left[(d_1 - 1)/2\right]$, $t_2 = \left[(d_2 - 1)/2\right]$ random errors can be corrected by C_1 and C_2 , respectively, $t = \left[(d_1 d_2 - 1)/2\right]$ random errors can be corrected by $C_1 \otimes C_2$. It can also prove that $b \le \max(n_1 t_2, n_2 t_1)$ burst errors can be corrected by $C_1 \otimes C_2$. So, the TPC is kind of code with strong error-correct abilities[5], and its encoding structure is depicted as follows.

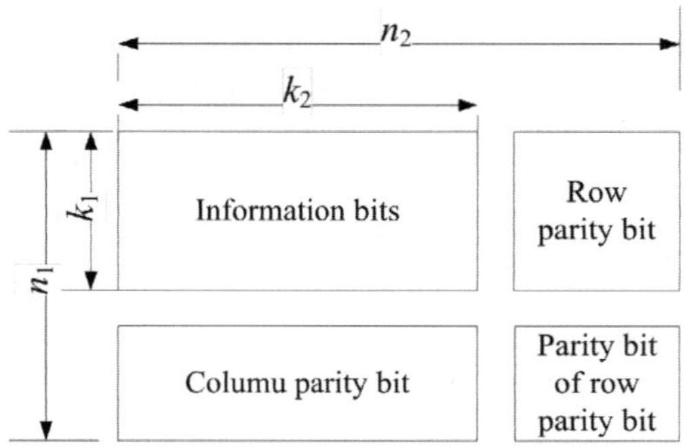

Figure 1. Encoding structure of TPC

2.2. Decoding Mechanism of TPC

2.2.1. TPC decoding structure. TPC is a serial concatenated code, it has good performance with an iterative decoding algorithm, among which, the unit-decoder is the most important part. The TPC unit-decoder is constituted with two parts, one is a soft input hard output decoder which is based on the Chase2 algorithm, and the other part is the calculation of extrinsic information which used to convert the hard input into soft output. Suppose the received code word matrix is r , extrinsic information matrix is $[W(m)]$, where m denotes the m th unit-decoder, and the soft input information is $[R(m)]$, soft output information is $[R'(m)]$. Based on the Chase2 decoding algorithm, the unit-decoder's work can be divided into four steps:

a) Calculate the soft input information $[R(m)]$ with the multipliers and adders.

b) Define the decoding subset Ω based on Chase 2 algorithm, get the decision code C_C , and find its competitive code C_D .

c) Calculate the soft output information $[R'(\mathrm{m})]$ and extrinsic information $W\left[(m+1)\right]$ with C_C and C_D .

d) Transfer the matrix $W[(m+1)]$ to next decoder to update the previous in it and the new extrinsic information $W[(m+1)]$ will be used as prior knowledge in decoding[5].

The soft input matrix of the m th decoder is denoted as $\left[R(m)\right] = r + \alpha(m)\left[W(m)\right]$, among which, $\alpha(m)$ is a factor to adjust the peak value of the m th decoder, and it can be changed flexibly based on different sub-code of TPC and different iteration decoding times. Input the $\left[R(m)\right]$ into the SISO decoder, and then decode it row by row (or column by column) to get the soft output matrix $\left[R'(m)\right]$. Subtract the received code word from $\left[R'(m)\right]$ and $\left[W(m+1)\right] = \left[R'(m)\right] - r$, and the result will be used as next decoder's prior information. Calculate the soft input matrix $\left[R(m+1)\right]$ of the next half iteration with $\left[W(m+1)\right]$, and repeat the process depicted above to decode column by column or row by row, than an iteration decoding process of a two-dimension TPC is completed. Concatenate multiple unit-decoders can make a concatenated decoding structure which is shows as follows.

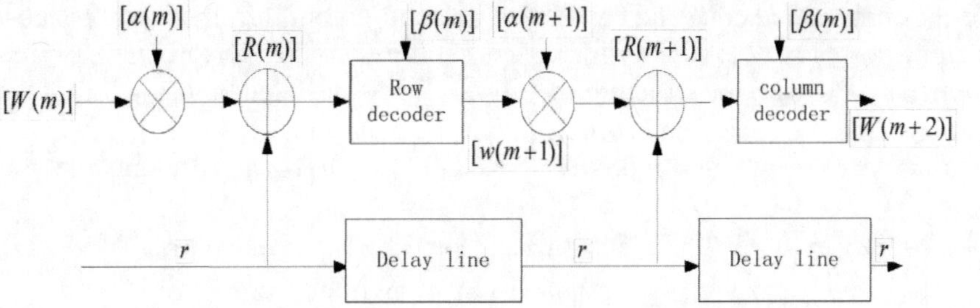

Figure 2. Concatenated iteration decoding structure

2.2.2. Chase2 Decoding Algorithm. The Chase 2 decoding algorithm is the most important and the most complex point of a unit-decoder. When a linear block code *C* is sent through a binary AWGN channel, suppose the received code word is *R,* and based on the maximum likelihood principle, the optimal decision word *D* is carried out by equation (1),

$$D = C^i \quad if \left|R - C^i\right|^2 \le \left|R - C^l\right|^2 \quad \forall l \in \left[1, 2^k\right] \quad \text{* MERGEFORMAT (1)}$$

Equation (2) shows the Euclid distance:

$$\left|R - C^i\right|^2 = \sum_{l=1}^{n} \left|r_l - c_l^i\right|^2 \quad \text{* MERGEFORMAT (2)}$$

We need to exhaust all possible code words to make a best decision, of which the complexity is very high. Chase 2 is a sub-optimal algorithm to decode the linear block code. When we decode the received information with the maximum likelihood principle, the produced word will drop into the sphere, whose centre is located at $Y = (y_1, \cdots, y_i, \cdots, y_n)$ and the radius is $(\delta - 1)$ with high probility. *Y* is the hard decision of the received signal. So, not all the code words should be considered, steps of Chase 2 is depicted as follows,

a) Select the most unreliable positions of *Y* and its number must be more than $p = \lfloor d/2 \rfloor$, the reliable is given by Equation (3):

$$\Lambda(d_j) = \ln\left(\frac{\Pr\{e_j = +1/d_j\}}{\Pr\{e_j = -1/d_j\}}\right) = (\frac{2}{\sigma^2})d_j \quad \text{* MERGEFORMAT (3)}$$

b) Construct the test pattern $T^q, q = 2^p$.
c) Construct test sequence Z^q according to $Z^q = Y \oplus T^q$.
d) Decode Z^q, and the result set is Ω.
e) Calculate the Euclid distance between code word C in Ω and the received word R, the word with the minimum distance is chosen as D.

As the square computation is hard to implement in hardware platform when calculates the Euclid distance. In practice, the Euclid distance calculating is simplified to correlation value calculating that showed in Equation (4). The correlation value calculating have advantages compared with the Euclid distance calculating, such as correlation value calculating only need the adder and subtractor, and the multiplier is not necessary.

$$COR\_V_{i+1} = \begin{cases} COR\_V_i + r_i & if\ z_i = 1 \\ COR\_V_i - r & if\ z_i = 0 \end{cases} \quad \text{* MERGEFORMAT (4)}$$

After getting the decision code word, in order to implement the iterative decoding, the Chase decoding result should be outputted as soft data, the data with reliability measure. The reliability measurement of the hard decision needs two code words, one is the hard decision result, and the other is the competitive code which has a great effect on the complexity and performance of the decoding

algorithm. For simplification, a factor β is directly used instead of looking for the competitive code. Simulation results shows that more than 1 dB is lost in performance. In this paper, a modified method for looking for the competitive code and calculating the soft output information is proposed, with which the decoding complexity is slightly increased but the decoding performance is greatly enhanced. For example, when a TPC with the structure of （64,57,4）is decoded by using 3 times iteration , 3 unreliable positions, and the code gain is 6.8dB under the condition of BER=10^{-6}.
Equation (5) shows how to find the competitive code, and $c\_se(i) = \text{-}1$ represents that the competitive code is not found.

$$c\_se(i) = \begin{cases} -1 & if\ z_i(k) = z_d(k) \\ k & if\ z_i(k) \neq z_d(k)\ \&\ COR\_V(k) > \max(COR\_V) \end{cases} \quad i = 1,\cdots,n;\ k = 1,\cdots,q \quad \text{* MERGEFORMAT}$$

In calculating the soft output information, From Equation (6), we can see that when the competitive code can not be found, the result is substituted by a factor β .

$$R' = \begin{cases} COR\_V(d\_se) - COR\_V(c\_se) & if\ c\_se(i) \neq \text{-}1\ \&\ if\ z_d(k) = 1 \\ COR\_V(c\_se) - COR\_V(d\_se) & if\ c\_se(i) \neq \text{-}1\ \&\ if\ z_d(k) = 0 \\ \beta & if\ c\_se(i) = \text{-}1\ \&\ R' > 0 \\ -\beta & if\ c\_se(i) = \text{-}1\ \&\ R' < 0 \end{cases} \quad \text{* MERGEFORMAT (6)}$$

$$i = 1,\cdots,n;\ k = 1,\cdots,q$$

When decoding a row or a column, if the competitive code can not be found, make a statistic as follows,

$$\begin{aligned} ex\_sum &= ex\_sum + |R' - R| \\ ex\_num &= ex\_num + 1 \end{aligned} \quad \text{* MERGEFORMAT (7)}$$

After decoding a row or a column, Equation (8) is implemented,

$$ex\_sum = ex\_sum / ex\_num \quad \text{* MERGEFORMAT (8)}$$

The soft output is finally achieved by Equation (9),

$$R' = \begin{cases} R' \times ex\_sum & if\ c\_se(i) = \text{-}1 \\ R' & if\ c\_se(i) \neq \text{-}1 \end{cases} \quad \text{* MERGEFORMAT (9)}$$

3. Performance Analysis of TPC Decoding

In order to simulate the performance of TPC, a two-dimension TPC code with the structure of （64,57,4）is used, and its columns and rows are all use the extended Hamming code.

3.1. Effect of the decision method and the quantization bits

In order to make best use of the information from every decoder, TPC use soft decision, and the output of the decoders is soft information. In AWGN, compared with hard decision, the system performance in terms of errors is enhanced around 2dB[6].

Compared with the unquantized method, the fixed point quantization can lose some code gain. In this work, we simulate the TPC with float point and fixed point quantization, respectively. The results shows that, in the AWGN channels, the code gain lose can be controlled under 0.2dB if more than 8bits is adopted in the linear quantization. But if the quantization bits are increased, the performance is not correspondingly increased.

3.2. Effect of the Iteration

The output of each decoding iteration includes the received signal of the decoder and the previous iteration, and more repeats can lead to a more reliable decision. In the SISO serialized decoding, increase the iteration time can increase the decoding performance, but when the iteration time is increased beyond a certain number, the performance changes little. The decoding complexity and the time delay are all increased along with the increase of the iteration time, and the decoding speed oppositely changes. Usually, 3-5 iterations is used to decode TPC. Simulations with 8bit quantization and 3 unreliable positions shows that the code gain can reach more than 6.8dB under the condition of BER=10^{-6}, and when the iteration time is increased, the code gain can also be increased but can not be increased in direct proportion.

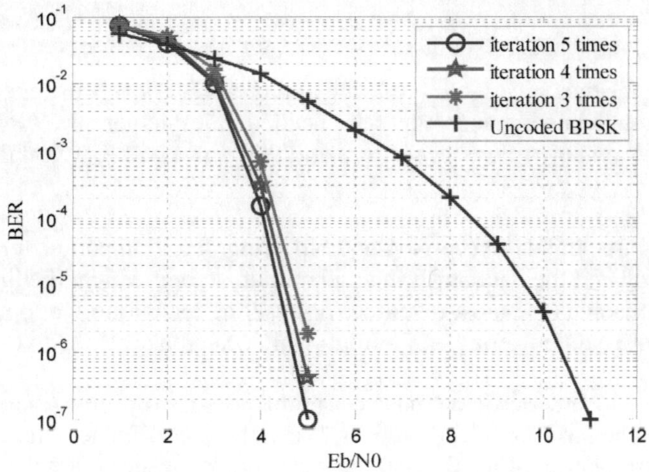

Figure 3. The influence of the iteration times in the TPC Decoding.

3.3. Effect of the Unreliable Positions

In the iterative decoder, in order to maximum likelihood decode the extended Hamming code (n,k,δ), the correlation value between the received code word R and 2^p possible code words should be calculated. So, the decoding complexity is increased exponentially with p. It is hard to physically realize when $p>6$. The following simulation shows the effect of p, when the quantization bits are constrained to 8 and the iteration time is set to 3. Simulation results indicated that p greatly affect the decoding performance, and increasing p can increase the decoding performance while the decoding complexity is also increased. When $p<3$, increasing it can greatly increase the decoding performance, and when $p>3$, increased p can not greatly improve the decoding performance.

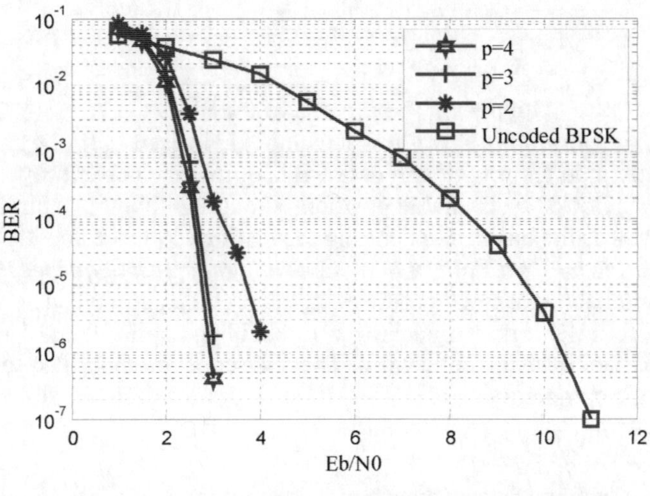

Figure 4. The influence of the numbers of unreliable positions in the TPC decoding.

Figure 4 shows the BER performance of the TPC decoding where 8 quantization bits, 3 iterations and 2 unreliable positions are used. When aiming at (64,57,4) TPC, simulation results confect that the code gain is more than 6.2dB at BER=10^{-6}, and if the unreliable positions is increased to 3 the code gain can be 6.8dB. The code gain can be improved if the iteration time and unreliable positions are increased, but it is too difficult to implement in engineering.

4. TPC in FPGA

4.1. TPC Encoding in FPGA

Four steps which include the data receiving, buffering and pre-processing, row encoding, column encoding, framing, data rate adjusting and output are needed to implement the TPC encoding in FPGA. The encoding is discussed in many references before and this paper will focus on the decoding.

4.2. TPC Decoding in FPGA

A decoder includes four parts: the data receiving and buffering, iteration parameters and data controlling, SISO decoding and the data buffering and output, among which, the SISO decoding is the most complex part and consumes the most resources. Ping-Pong technique is adopted between data receiving and buffering and data buffering and output, which can makes these two parts parallel the SISO decoding.

In the decoding process, the pre-decoding data sent to the decoder is 8 bits quantized in frame. Firstly, the received data is Ping-Pong buffered by the decoder. Secondly, once a frame of data is filled, the decoder begins to decode this frame. Thirdly, when the iterative decoding is finished, hard decision is implemented, and the data is Ping-Pong outputted. The first frame is decoded completely. In the process of the row decoding and the column decoding, the main flow is almost similar, the difference is only exist in some parameters.

4.2.1. SISO Decoding Module. SISO is the key part of the TPC decoder, and it is used to decode the receiving data with soft-input and soft-output and to calculate and out the extrinsic information $[W(m)]$. A whole SISO is depicted in figure 5. The purpose of decomposing the SISO decoder into several modules is to increase the decoding efficiency by using the pipeline technologies.

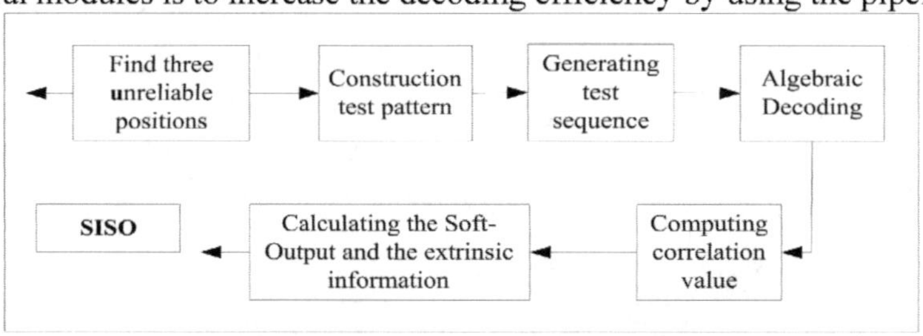

Figure 5. The process of the SISO

a) Module of Finding the Unreliable Positions. This module is used to find p data with minimum absolute values in a serially inputted row of soft data while output the current row of data after a hard decision. In order to saving clocks, p unreliable positions should be found in a circle. Finding the p unreliable positions in a row is the basis of generating the testing sequence and algebraic decoding.

b) Module of Constructing the Test Pattern and Generating the Test Sequence. Construct q test patterns based on the p unreliable positions produced in the previous module. The way of construct the test pattern is depicted as follows. Firstly, make the p unreliable positions 0 and 1 respectively, and then exclusive or with the hard decision code word. The test sequence should be saved until the decision code C_C and the competitive code C_D is get and the middle variable related to the soft output and the extrinsic information.

c) Module of Algebraic Decoding

Based on the algebraic decoding rule of the extended Hamming codes, this module finds the error bit according to the value of the check bit. Then, the error bit is reversed.

d) Module of Computing Correlation Value

Hardware resources, such as multipliers, are used to calculate the Euclid distance of different test sequences and received sequences. The calculating of Euclid distance is simplified to calculating of correlation value. Both the cycle accumulation and the binary trees adder can be used to calculate the correlation value, the binary trees adder saves the resource of clock but lots of middle value will be imported and then consumes lots of hardware resources. Thus, cycle accumulation is used to calculate the correlation value in our work.

e) Calculating of the C_C *and* C_D. The sequence corresponding to the maximum correlation value is used as the C_C, and then the C_D is achieved. The correlation values C_C and C_D will be used in the subsequent calculating of the soft output and extrinsic information. Looking for the C_D is the hardest part in hardware implementation since it consumes lots of hardware resources due to the pipeline technique. Figure 6 shows the pipeline simulation.

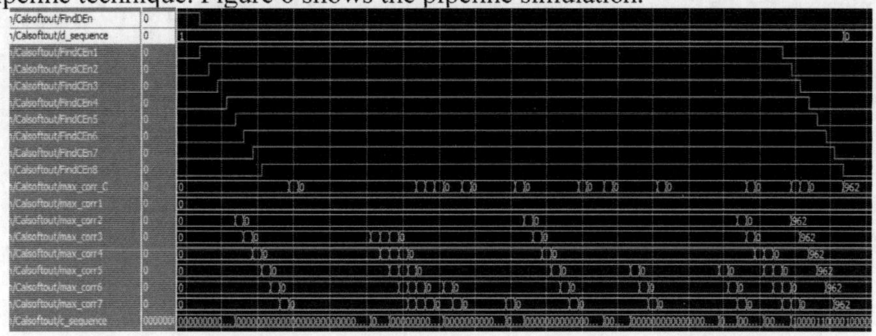

Figure 6. Pipeline simulation figure of the competitive code calculating.

f) Module of Calculating the Soft-Output and the Extrinsic Information. While calculating the soft-out, it must firstly look up the competitive code corresponding to each soft-out. If the competitive code exists, then compute the soft-out based on the correlation value, extrinsic information value, and if the competitive code doesn't exist, substitute the soft-out with the reliable factor β.

4.2.2. Iteration Control. Iteration Control, which is used to implement the data transfer and iteration parameter control, is the most important part of the TPC decoder. In decoding, the α is decided by the iteration time, and get the $r+\alpha[W(m)]$ by the internal signal composition module, and the composited signal is sent to the SISO module. Because there are row SISO and column SISO in decoding, in the next half iteration, the data should be read according to the row address and the column address. When the row decoding is finished, the memory should be written in row while the memory should be read in column in column decoding, and vice versa.

The module of iteration control is an opened interface in the program, where the iteration time can be easily modified by changing a few parameters. The iteration is a serialized process, and multi-iteration is realized by multiple calling of the same module in different times. So, in theory, changing the iteration time will not change the resources used by the program, and only the input data rate will be affected. Generally, the less iteration time the much higher decoding rate, and the less coding gain. If the iteration time is reduced, the pre-decoding data rate will be reduced accordingly, and the code gain will be enhanced.

5. Conclusions

Soft-input and soft-output is adopted by the TPC, the code gain can achieve 6.8dB when the BER=10^{-6}, and it can be implemented in satellite communications and deep space communications. The Ping-Pong technique and pipeline technique in this work can greatly alleviate the time delay and improve the decoding rate. In practice, a decoding rate of 20Mbps can be achieved by using a Xinlinx XC5VLX85-3FF676 FPGA chip. In summary, the TPC has a good prospect in engineering.

6. References

[1] X M Wang, G Z Xiao 2001 *Error Correction Code--Principle and method [M].* (Xian : Xidian University Publisher House)

[2] D H Liu, G M Liang 2011 *Turbo Code Design and Application [M]*. (Beijing: Electronic industry Publisher House)
[3] S Lin, M Fossorier 1998 *Soft input soft output decoding of linear block codes based on ordered statistics [J]*.(Global Telecom. Conference,2828-2833)
[4] X F Wd, A N Akansu 2001 *On parallel iterative decoding of product code [J]*. (VIC 2001. IEEE VTS 54 7-11, Oct.2001)
[5] F P Wen. 2010 *Research and Application of Turbo Product Code Decoding Algorithm Based on Chase Algorithm [D]*, (Chengdu： University of Electronic Science and Technology of China 2010.5)
[6] X Luo. 2007 *Turbo Product Code Technology Research and its FPGA Implementation [D]*. (Xian： Northwestern Polytechnical University 2007.3)

Journal of Physics: Conference Series **679** (2016) 012024 doi:10.1088/1742-6596/679/1/012024

The design of an improved matched filter in DSSS-GMSK system

Mao Wei-tong, Zheng Lin-hua, Xiang Liang-jun, Wang Tan

College of Electronic Science and Engineering,
National University of Defense Technology,Changsha, China,410073

E-mail address : 447535556@qq.com

Abstract. This paper introduces the principle of DSSS-GMSK system, analyses the superiority of GMSK modulation over MSK modulation. Accord that the method of de-spread before demodulation can effectively improve the capability of the system with spread spectrum gain, this paper researches an improved method with matched filter to de-spread and demodulate the DSSS signals. The local PN code is modulated with GMSK modulation before being correlated with received signal, then we can get the synchronous PN code, de-spread and demodulate the signal. MATLAB simulation shows that this method is more efficient than the method of demodulation before de-spread in low SNR environment.

1. Introduction

Spread-spectrum system has been widely used with the advantage of low probability of being captured, strong ability on anti-jamming, high resolving power of distance and the power of random stochastic location by multi-user. As a working mechanism of spread-spectrum system, direct sequence spread-spectrum (DSSS) uses BPSK/QPSK/MSK as its common mode of modulation. GMSK modulation, which is an improvement on MSK modulation, uses a Gaussian low pass filter before MSK. GMSK modulated signals have more tightly band limit than other constant envelope signals, so that they are more suitable.

The advantage of DSSS system is the spread spectrum gain to the signal. To get the gain, the signals must be de-spread before demodulation. DSSS-BPSK/QPSK/MSK systems all use the method of de-spread before demodulation to manage the signals. The local PN code is made correlation calculation with received signals by matched filter, then we can get the correlation peak , which could help us get the synchro PN code and demodulate the signals. But in GMSK modulated system, we could not get the correlation peak by using the same method. To solve this problem, literature [4] and literature [5] both use the method of demodulation before de-spread to process the received signals. But this method could not efficiently use the spread-spectrum gain. The system could not work well in low SNR environment. In order to efficiently use the spread-spectrum gain, this paper researches an improved method with zero IF direct correlator to demodulate the DSSS signals, analyses and emulates the principle and capability of the system.

2. Principle and capability of DSSS-GMSK modulation

Journal of Physics: Conference Series **679** (2016) 012024 doi:10.1088/1742-6596/679/1/012024

2.1. Principle of DSSS-GMSK modulation

GMSK（Gaussian-filtered minimum-shift keying） is a kind of modulation scheme with constant envelope, which develops from MSK. The principle of GMSK modulation is that the signals go through a Gaussian low-pass filter before MSK modulation. The impulse response of the Gaussian filter is defined as follows:

$$h(t) = \frac{\sqrt{\pi}}{\alpha}\exp[(-\frac{\pi}{\alpha}t)^2] \tag{1}$$

Where $\alpha = \sqrt{\ln 2/2}/B$, B is the 3dB bandwidth of the Gaussian filter. If the width of a pulse is T, then BT is the normalized 3dB bandwidth. We can obtain the frequency impulse signal of the Gaussian filter:

$$g(t) = Q\left(\frac{2\pi B_b}{\sqrt{\ln 2}}\left(t-\frac{T}{2}\right)\right) - Q\left(\frac{2\pi B_b}{\sqrt{\ln 2}}\left(t+\frac{T}{2}\right)\right) \tag{2}$$

Where $Q(x) = \int_x^{\infty} \frac{1}{\sqrt{2\pi}}\exp\left(-x^2/2\right)dx$.

Suppose that the transmitting data multiplied by PN sequence is $a_k (a_k \in \{-1,1\})$, then the phase of the modulated signals is shown as follows:

$$\phi(t_k) = \frac{\pi}{2T}\int_{-\infty}^{(k+1)T}[\sum_k a_k g(\tau - nT - \frac{T}{2})]d\tau \tag{3}$$

Where $kT \le \tau \le (k+1)T$. The frequency of the carrier wave is f_c , then the GMSK modulated signal is produced after IF modulation:

$$s(t_k) = \cos\left(2\pi f_c t_k + \phi(t_k)\right) = \cos\left(2\pi f_c t_k\right)\cos\phi(t_k) - \sin\left(2\pi f_c t_k\right)\sin\phi(t_k) \tag{4}$$

2.2. Capability analysis of GMSK modulated signals

GMSK is an improvement on MSK. Figure 1 gives the BER performances and frequency spectrum for GMSK and MSK modulated signals. Although the error rate performance of GMSK modulated signal is less than MSK modulated signal, but the gap is quite small. Figure 2 shows that with respect to MSK modulated signals, GMSK modulated signals have a more compact spectrum and faster band attenuation. So GMSK has a broader development prospect.

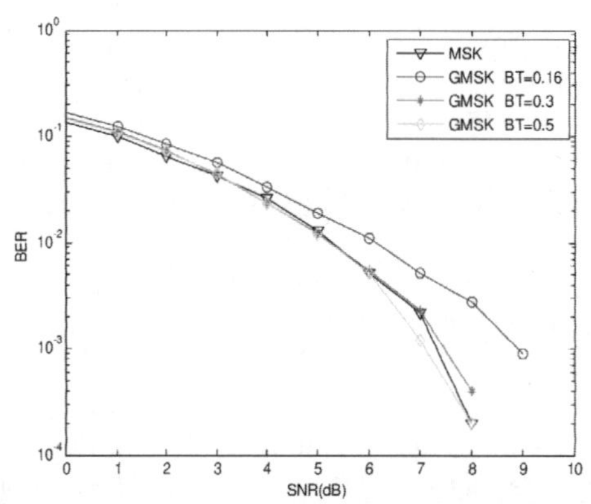

Figure 1. BER comparison of GMSK and MSK.

Normalized 3dB bandwidth is a very important parameter of GMSK modulated signals. From figure 2 we can see that the spectrum structure of GMSK modulated signals is closely related to BT value. The smaller the BT value, the more compact the frequency spectrum. When BT $\rightarrow +\infty$, the GMSK modulated signals change into MSK modulated signals. BT value will also have an impact on the error rate of GMSK modulated signals. Figure 1 also shows that the BER performance will be getting worse as BT value decreases. So, BT value usually is 0.3 in GSM system.

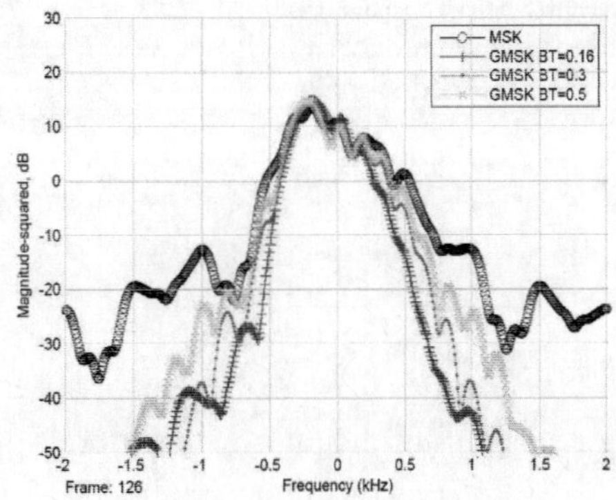

Figure 2. Frequency spectrum of GMSK and MSK.

3. Improved method with matched filter to de-spread and demodulate the DSSS signals

Two methods are used for the signal processing at the receiving terminal of the DSSS system: de-spread before demodulation and demodulation before de-spread. In high SNR environment, the two methods both have good BER performances. But in low SNR environment, the method of de-spread before demodulation has a better BER performance because this method could effectively use the spread spectrum gain. At the receiving terminal, we must get the the synchronous PN code before despread the received signals. Common methods to get synchronous PN code are sliding correlation and matched filter. The principle of matched filter is that we use the PN code as the coefficients of the filter to filter the received signals. The result will be sent to the threshold decider. If the result is larger than the threshold value, the PN code is synchronous.

In DSSS-BPSK system, signals at the receiving terminal is shown as follows:

$$s(t) = \sqrt{2P}c(t-\tau)\cos(\omega_0^{'}t + \phi^{'}) \tag{5}$$

Where $c(t)$ is the PN code multiplied by transmitted data. τ is the time delay. $\omega_0^{'}$ is the frequency of the received signal， $\phi^{'}$ is the initial phase which can be set to zero. The impulse response of the matched filter is: $h(t) = c(NT_c - t)$， where $0 \le t \le NT_c$， NT_c is the period of PN code. After DDC, the signals will be sent to the matched filter. The result is that:

$$\begin{aligned} y(t) &= s(t) * h(t) \\ &= \int_{-\infty}^{t} s(\lambda)h(t-\lambda)d\lambda \\ &= \int_{t-NT_c}^{t} \sqrt{P}c(\lambda-\tau)c(NT_c - t + \lambda)\cos(\Delta\omega\lambda)d\lambda \end{aligned} \tag{6}$$

Where $\Delta\omega$ is the variation of the carrier frequency. The correlation peak could be found from the result. However, the GMSK modulated signals are shown as formula(4), the result of the signals correlated with PN code is that:

$$\begin{aligned} z(t) &= \cos\left(\Delta\omega t + \phi(t)\right) * h(t) \\ &= \int_{-\infty}^{t} \cos\left(\Delta\omega\lambda + \phi(\lambda)\right) h(t-\lambda)d\lambda \\ &= \int_{t-NT_c}^{t} \cos\left(\Delta\omega\lambda + \phi(\lambda)\right)\cdot c(NT_c - t + \lambda)d\lambda \end{aligned} \tag{7}$$

We can not find the correlation peak from formula(8), then we can not get the synchronous PN code and

Journal of Physics: Conference Series **679** (2016) 012024 doi:10.1088/1742-6596/679/1/012024

despread the received signals. To solve this problem, this paper researches an improved method with matched filter to de-spread and demodulate the DSSS signals. Figure 3 shows the process of the method. The local PN code, which modulated by GMSK mode first, is set as the coefficient of the matched filter. Then the matched filter is used to filter the received signals.

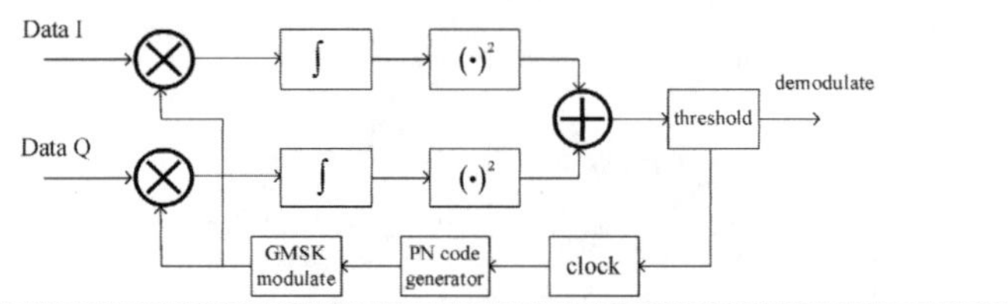

Figure 3. Process of the improved method.

Assume that the carrier is synchronous, after DDC, the GMSK In-phase data and Quadrature data of the signals are that:

$$s_i(t) = A\cos(\phi(t)) \tag{9}$$

$$s_q(t) = A\sin(\phi(t)) \tag{10}$$

Where $\phi(t) = \frac{\pi}{2T}\int_{-\infty}^{t}\left[\sum_n a_n g(\tau - nT - \frac{T}{2})\right]d\tau$. Assume the period of the PN code is k, then the local GMSK modulated signals are that:

$$r(t) = \cos(\phi(t)), 0 \le t \le kT \tag{11}$$

The output of the matched filter is that:

$$\begin{aligned} y_i(t) &= s(t) * r_i(t) = \int_{-\infty}^{t} s(\lambda) r_i(t-\lambda) d\lambda \\ &= A\int_{t-kT_c}^{t} \cos(\phi(\lambda))\cos(\phi(t-\lambda))d\lambda \end{aligned} \tag{12}$$

$$\begin{aligned} y_q(t) &= s(t) * r_q(t) = \int_{-\infty}^{t} s(\lambda) r_q(t-\lambda) d\lambda \\ &= A\int_{t-kT_c}^{t} \cos(\phi(\lambda))\sin(\phi(t-\lambda))d\lambda \end{aligned} \tag{13}$$

Then the peak value is that:

$$z(t) = y_i(t)^2 + y_q(t)^2 \tag{14}$$

If $z(t)$ is larger than threshold value, then the local PN code is synchronous. According to the correlation peak values in each symbol period, we can demodulate the received signals.

4. Simulation results

Assume that signal transmission rate $R_b = 80kHz$, PN code period $k = 128$, then PN code transmission rate is $10.24MHz$ 。The sampling rate is 12 times the code transmission rate. Time delay is set to be 48 sampling duration. Carrier is synchronous at the receiving terminal. We can get the simulation results with the input SNR of $-6dB$:

（1）The output of the matched filter is shown as Figure 4.

From Figure 4 we can see the correlation peak clearly. Then we can get the synchronous PN code by setting a suitable threshold value.

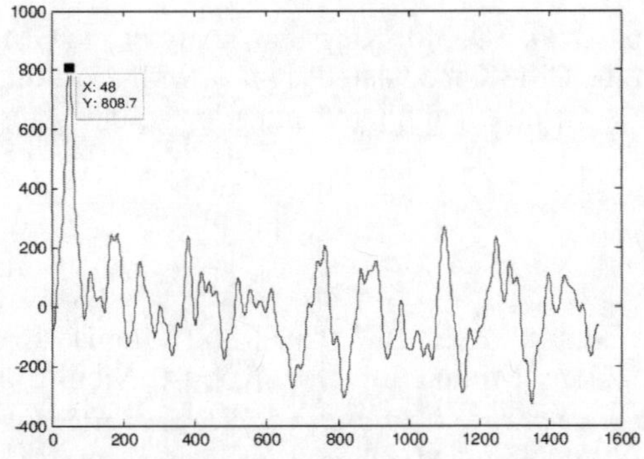

Figure 4. Output of the matched filter.

(2) Figure 5 shows the correlation peak after de-spread. Since every symbol has 1536 sampling points, the correlation peak will appear every 1536 sampling points. We can get the sending signals through the positive and negative of every peak value.

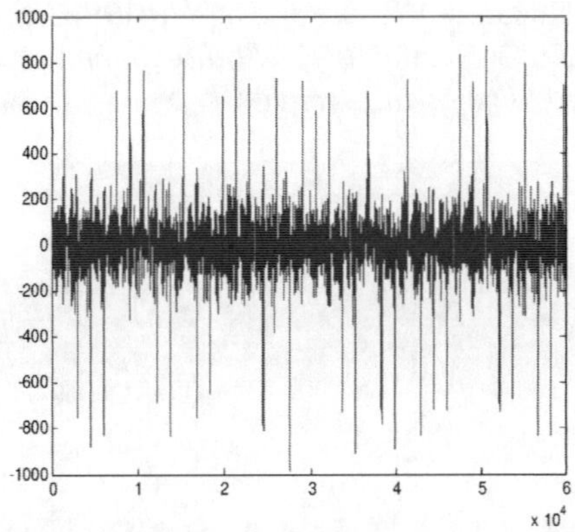

Figure 5. Correlation peak after de-spread.

(3) Figure 6 shows the BER performances of the two methods. The method this paper researched has a better BER performance than the method of demodulation before de-spread in low SNR environment.

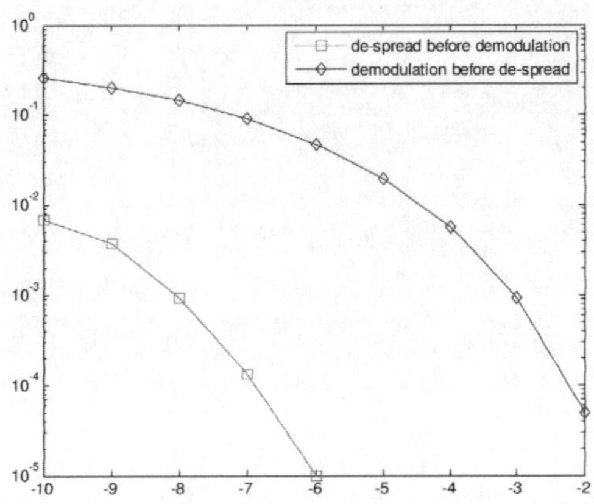

Figure 6. BER performances of the two methods.

Journal of Physics: Conference Series **679** (2016) 012024 doi:10.1088/1742-6596/679/1/012024

5. Summary

This paper introduces the principle of DSSS-GMSK system, analyses the GMSK modulated signals. According to the special form of the GMSK modulated signals, with the thought of de-spread before demodulation, this paper improves the matched filter. Simulation results show that we can get the correlation peak from the filter output and despread the signals. When input SNR is $-6dB$, the BER can be 10^{-5}.

References

[1] Hanwen Cao, Friederike Maier, Andreas Wilzeck and Thomas Kaiser 2012 DSSS Differential GMSK With Space-Time Modulation for Robust Mobile Data Links *IEEE Wireless Communications and Networking Conference: PHY and Fundamentals*

[2] SHIGA.S, IKEGAHI.T 2000 Digital Matched Filter Detection of GMSK/DSSS Signal *IEEE Int. Symp. an Personal, Indoor and Mobile Radio Communication* pp 1281-1285

[3] LOU Li 1994 The Emulation and Analysis of GMSK Digital Modulation *Modern Electronics Technique* Vol 12 pp 11-13

[4] Liang Xianming 2010 Realization of Demodulation and Despread Technology for GMSK Spread Spectrum Signal *Telecommunication Engineering* Vol 5 pp 51-55

[5] Yasuhiro yano, Hisao Tachika and Tadashi Fujino 1996 Performance of DS/GMSK/PSK Modern Using Four-phase Correlator *IEEE.Radio Communication Department, Information Technology R&D Center, Mitsubishi Electric Corp*

8th International Conference on Advanced Infocomm Technology (ICAIT 2015) IOP Publishing
Journal of Physics: Conference Series **679** (2016) 012025 doi:10.1088/1742-6596/679/1/012025

Architecture Design and Experimental Platform Demonstration of Optical Network based on OpenFlow Protocol

Fangyuan Xing[1], Honghuan Wang[1], Hongxi Yin[1,*], Ming Li[1], Shenzi Luo[1], Chenguang Wu[2]
[1]Lab of Optical Communications and Photonic Technology, School of Information and Communication Engineering,
Dalian University of Technology, Dalian 116024, Liaoning, China
[2]HAEPC Information and Telecommunication Company, Zhengzhou 450052, Henan, China

E-mail: hxyin@dlut.edu.cn

Abstract. With the extensive application of cloud computing and data centres, as well as the constantly emerging services, the big data with the burst characteristic has brought huge challenges to optical networks. Consequently, the software defined optical network (SDON) that combines optical networks with software defined network (SDN), has attracted much attention. In this paper, an OpenFlow-enabled optical node employed in optical cross-connect (OXC) and reconfigurable optical add/drop multiplexer (ROADM), is proposed. An open source OpenFlow controller is extended on routing strategies. In addition, the experiment platform based on OpenFlow protocol for software defined optical network, is designed. The feasibility and availability of the OpenFlow-enabled optical nodes and the extended OpenFlow controller are validated by the connectivity test, protection switching and load balancing experiments in this test platform.

1. Introduction

Optical networks have increasingly become important information infrastructure because of having many advantages such as the ultra-high speed rate, the ultra-wide bandwidth, the ultra-great capacity and the long transmission distance. The software defined optical network (SDON) is a novel optical network architecture, in which the users and operators can dynamically design the structures and functions through software programming in order to respond to the requests rapidly, utilize the resources effectively and provide the services flexibly [1].

In 2008, the concept of software defined network (SDN) is firstly proposed by Professor Nick Mckeown in Stanford University [2]. SDN, a novel network architecture, enables the network control to become programmable and the underlying infrastructure to be abstracted from applications and network services through decoupling network control and forwarding [3]. By the year 2011, KDDI R&D Laboratory in Japan and Beijing University of Posts and Telecommunications firstly introduced SDN into optical networks [4]. In 2013, High Performance Networks Group (HPNG) in UK applied SDON to the cloud service architecture [5]. Though the researches on the SDON have made some progresses recently, the layered architecture of the SDON optical nodes and the experimental verification on the protocols, the strategies and the functions of the more complex network topology

Journal of Physics: Conference Series **679** (2016) 012025 doi:10.1088/1742-6596/679/1/012025

still need further exploring. Consequently, a novel OpenFlow-enabled optical node is proposed and an open source OpenFlow controller is extended on routing strategies in this paper. Moreover, the experiment platform based on OpenFlow protocol for software defined optical network is designed and the performance of the OpenFlow-enabled optical nodes and the extended OpenFlow controller are experimental evaluated.

2. Architecture design of OpenFlow-enabled optical node

2.1. Architecture of optical node

The architecture of OpenFlow-enabled optical node is shown in Figure 1.

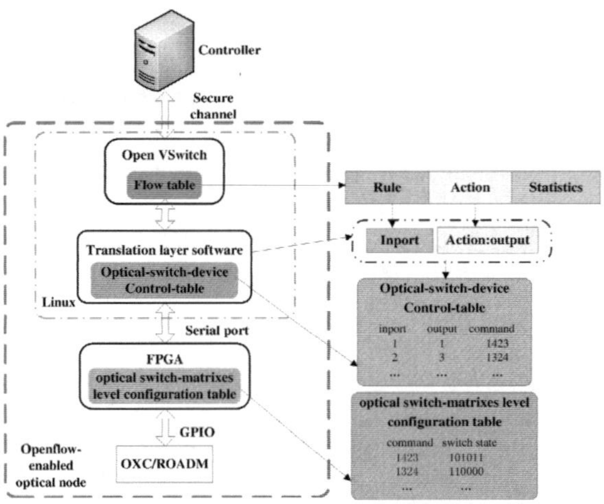

Figure 1. Architecture of OpenFlow-enabled optical node.

The OpenFlow-enabled optical node consists of an Open vSwitch software, a translation layer software, an FPGA and an OXC/ROADM device. The Open vSwitch is a multilayer virtual switch in accord with the open source Apache 2.0 [6]. Open vSwitch software comprises a secure channel, a flow table and an OpenFlow protocol [7]. The OpenFlow controller communicates with the OpenFlow-enabled optical nodes via the secure channel of the Open vSwitch. The flow table is used to store commands from the controller while the OpenFlow protocol acts as the signaling between the OpenFlow controller and optical nodes [8]. The translation layer software composed of optical-switch-device control-table and serial communication module, is responsible for reading out the input, output and wavelength information from the flow table. The FPGA consisting of optical switch-matrixes level configuration table and serial communication module, communicates with the translation layer software via serial ports and controls the OXC/ROADM devices according to the level configuration table.

The signaling proceeding of the OpenFlow-enabled optical node is shown as below. When the OpenFlow-enabled optical node receives a packet that does not match with any flow entry in the flow table, it sends the packet to the OpenFlow controller. The OpenFlow controller performs routing strategies based on its knowledge of the whole network and then, inserts a new flow entry into the flow table of the Open vSwitch via the secure channel. The translation layer software reads out the input, output and wavelength information from the flow entry and acquires commands through looking up the optical-switch-device control-table. Afterwards, the command is passed to the FPGA via serial ports. According to the command, FPGA obtains the states of optical switch matrixes in the OXC by searching the level configuration table. Finally, the FPGA sends the states of optical switch matrixes to OXC/ROADM device via general purpose input and output ports (GPIO).

2.2. Extension of controller

An open source floodlight [9] extended on the routing strategies, is utilized as an OpenFlow controller. The routing and wavelength assignment (RWA) algorithm is added into the routing and forwarding modules, in which the Dijkstra Algorithm to find the shortest path between each node pair and the First-Fit Algorithm to solve wavelength assignment issues are deployed.

The extended OpenFlow controller workflow is shown in Figure 2. The controller discovers the link resources through sending Link Layer Discovery Protocol (LLDP) periodically. When the network adds or deletes links, the link discovery manager calls the topology manager. The topology manager computes the network routing resources to create routing list based on the Dijkstra Algorithm. When a service request is produced from users, the forwarding module receives the request, chooses the shortest path according to the routing list, assigns the wavelength utilizing the First-Fit Algorithm and passes a flow entry to the OpenFlow-enabled optical nodes.

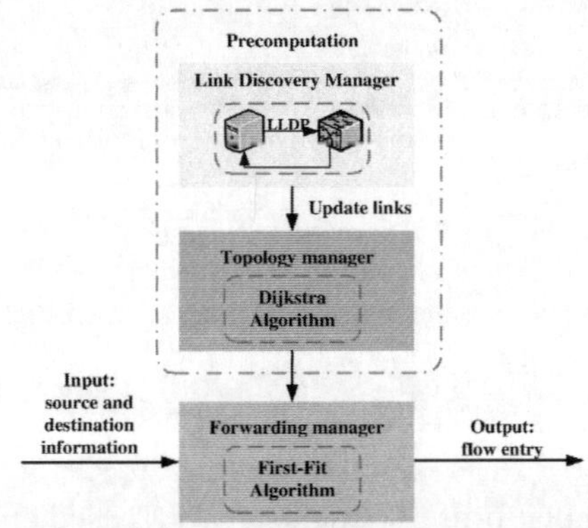

Figure 2. Extended OpenFlow controller workflow.

In the flow entry, the cookie field comprising 56 bits binary numbers, stores the statistic information of the flow table. After numerous tests, the 16-19 bits of the cookie field are found to be empty. Consequently, the wavelength information is stored in the 16-19 bit. For instance, in Figure 3, the flow entry shows that the input port is No. 4, the output port is No. 2 and the wavelength is 1.

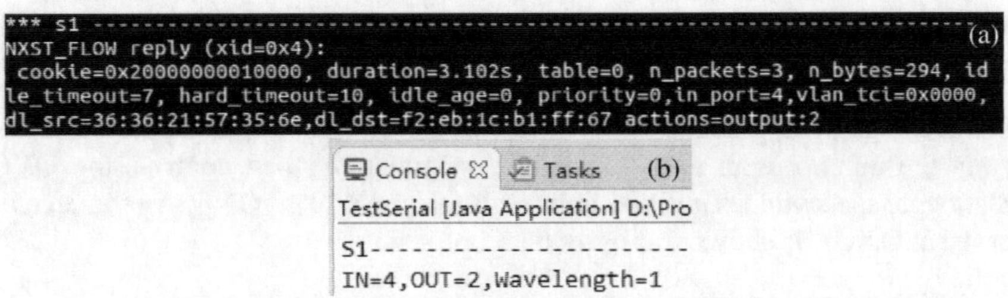

Figure 3. (a) Flow entry of virtual switch, (b) Information after translating by the translation layer software.

3. Experimental platform demonstration of optical network based on OpenFlow protocol

3.1. Style and spacing

The experimental platform of optical network based on OpenFlow protocol is shown in Figure 4.

The experiment platform consists of a control plane and a data plane. The control plane comprising an extended OpenFlow controller, an Open vSwitch software and a translation layer software, is responsible for the routing decision, the wavelength assignment and the configuration of optical switch states. In this control plane, the controller runs on the Linux operation system embedded in the IBM server while the Open vSwitch software and the translation layer software run on the Linux operation system embedded in the computer. The Open vSwitch software creates several virtual switches corresponding to OXC/ROADMs in the data plane. The topology of the virtual switches is consistent with the data plane, as shown in the virtual plane of the inset in the Figure 4. The data plane composed of FPGA and several optical nodes such as OXC and ROADM, is employed to implement the packet forwarding according to the optical switch-matrix level configuration table. The topology of the data plane includes ring, mesh and star networks. The performance of several network protocols and algorithms can be evaluated in this experiment platform, and many functions such as the multicast, the multi-granularity switching, the load balancing, the protection switching and etc., are verified.

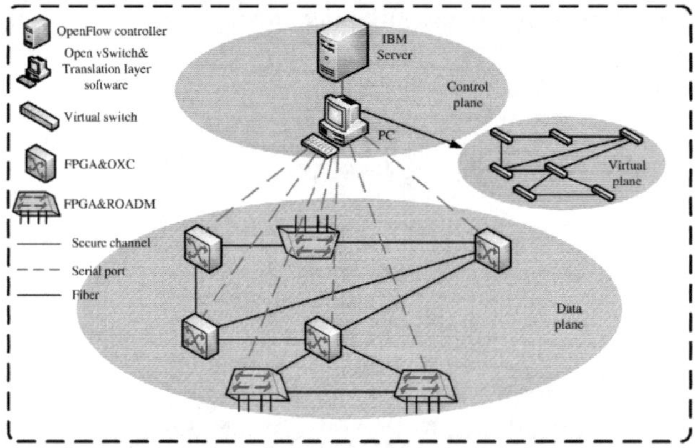

Figure 4. Experimental platform of optical network based on OpenFlow protocol.

3.2. Experimental demonstration

3.2.1. Connectivity test. The structure of the data plane under the connectivity test is shown in Figure 5. Two hosts (h1 and h4) are connected to the virtual switches (S1 and S3) corresponding to OXC1 and OXC3. The IP address of h1 and h4 is 10.0.0.1 and 10.0.0.4 respectively. On the host h1, ping 10.0.0.4 command is utilized to test the network connectivity. The virtual switch S1 receives the request information and looks up the flow table. If the matched flow entry is not found, S1 passes the request to the OpenFlow controller. The controller preforms RWA Algorithm, chooses the shortest path OXC1-OXC3, assigns wavelength 1 and sends flow entries to S1 and S3. Afterwards, the translation layer software reads out the input, output and wavelength information, then controls the OXCs via serial ports. The OXCs toggle the switches and the light path from OXC1 to OXC3 is connected. The output waveform observed at OXC3, is shown in Figure 6.

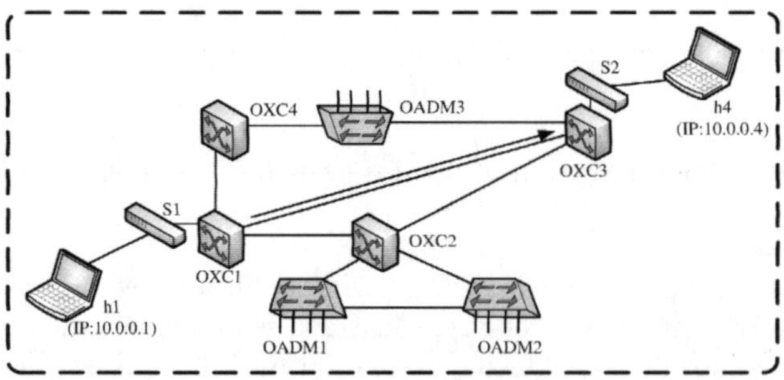

Figure 5. The structure of the data plane under the connectivity test.

Journal of Physics: Conference Series **679** (2016) 012025 doi:10.1088/1742-6596/679/1/012025

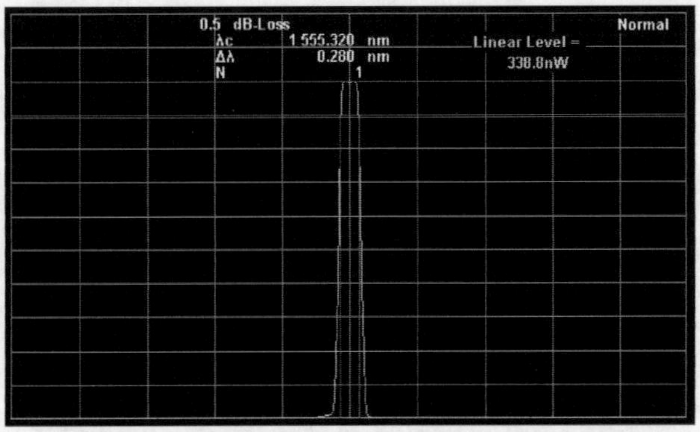

Figure 6. The output waveform at OXC3.

3.2.2. Protection switching experiment. The structure of the data plane under the protection switching experiment is shown in Figure 7. On the basis of the connectivity test above, the link between OXC1 and OXC3 is disconnected. When the topology of the network is changed, the controller performs the routing strategy again and sends the flow entries. The link is switched as OXC1-OXC2-OXC3, and the service continues to be transmitted.

The experimental result of the protection switching experiment is shown in Figure 8. When the link between OXC1-OXC3 is broken, the network loses packets. After 6.98 ms, the network reestablishes the links and switches the service.

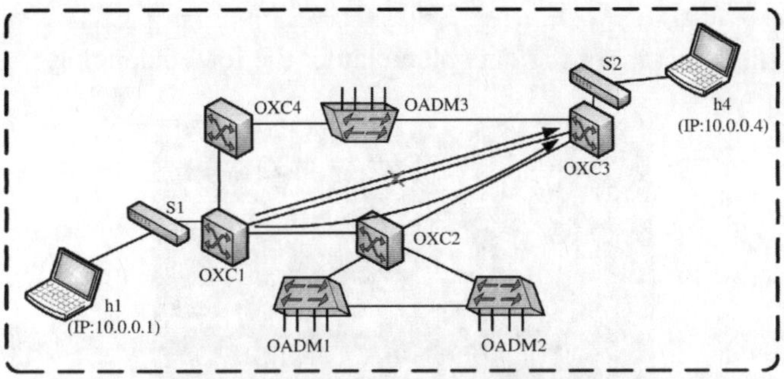

Figure 7. The structure of the data plane under the protection switching experiment.

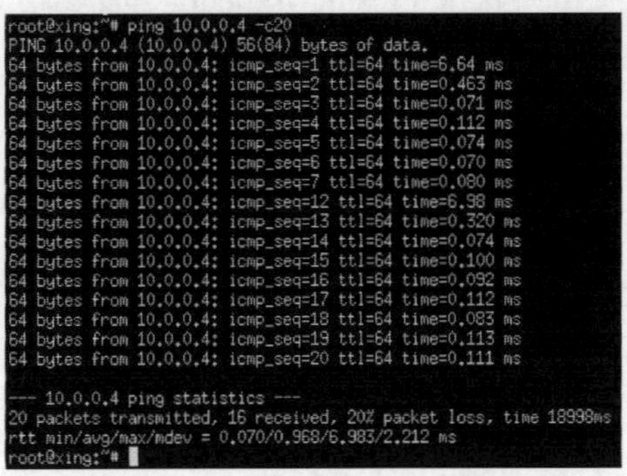

Figure 8. The experimental result of the protection switching experiment.

Journal of Physics: Conference Series **679** (2016) 012025 doi:10.1088/1742-6596/679/1/012025

3.2.3. Load balancing experiment. The structure of the data plane under the load balancing experiment is shown in Figure 9. The hosts h8 and h9 are both connected to the virtual network of which IP address is 10.0.0.100.

The experimental result of the load balancing experiment is shown in Figure 10. On the host h1, the command ping 10.0.0.100 is executed. The h1 receives the replies from the host h9 and the output waveform is observed at OADM1. This result illustrates that the host h1 is connected to 10.0.0.100 via the link OXC1-OXC2-OADM1. However, when the command ping 10.0.0.100 is executed on the host h4, the h4 receives the replies from the host h8 and the output waveform is observed at OADM2. This result illustrates that the host h4 is connected to 10.0.0.100 via the link OXC3-OXC2-OADM2. The experiment above shows that the network can choose different hosts and different switches to process services when the same services are received.

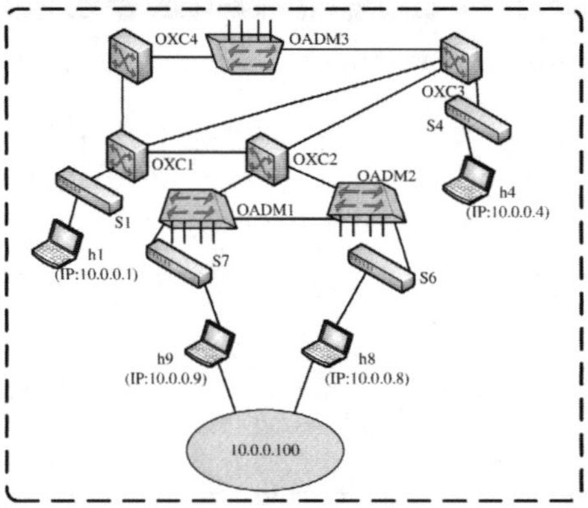

Figure 9. The structure of the data plane under the load balancing experiment.

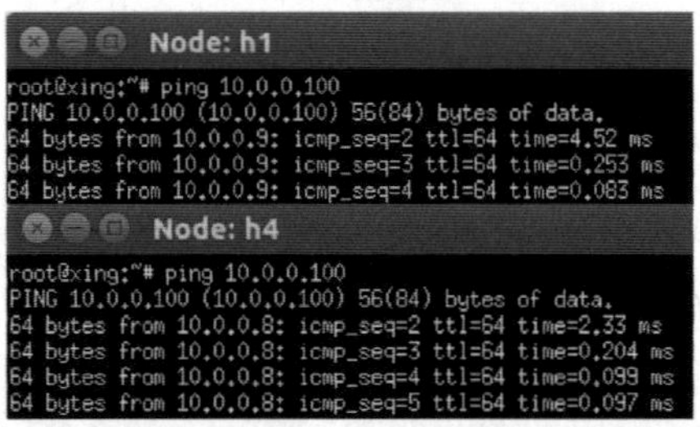

Figure 10. The experimental result of the load balancing experiment.

4. Conclusion

In this paper, a novel OpenFlow-enabled optical node is proposed. An open source OpenFlow controller is extended on the routing strategies and the experiment platform based on OpenFlow protocol for software defined optical network is designed. The feasibility and availability of the OpenFlow-enabled optical nodes and the extended OpenFlow controller are validated by the connectivity test, protection switching and load balancing experiments in this test platform. The structure of the OpenFlow-enabled optical node is clear and the function extension is flexible. The OpenFlow-enabled optical node can connect to several different optical devices via serial ports and the versatility is enhanced. The experiment platform includes different topologies and can be utilized to verify numerous protocols, algorithms and functions.

5. References

[1] Simeonidou D, Nejabati R and Azodolmolky S 2011 Enabling the Future Optical Internet with OpenFlow: A Paradigm Shift in Providing Intelligent Optical Network Services *In Proceedings*

of 13th International Conference on Transparent Optical Networks (ICTON) (Stockholm, Sweden) pp 1-4

[2] McKeown N et al. 2008 OpenFlow: enabling innovation in campus networks *ACM SIGCOMM Computer Communication Review* **38** pp 69-74

[3] https://www.opennetworking.org/sdn-resources/sdn-definition.

[4] Liu L Tsuritani T Morita I Guo H and Wu J 2011 Experimental validation and performance evaluation of OpenFlow-based wavelength path control in transparent optical networks *Optics Express* **19** pp 26578-26593

[5] Channegowda M Nejabati R and Simeonidou D 2013 Software-Defined Optical Networks Technology and Infrastructure: Enabling Software-Defined Optical Network Operations [Invited] *Journal of Optical Communications and Networking* **5** pp A274-A282

[6] http://openvswitch.org/.

[7] http://archive.openflow.org/.

[8] https://www.opennetworking.org/images/stories/downloads/specification/openflow-spec-v1.3.1.pdf

[9] http://www.projectfloodlight.org/.

Acknowledgments

This work is supported in part by the National Natural Science Foundation of China (NSFC) under Grant 61071123, the First HAEPC Science and Technology Project in 2015 (5217Q014006V).

Journal of Physics: Conference Series **679** (2016) 012026 doi:10.1088/1742-6596/679/1/012026

Laser Illumination Modality of Photoacoustic Imaging Technique for Prostate Cancer

Dong-qing Peng[1,2], Yuan-yuan Peng[1], Jian Guo[1] and Hui Li[1]*

1.Key Laboratory of Optoelectronic Science and Technology for Medicine (Ministry of Education of China), College of Photonic and Electronic Engineering, Fujian Normal University, Fuzhou, Fujian, 350007

2.School of Science, Jimei University, Xiamen, Fujian, 361021

Corresponding author's e-mail:hli@fjnu.edu.cn

Abstract. Photoacoustic imaging (PAI) has recently emerged as a promising imaging technique for prostate cancer. But there was still a lot of challenge in the PAI for prostate cancer detection, such as laser illumination modality. Knowledge of absorbed light distribution in prostate tissue was essential since the distribution characteristic of absorbed light energy would influence the imaging depth and range of PAI. In order to make a comparison of different laser illumination modality of photoacoustic imaging technique for prostate cancer, optical model of human prostate was established and combined with Monte Carlo simulation method to calculate the light absorption distribution in the prostate tissue. Characteristic of light absorption distribution of transurethral and trans-rectal illumination case, and of tumor at different location was compared with each other.The relevant conclusions would be significant for optimizing the light illumination in a PAI system for prostate cancer detection.

1.Introduction

In recent years, prostate cancer has been a serious health concern around the world, especially in the United States[1-4]. As a new biomedical imaging modality for various biomedical applications which combined contrast capability of optical imaging with the resolution of ultrasound imaging, Photoacoustic imaging (PAI) has recently emerged as a promising imaging technique for imaging prostate lesions [5-14]. But there was still a lot of challenge in the PAI for prostate cancer detection[13]. Some critical issues were required to be considered, such as a light delivery to the prostate with a sufficient penetrating depth and a minimal invasiveness. Nowadays, trans-urethral and trans-rectal light illumination had both been used as light irradiation model in the past PAI studies[11,14]. As we known,the temporal amplitude and profile of photoacoustic signals were spatial mappings of the absorbed optical energy.Therefore,knowledge of absorbed light distribution in prostate tissue was essential since that the distribution characteristic of absorbed light energy would influence the imaging depth and range of PAI.

According to the structural characteristic of prostate tissue, in this paper, 3D optical model of human prostate was established to calculate the light absorption distribution in the prostate tissue at transurethral and trans-rectal illumination modality. Followingly, a three-tumors-embedded prostate optical model with trans-urethral laser illumination was also employed. Characteristic of light absorption distribution of transurethral and trans-rectal case was compared with each other. Comparison even be made between the light absorption of the three tumors. The relevant conclusions would be significant for optimizing the light illumination in a PAI system for prostate cancer detection.

Journal of Physics: Conference Series **679** (2016) 012026 doi:10.1088/1742-6596/679/1/012026

2. Model and Method

A 3D triangular meshes prostate optical model was established based on human prostate morphology through programming, as shown in Fig.1,(a) by trans-rectal and (b) trans-urethral laser illumination. The prostate phantom's transverse diameter at *x*–axis, *y*-axis and the height were 2.0cm,3.8cm,3.18cm,respectively. In order to meet to the actual surrounding situation, here, a cube tissue with the diameter at *x*–axis, *y*-axis and the height 3cm,5cm,5cm,respectively, was used to surrounding around the prostate model. Then, an isotropic cylindrical diffusing light with diameter of 0.6mm,and effective light length 16mm,the center of light source was (x=0,y=0,z=15)mm, was employed in the transurethral light irradiation.and the center of light source for trans-rectal light illumination was (x=10,y=0,z=15)mm.

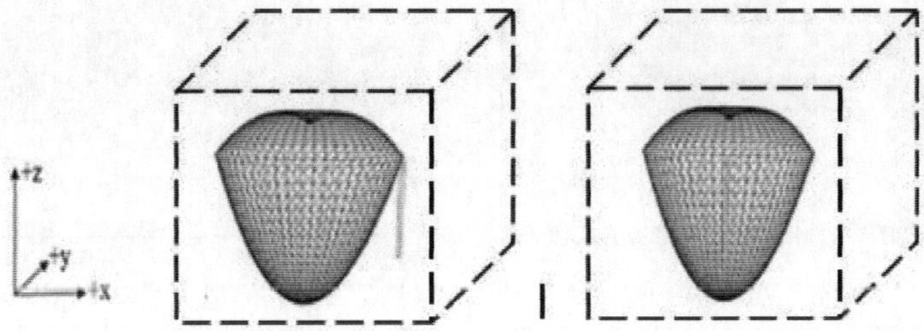

(a) (b)

Fig.1 3D prostate model. (a)by trans-rectal light illumination; (b)by trans-urethral laser illumination

In this section, the optical model was combined with Monte Carlo method[15-16]to investigate light energy absorption distribution in prostate tissue by different laser illumination. Its simulation process can be found in reference[16].The wavelength of these two light source was both set to 732nm, and the total energy of incident light was 1J. The recorded range of light distribution in this simulation was always keep as: *x*-axis (-15,15)mm; *y*-axis (-25,25)mm; z-axis(-10,40)mm; and the record step was set to 0.3mm,0.5mm and 0.5mm,respectively. And the total incident photon number was 500,000. A Cartesian coordinates was used for the simulation. According to optical properties of human prostate from foreign groups [17], main optical properties, i.e.absorption coefficient μ a, reduced scattering coefficient μ ′ s and refractive index n at 732nm wavelength for prostate optical model and surrounding cube tissue were specified as Table 1.

Table 1 Optical parameters of the simulation model

Tissue	μ_a (mm^{-1})	μ'_s(mm^{-1})	n
Prostate	0.03	0.19	1.43
Surrounding tissue	0.01	1.0	1.41

3. Simulation and Results

Using model as shown in Fig.1(a) and (b), the light absorption distribution patterns in the tissue model by trans-rectal light illumination and trans-urethral laser illumination were obtained, which was demonstrated in Fig.3.(a) and (b).

Comparison results between Fig.3(a) and Fig.3(b) showed that there was a full light absorption within the prostate tissue through trans-urethral laser illumination. It indicated that laser illumination from urethral would allow the prostate tissue to obtain a more efficient light absorption than trans-rectal light illumination, most of the light energy was absorbed within the prostate. In order to give a quantitative comparison of irradiance range of two kinds of light illumination model, contour line of the absorbed light energy was marked in the *xz*−plane where y=0. Value of contour line -9.4J/cm2 was logarithm of absorbed energy,i.e,log(A) .It demonstrated furtherly that most the light absorption can be found in the whole volume of prostate for transurethral case. However, lots of the light absorption were happened outside the prostate, there was absorption only in the vicinity of prostate since of the attenuation of surrounding tissue. In the Fig.4, profile of relative absorbed light energy distribution along z direction and contour line of the relative light absorption energy in the xz plane were further drawn, (a) along z direction in the xz-plane where x=-1.5mm. and (b) along x direction in the xz-plane where z=15mm.Two dash lines were used to show the boundary of model. From fig.4(a)and fig.(b).we could find that the relative absorbed light energy along z direction or x direction at internal illumination case was almost larger than that of external illumination, with the except of vicinity of prostate. Note that absorption at two sides of light source was similar at trans-urethral laser illumination case,which was good for analysis of photoacoustic image.This hint there was a strong light energy absorption in the deep position for internal illumination model considering the transrectal photoacoustic signal detection model.

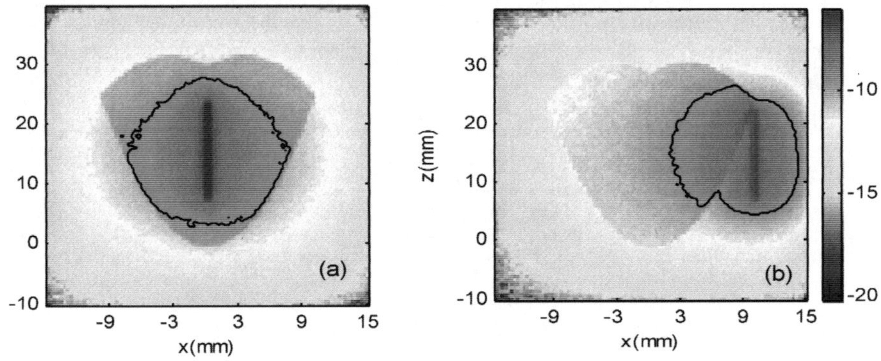

Fig.3 Light absorption pattern and contour line of the absorbed light energy distribution in the xz–plane where y=0.Value of contour line -9.4J/cm2 was logarithm of absorbed energy. (a)by trans-rectal light illumination;(b)by trans-urethral laser illumination.

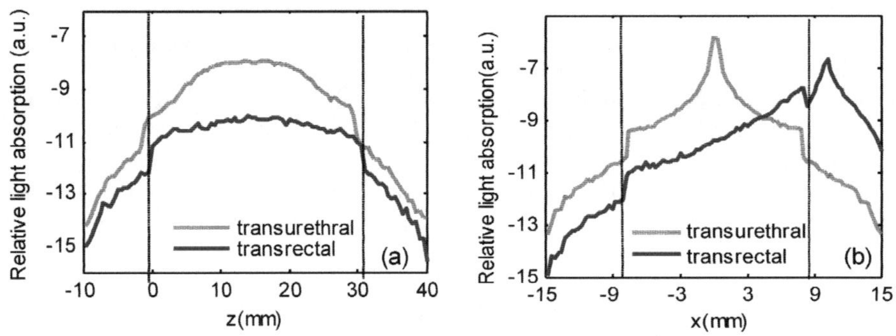

Fig.4 Profile of relative absorbed light energy distribution in the xz–plane.(a) along z direction where x=-1.5mm.(b) along x direction where z=15mm.

Next, a three-tumors-embedded prostate optical model with trans-urethral laser illumination was also established, as shown in Fig.5(a). Diameters of the three tumors were all set to 2 mm. The center of three tumors was located at (2,0,19)mm.(-2,0,19)mm,and (2,0,12)mm,respectively. The center of light source was (x=0,y=0,z=15.9)mm Considering tumor has a strong light absorption capacity, here, the absorption coefficient of tumor was set to be 10 times of the background prostate tissue. Main optical properties at 732nm wavelength for prostate optical model were specified as the absorption coefficient of μ_{ap}=0.03 mm^{-1}, scattering coefficient μ'_s =0.19mm^{-1}, refractive index n=1.43. and for tumor, the absorption coefficient of μ_{ac}=0.30 mm^{-1}, scattering coefficient μ'_s =0.135mm^{-1}, refractive index n=1.40.

The light absorption results for the tumours-embedded prostate model were shown in Fig.5(b) for xz-plane. It can be seen that light absorption was happened everywhere of the prostate tissue model. It was worth noting that there was a symmetrical light absorption property around the light source. Simulation results revealed that the light absorption of tumor1 and tumor3 irradiated by the cylindrical diffusing light was roughly alike as shown in Fig.6(a), which illustrates a better uniform lateral illumination range of cylindrical diffusing light source. These hinted that the cylindrical diffusing light would be suitable to the 3D scanning of PAI and was a proper light source for PAI.

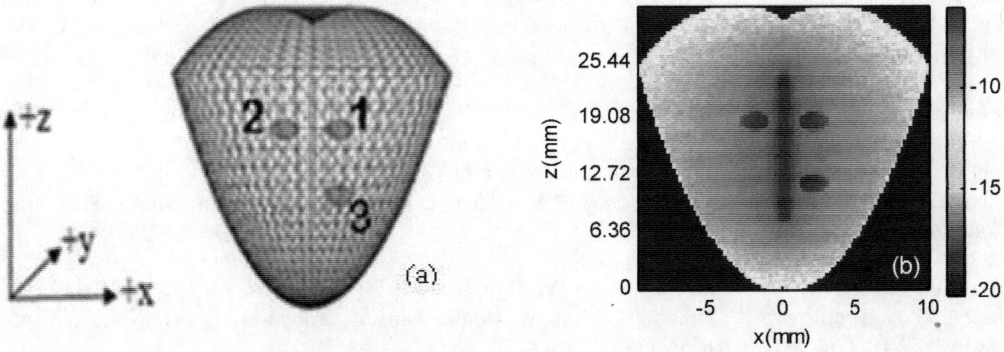

Fig.5 Embedded tumors optical model (a) and its light absorption distribution (b)

Meanwhile, it can be seen from Fig.6(b) that the laser absorption energy in tumor-1 was similar to that of tumor-2 which was more far from the light source. Results demonstrated the light absorption of tumor at left side of the light source was the same to that of tumor at right side because of the symmetrical light emitting characteristic of cylindrical diffusing light through internal irradiation model via urethra. Therefore a further imaging depth could be obtained since a ultrasound transducer was located outside of prostate tissue.

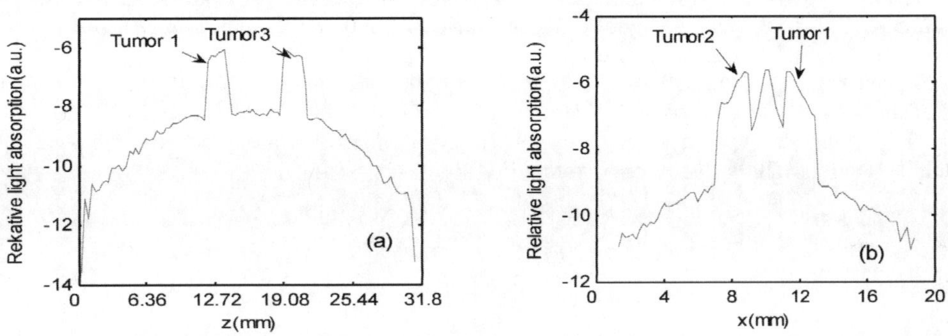

Fig.6 Profile of absorbed light energy distribution through center of tumors (a)along z direction (b)along x direction

4.Conclusion

In a word, a numerical modeling was still an important and effective way of gaining a deeper understanding of the complex interactions of the laser-tissue interaction process. In this paper, a 3D optical model of human prostate was established and was combined with Monte Carlo simulation method to calculate the light absorption distribution in the prostate tissue. A three-tumors-embedded prostate optical model with trans-urethral laser illumination was also employed. Characteristic of light absorption distribution of transurethral and trans-rectal case was compared with each other. Results show there was a more efficient light absorption in the whole volume of prostate using transurethral light illumination, and there was a strong light energy absorption in the deep position for internal illumination. Comparation results also demonstrated the light absorption of tumor at two side of the light source was the same because of the symmetrical light emitting characteristic of cylindrical diffusing light through transurethra irradiation model. Therefore a further imaging depth could be obtained since a ultrasound transducer was located outside of prostate tissue. In contrast, there was absorption only in the vicinity of prostate boundary since of the attenuation of surrounding tissue, and light absorption was quite small at the other side of urethral. Therefore, only half of the volume of the prostate has the possibly of imaging. The relevant conclusions would be significant for optimizing the light illumination in a PAI system for prostate cancer detection, and be helpful to the choice and the placement of laser source in the system.

Acknowledgment

This work has been sponsored by Natural Science Foundation of China (No.61178089), Natural Science Foundation of China (No.81201124), Fujian provincial key program of science and technology (No.2011Y0019), Fujian provincial education science research project(JA15125). Natural Science Foundation of Fujian

Province(No.2014J01226).and supported by the Key Laboratory of OptoElectronic Science and Technology for Medicine of Ministry of Education, Fujian Provincial Key Laboratory for Photonics Technology, Fujian Normal University,China(Grant No.JYG1417)

References

[1]Siegel R, Naishadham D, Jemal A 2013 *CA-Cancer J.Clin.***63** 11.
[2] Siegel R, Ma J M, Zou Z H, Jemal A 2014 *CA-Cancer J.Clin.***64** 9.
[3] Dogra V S, Chinni B K, Valluru K S, Joseph J V, Ghazi A, Yao J L, Evans K, Messing E M, Rao N A 2013 *J.Clin.Imaging Sci.***3** 41
[4] Kuo N, Kang H J, Song D Y, Kang J U, Boctor E M 2012 *J.Biomed.Opt.***17** 066005.
[5] Wang X D, Roberts W W, Carson P L, Wood D P, Fowlkes J B 2010 *Biomed.Opt.Express.* **1** 1117.
[6] Yaseen M A, Ermilov S A, Brecht H P, Su R, Conjusteau A, Fronheiser M, Bell B A, Motamedi M, Oraevsky A A 2010 *J.Biomed.Opt.***15** 021310.
[7] Spirou G M, Vitkin I A, Wilson B C, Whelan W M, Henrichs P M, Mehta K, Miller T, Yee A, Meador J, Oraevsky A A 2004 Proc.SPIE 5320 44.
[8] Bauer D R, Olafsson R, Montilla L G, Witte R S 2011 *J.Biomed.Opt.***16** 026012.
[9] Harrison T, Zemp R J 2011 *J.Biomed.Opt.***16** 080502.
[10] Su J L, Bouchard R R, Karpiouk A B, Hazle J D, Emelianov S Y 2011 *Biomed.Opt.Express* **2** 2243.
[11] Bell M A L, Kuo N P, Song D Y, Kang J U, Boctor E M 2014 *J.Biomed.Opt.***19** 126011.
[12] Xie W M, Li L, Li Z F, Li H 2012 *Proc.SPIE* 8553 85532V.
[13] El-Gohary S H, Metwally M K,Eom S,Jeon S H,Byun K M,Kim T S 2014 *Biomed.Eng. Lett.***4** 250.
[14] Valluru K, Chinni B,Bhatt S,Dogra V,Rao N,Akata D 2010 *IEEE International Conference on Imaging Systems and Techniques* Thessaloniki,Greece,July 1-2,2010,p121.
[15] Ren S,Chen X,Wang H,Qu X,Wang G,Liang J,Tian J 2013 *Plos One* **8** 1.
[16] Li H,Tian J, Zhu F,Cong W,Wang L V,Hoffman E A,Wang G 2004 *Acad. Radio.l* **11** 1029.
[17] Zhu T C, Dimofte A,Finlay J C, Stripp D,Busch T,Miles J,Whittington R, Malkowicz S B,Tochner Z,Glatstein E,Hahn S M 2005 *Photochem photobiol.* **81** 96.

Journal of Physics: Conference Series **679** (2016) 012027 doi:10.1088/1742-6596/679/1/012027

Design, Fabrication and Demonstration of a 1×20 Multimode Interference Splitter for Parallel Biosensing Applications

***N Najeeb [1,3], Y Zhang [1], C J Mellor [2] and T M Benson[3]**

[1]Department of Electrical and Electronic Engineering, University of Nottingham Ningbo China, 199 Taikang East Road, Ningbo, 315100, China.
[2]School of Physics and Astronomy, University of Nottingham, University Park, Nottingham NG7 2RD, UK
[3]Department of Electrical and Electronic Engineering, University of Nottingham, University Park, Nottingham NG7 2RD, UK.
*najla.najeeb@nottingham.edu.cn

Abstract. This paper presents the experimental achievement of a silicon-on-insulator 1×20 MMI splitter andsimulation evaluations of the TE-like and TM-like mode MMI splitters for parallel biosensing applications. Device fabrication technology and optical characterisation results are provided.

1. Introduction

In 1973,OlofBrynghal used Fresnel image planes and a pinhole array to demonstrate the formation of multiple images of an object based on self-imaging techniques [1]. In 1975 R. Ulrich reviewed this phenomenon on multimode planer slab waveguides [2, 3] which was explained in more detail by Heaton et al. [4] in 1992. In 1995 Soldano and Pennings[5] stated the principle as "*Self-imaging is a property of multimode waveguides by which an input field profile is reproduced in single or multiple images at periodic intervals along the propagation direction of the guide*".

Multimode Interference (MMI) devices are based on this self-imaging principle where the input image is reproduced into multiple images at the output with high uniformity [5]. Compared to Y-branches, MMI splitters can be made very compact and sharply increase tolerance to fabrication errors, as sharp edges near the branching points are avoided. MMI devices also have wide optical bandwidths, low cross talk and polarisation independence and are widely used in integrated optics as power splitters [6, 7], modulators [8], multiplexers [9], switches [10] and Mach-Zehnder interferometers [7].

1×N MMI splitters (where N is the number of outputs) have been theoretically and experimentally investigated widely especially for SOI based devices because of their compactness due to high index contrast, low losses, and compatibility with Si electronics and biological samples [11-16]. The highest number of outputs presented theoretically is 64 [17] and experimentally is 12 [13].This paper reports on the experimental achievement of 20 outputs.

Although MMI splitters are widely used in telecommunication, the devices presented in this paper are developed for biosensors to enable parallel detection and identification of multiple analytes in multiple samples with extremely high sensitivity of detection[18]. This is achieved by connecting a Mach-Zehnder Interferometer (MZI) via S-bends to each output of the MMI splitter. The S-bends are used to space-out the MZI on the chip. For the device to work as an evanescent field sensor it should be designed to operate in TM polarisation mode to give enhanced sensitivity.

2. Design of a MMI Device

Figure 1 shows the schematic diagram of a 1×2 MMI structure.

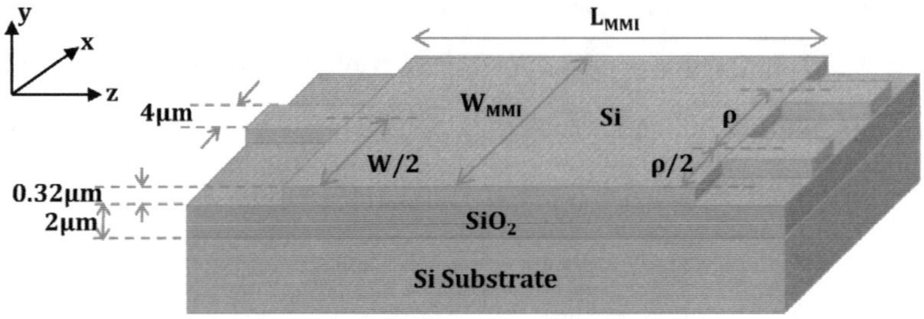

Figure 1: Schematic diagram of a1x2 MMI structure.

Inside the MMI cavity the beam is split into numerous modes and the symmetric intensity pattern is repeated periodically at the intervals of Λ along the guide:

$$\Lambda = \frac{n_{eff} {W_{MMI}}^2}{\lambda_0} \tag{1}$$

where$n_{eff}$and $W_{MMI}$are the effective refractive index and width of the MMI cavity respectively, and $\lambda_0$is the operating wavelength.

The N images formed are equally spaced across the multimode guide. The distance between the images is known as the Pitch, $\rho = W_{MMI}/N$. In the present work the pitch is fixed at 15μm to prevent interference between the output waveguides. W_{MMI} is determined by the number of output waveguides, N, where $W_{MMI}=N\rho$.

Using these values the approximate MMI length, L_{MMI} is calculated by Equation 2:

$$L_{MMI} = \frac{\Lambda}{N} = \frac{n_{eff} {W_{MMI}}^2}{N\lambda_0} \tag{2}$$

Equation 2 gives only an approximate length as the modes are not strictly confined to the multimode MMI region [15]. The optimum cavity length is found using a 3D BPM, and attained when the power of all the outputs isequal or almost the same. The MMI device has the lowest insertion loss and highest uniformity at this cavity length.

Figure 2 shows a 1×20 MMI simulated on RSoft Beam Prop. The waveguide width and height are kept at 4μm and 0.32μm respectively in this paper and theoperating wavelength is 1.55μm.

Journal of Physics: Conference Series **679** (2016) 012027 doi:10.1088/1742-6596/679/1/012027

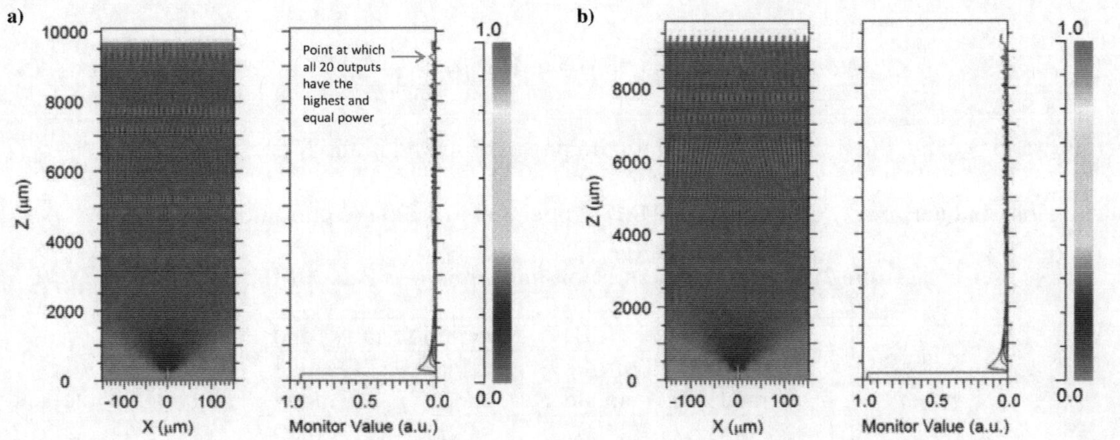

Figure 2: Illustrating TE-like mode simulation for 1x2 MMI with a) longer cavity length and b) optimum cavity length.

3. Simulation Evaluation

3.1. Polarization Dependence

From Table 1 it can be seen that the optimum lengths for theTM-like mode is significantly shorter than for theTE-like mode making the device polarization sensitive. This is an advantage for evanescent field based biosensors which operate in TM-like mode to obtain high sensitivity.

Table 1: Optimum 1xN MMI cavity length for TE-like and TM-like modes.

N	W_{MMI} (µm)	L_{MMI} calculated (µm)	L_{MMI} Optimum (µm)	
			TE-like mode	TM-like mode
2	30	889.7	910.1	802.6
3	45	1334.5	1361.2	1200.4
5	75	2224.3	2269.7	2008.3
8	120	3558.9	3630.4	3205.1
12	180	5338.3	5443.4	4804.2
15	225	6672.9	6803.0	6006.2
20	300	8897.2	9063.5	8012.2

3.2. Insertion Loss and Uniformity

Insertion loss and uniformity are the two most commonly used Figures of Merits to compare the efficiency of MMI devices [19, 20]. Insertion loss is defined as:

$$Loss\ insertion(dB) = -10log_{10}\left(\sum_{n=1}^{N}\frac{P_n}{P_{in}}\right) \tag{3}$$

where P_{in} is the input power and P_n is the power of the nth port.

The Non-uniformity is defined as:

$$Non\text{-}uniformity(dB) = -10log_{10}\left(\frac{P_{min}}{P_{max}}\right) \quad (4)$$

whereP_{min} and P_{max} are the maximum and minimum power observed in the *N* ports.

Insertion loss and uniformity calculated for MMI splitters up to 1×20 are presented in Table 2.

Table 2: Insertion loss and Non-uniformity for 1×N MMIs.

N	Insertion Loss (dB)		Non-uniformity (dB)	
	TE-like mode	TM-like mode	TE-like mode	TM-like mode
2	0.11	0.07	0.00	0.00
3	0.12	0.12	0.00	0.00
5	0.16	0.17	0.01	0.01
8	0.19	0.28	0.02	0.01
12	0.21	0.40	0.04	0.03
15	0.23	0.49	0.04	0.03
20	0.23	0.73	0.06	0.08

In theory it is assumed that all guided waves are confined to the MMI cavity whereas in reality there will be leaky modes and radiation modes [4]. Therefore, losses increase as the MMI length increases when N increases [16].

3.3. Optical Bandwidth and Fabrication Tolerance

The wavelength dependence or optical bandwidth is evaluated at the cost of a maximum additional loss of 1dB. Theoretically the optical bandwidth is inversely proportional to the number of outputs [21]. Fabrication tolerance is defined as the 1×N MMI cavity width (W_{MMI}) variation thatproduces an additional loss of 1dB. Table 3 shows that the bandwidth reduces as the number of outputs increases, however, the fabrication tolerance stays the same.

Table 3: Bandwidth and Fabrication Tolerance for 1×N MMIs.

N	Optical Bandwidth (nm)		Fabrication Tolerance (µm)
	TE-like mode	TM-like mode	
2	81	63	±1
3	60	41	±1
5	36	26	±1
8	23	16	±1
12	15	12	±1
15	12	9	±1
20	9	8	±1

3.4. Wafer Core Thickness Tolerance

The core height of the SOI devices depends on the wafer purchased. The wafer used for this experiment has a standard deviation of 19.8 nm on a silicon core thickness of 320nm and buried oxide of 2µm. Therefore MMIs were designed to have +/- 20nm of silicon core height tolerance.

Journal of Physics: Conference Series **679** (2016) 012027 doi:10.1088/1742-6596/679/1/012027

Figure 3 shows that the length of the MMI cavity increases linearly as the number of outputs increases; while the number of outputs increases, the tolerance for core thickness decreases.

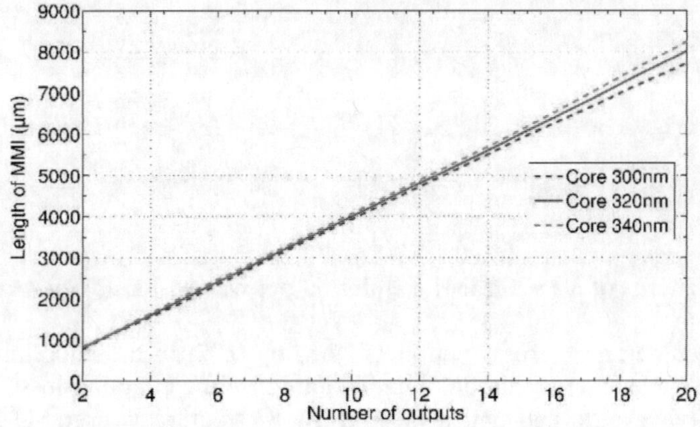

Figure 3: Optimum MMI cavity length for silicon core thickness 0.32μm and +/-20nm for 1×N MMI.

4. Device Fabrication

A 1×20 MMI was fabricated using photolithography and etch process as schematically illustrated in Figure 4. A commercially available SOI from SOITEC was used with a 0.32μm silicon layer on top of 2μm buried oxide layer.

The wafer was thoroughly cleaned by sonicating in solvents then dried with Nitrogen and dehydrated at 150°C for 30 minutes. Positive resist AZ6612 was spin coated at 4000rpm for 30 seconds to obtain a 1.2μm thick resist layer and soft baked at 110°C for 50 seconds. The resist was then patterned by exposing UV light for 6 seconds at 9mJ/s in a mask aligner (Karl Suss MJB-3). The exposed sample was developed for 30 seconds using AZ726MIF developer, rinsed with water and dried with nitrogen.

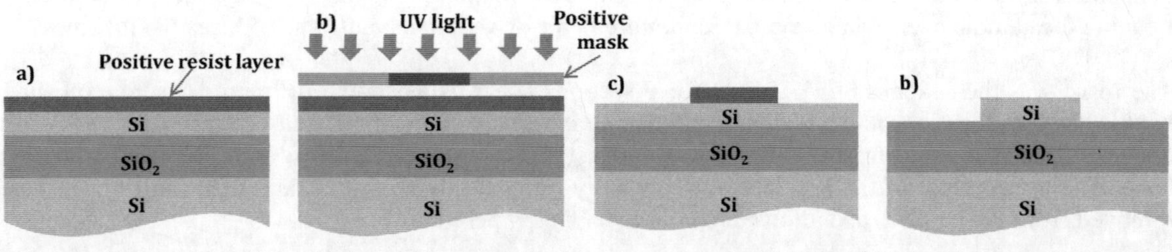

Figure 4: Schematic diagram of photolithography and etch process; a)Spin coating positive resist, b)UV exposure through a mask, c)Developed photoresist patterns and d)Etched silicon patterns.

The silicon was etched for 110 seconds by inductively coupled plasma (ICP) etching using CORIAL 200L plasma etcher and the resist was removed with oxygen plasma for 105 seconds. In both processes the etching time included an extra 20 seconds to compensate for different etch rates at different point of the sample. This may result in etching in to the oxide layer slightly, but this will not affect the function of the product. Figure 5 shows the microscopic image of the outputs of an etched MMI.

Journal of Physics: Conference Series **679** (2016) 012027 doi:10.1088/1742-6596/679/1/012027

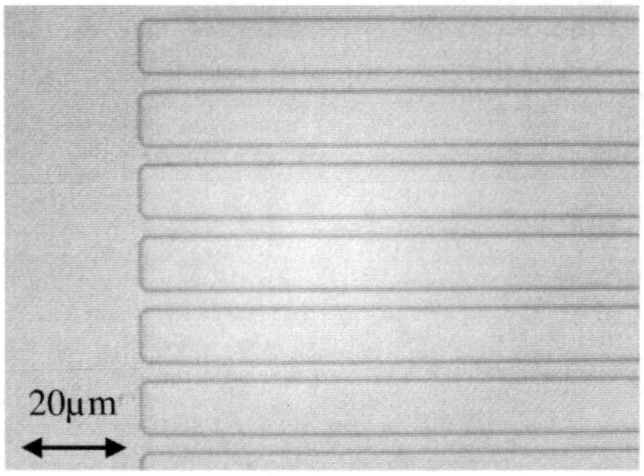

Figure 5: Etched Si patterns of a 1×20 MMI coupler output waveguides(View from microscope x100).

Since the input waveguide has a cross section of 4μm by 0.32μm the coupling loss from an input tapered fibre to the Si waveguide would be high. To minimise the coupling loss, and more effectively couple light into the waveguide and out, a tapered SU-8 structure is used.SU-8 is an epoxy-based negative photoresist that contains eight epoxy groups per molecule giving the polymer very high functionality. Regions exposed to UV light polymerise resulting in an extremely high crosslink density [22] and a high degree of thermal stability. SU-8 has been used as optical waveguide in biosensing applications as it provides good transmission properties and low propagation loss [23-25]. The tapered SU-8 structure is 2μm in height and a width that tapers from 10μm to 4μm.

The SU-8 photolithography process as schematically illustrated in figure 6 was altered slightly from the standard protocol to prevent crack formation and poor adhesion between the resist and silicon.

1ml of resist per 25ml of substrate diameter was dispensed on to the surface of the etched sample. To obtain 2μm thick layer the SU-8 was spread at 500rpm for 5 seconds and Spin coated at 2000rpm for 30 seconds as shown in Figure. The sample was then soft baked at 65°C for 15 minutes and 95°C for 1 minute on two different hotplates. The sample needs to be gradually heated and cooled to room temperature to reduce thermal shock which causes the formation of waves on the surface of the resist. Leaving the sample overnight at room temperature helps drive out solvents and reduces the tackiness.

The resist was then exposed to UV light for 6 seconds at 9mJ/s as shown in figure. A post exposure bake was then done in a hotplate where the temperature was gradually increased from 65°C to 95°C during a time span of 20 minutes. The exposed resist cross links and harden during the post exposure bake. The temperature of the hotplate was gradually ramped down to 40°C and then the sample was removed from the hotplate and allowed to cool to room temperature.

Once the sample was cooled down it was developed with EC solvent for 40 seconds and then washed with Isopropanol and dried with Nitrogen to obtain tapered structures as shown in Figure 7.

To make the SU-8 mechanically and thermally stable a hard bake was incorporated into the process. The sample was hard baked in an oven where the temperature was gradually ramped up from 65°C to 175°C during a 35 minute time period and then allowed to cool naturally at room temperature.

Journal of Physics: Conference Series **679** (2016) 012027 doi:10.1088/1742-6596/679/1/012027

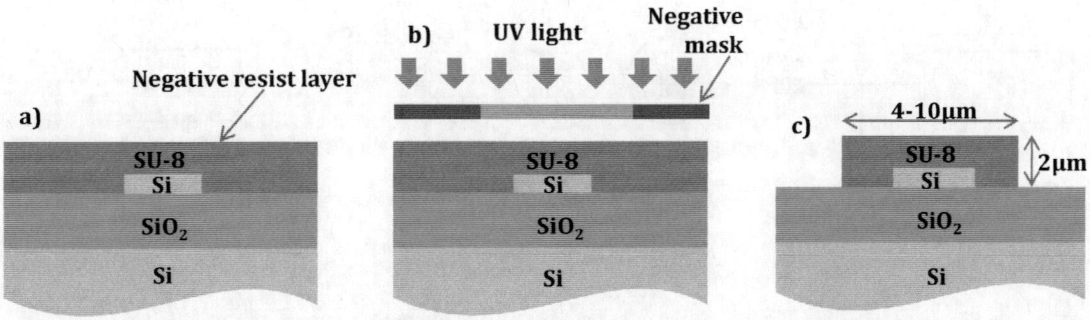

Figure 6: Schematic diagram of SU-8 photolithography process a) 2μm thick SU-8 resist later spin coated, b) UV exposure over negative mask and c) Final structure with SU-8 on Si.

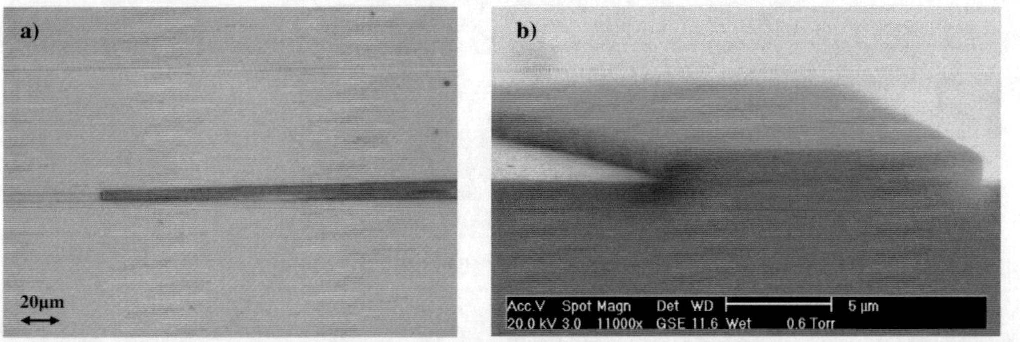

Figure 7:a) SU-8 taper well aligned with the waveguide (view from microscope x20) and b) Wafer cleaved through the SU-8(view from ESEM).

After fabrication the samples were cleaved perpendicular to the SU-8 tapers to obtain good quality end facets. A smooth facet is needed to minimise light scatter and maximise coupling efficiency. First a layer of positive resist was spin coated and soft baked for 30 seconds at 110°C in order to protect the device features. Two lines were scribed on either side of the tapered waveguide on a manual diamond scriber and then placed on a rubber mat with the sample features facing down. A small pressure was applied with a pen just enough to crack the wafer. Figure 7 b shows an ESEM image of a wafer cleaved smoothly through the SU-8.

5. Experimental Characterisation Results

An optical bench set up was used as shown in figure 8 to evaluate the 1×20 MMI. A tuneable laser (Agilent 8164B Lightwave measurement system from 1465μm-1575μm) and a red diode was connected to a tapered fibre with a spot diameter of 2.5μm was used to couple light into the input SU-8 tapered waveguide of the MMI. A power meter was used to measure the light intensity of the light in front of the camera.

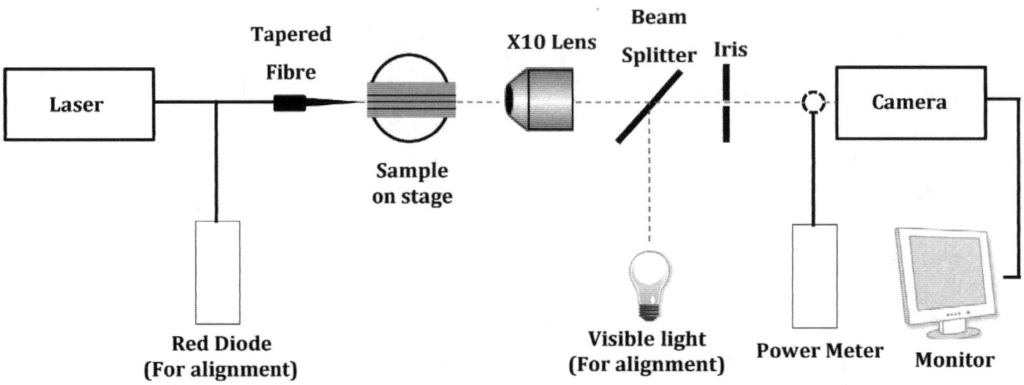

Figure 8: Schematic representation of Optical Bench set-up.

Figure 9 shows the successful operation of a 1×20 MMI splitter as seen from the infrared camera. It can be seen that the outputs are multimoded as expected.

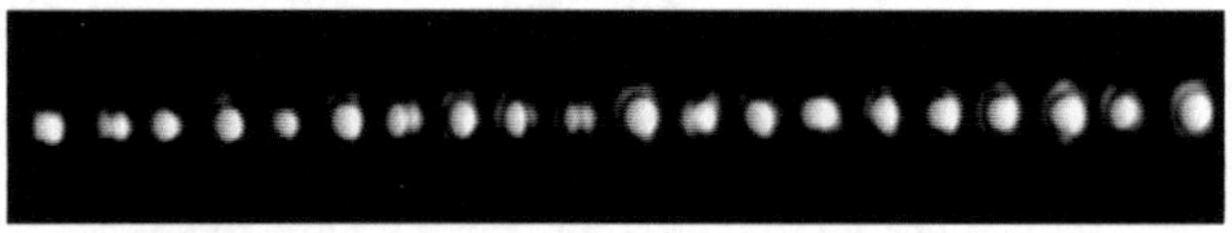

Figure 9: Outputs of a 1×20 MMI splitter.

6. Conclusion

A 1×20 MMI splitter has been designed, fabricated and tested in the lab. This device can be further improved by reducing the silicon core thickness from 0.32 μm to 0.22 μm which will greatly reduce the footprint of the MMI and will be more suitable for standard SOI fabrication. By using narrow single mode input and output waveguides and connecting N individual Mach-Zehnder interferometers to the outputs, this device can be used for evanescent field biosensing applications [18, 26].

7. References

[1] O. Bryngdahl, "Image formation using self-imaging techniques," *J. Opt. Soc. Am.,* vol. 63, pp. 416-419, 1973.

[2] R. Ulrich, "Light-propagation and imaging in planar optical waveguides," *Nouvelle Revue d'Optique,* vol. 6, p. 253, 1975.

[3] R. Ulrich, "Image formation by phase coincidences in optical waveguides," *Optics Communications,* vol. 13, pp. 259-264, 1975.

[4] J. Heaton, R. Jenkins, D. Wight, J. Parker, J. Birbeck, and K. Hilton, "Novel 1-to-N way integrated optical beam splitters using symmetric mode mixing in GaAs/AlGaAs multimode waveguides," *Applied Physics Letters,* vol. 61, pp. 1754-1756, 1992.

[5] L. B. Soldano and E. C. M. Pennings, "Optical multi-mode interference devices based on self-imaging: principles and applications," *Lightwave Technology, Journal of,* vol. 13, pp. 615-627, 1995.

[6] D. S. Levy, P. Kyung Hyun, R. Scarmozzino, R. M. Osgood, C. Dries, P. Studenkov, *et al.*, "Fabrication of ultracompact 3-dB 2 x 2 MMI power splitters," *Photonics Technology Letters, IEEE,* vol. 11, pp. 1009-1011, 1999.

[7] L. Zhuang, P. Wang, Z. Cao, G. Cao, Z. Wang, and F. Qiu, "A novel MMI–MZ all-optical switch," *Optik - International Journal for Light and Electron Optics*.

[8] P. Yue, B. M. Mao, X. Yi, Q. N. Li, and Z. J. Liu, "A MMI-based ultra-linear high-gain modulator and its performance analysis," *Optics & Laser Technology,* vol. 53, pp. 33-39, 2013.

[9] M. R. Paiam and R. I. MacDonald, "A 12-channel phased-array wavelength multiplexer with multimode interference couplers," *Photonics Technology Letters, IEEE,* vol. 10, pp. 241-243, 1998.

[10] W. Fan, Y. Jianyi, C. Limei, J. Xiaoqing, and W. Minghua, "Optical switch based on multimode interference coupler," *Photonics Technology Letters, IEEE,* vol. 18, pp. 421-423, 2006.

[11] A. Z. Chowdhury, "Performance study of silica-on-silicon based multimode interference (MMI) optical coupler," *Photonic Sensors,* pp. 1-9, 2013/06/07 2013.

[12] P. Trinh, S. Yegnanarayanan, F. Coppinger, and B. Jalali, "Compact multimode interference couplers in silicon-on-insulator technology," in *Lasers and Electro-Optics, 1997. CLEO'97., Summaries of Papers Presented at the Conference on*, 1997, pp. 441-441.

[13] D. Kwong, Y. Zhang, A. Hosseini, Y. Liu, and R. T. Chen, "Demonstration of Rib Waveguide Based 1x12 Multimode Interference Optical Beam Splitter on Silicon-on-Insulator," in *Photonics Society Summer Topical Meeting Series, 2010 IEEE*, 2010, pp. 221-222.

[14] A. Maese-Novo, R. Halir, S. Romero-García, D. Pérez-Galacho, L. Zavargo-Peche, A. Ortega-Moñux, *et al.*, "Wavelength independent multimode interference coupler," *Optics express,* vol. 21, pp. 7033-7040, 2013.

[15] A. Cleary, S. Garcia-Blanco, A. Glidle, J. S. Aitchison, P. Laybourn, and J. M. Cooper, "An integrated fluorescence array as a platform for lab-on-a-chip technology using multimode interference splitters," *Sensors Journal, IEEE,* vol. 5, pp. 1315-1320, 2005.

[16] Q. Lai, M. Bachmann, and H. Melchior, "Low-loss 1× N multimode interference couplers with homogeneous output power distributions realised in silica on Si material," *Electronics letters,* vol. 33, pp. 1699-1700, 1997.

[17] T. Rasmussen, J. K. Rasmussen, and J. H. Povlsen, "Design and performance evaluation of 1-by-64 multimode interference power splitter for optical communications," *Lightwave Technology, Journal of,* vol. 13, pp. 2069-2074, 1995.

[18] N. Najeeb, Z. Yaping, C. Mellor, and T. Benson, "Photonic biosensor chip for early-stage cancer diagnosis," in *Transparent Optical Networks (ICTON), 2015 17th International Conference on*, 2015, pp. 1-4.

[19] R. Yin, J. Yang, X. Jiang, J. Li, and M. Wang, "Improved approach to low-loss and high-uniformity MMI devices," *Optics Communications,* vol. 181, pp. 317-321, 2000.

[20] Y. Shi, D. Dai, and S. He, "Improved performance of a silicon-on-insulator-based multimode interference coupler by using taper structures," *Optics Communications,* vol. 253, pp. 276-282, 2005.

[21] P. A. Besse, M. Bachmann, H. Melchior, L. B. Soldano, and M. K. Smit, "Optical bandwidth and fabrication tolerances of multimode interference couplers," *Lightwave Technology, Journal of,* vol. 12, pp. 1004-1009, 1994.

[22] Available: http://www.nanotech.dtu.dk

[23] D. Esinenco, S. Psoma, M. Kusko, A. Schneider, and R. Muller, "SU-8 micro-biosensor based on Mach-Zehnder interferometer," *Rev. Adv. Mater. Sci,* vol. 10, pp. 295-299, 2005.

[24] B. Shew, C. Kuo, Y. Huang, and Y. Tsai, "UV-LIGA interferometer biosensor based on the SU-8 optical waveguide," *Sensors and Actuators A: Physical,* vol. 120, pp. 383-389, 2005.

[25] B. Y. Shew, Y. C. Cheng, and Y. H. Tsai, "Monolithic SU-8 micro-interferometer for biochemical detections," *Sensors and Actuators A: Physical,* vol. 141, pp. 299-306, 2008.

[26] Z. Yaping, "Multi-channelled waveguide chemical and biochemical optical sensing device," Patent, Great Britain Patent, 2010.

Application of a colorimeter for turbidity measurement

Yizhang Wen, Yingtian Hu, and Xiaoping Wang*

State Key Laboratory of Modern Optical Instrumentation, Department of Optical Engineering, Zhejiang University, Hangzhou, China

E-mail address: xpwang@zju.edu.cn

Abstract. This paper describes a new turbidity transducer based on color measurement. The absorbance of solutions reflects the absorption and scattering of suspended particle for incident light which could determine the turbidity of solutions. The experimental results indicate that there are good linear relationships between chromaticity and turbidity. The new way is suitable for continuous monitoring of water turbidity in the wide range.

1. Introduction

Accurate determination of turbidity is of great importance for many aspects of water resource development and river basin management [1]. Turbidity is an important parameter in monitoring water pollution and water eutrophication because spread of protozoa can be greatly reduced when the turbidity is very low [2]. Regular monitoring of turbidity can help to detect trends that might indicate increasing waste discharge or excessive algal growth [3].

There are two main types of turbidity transducers. The first kind of common turbidity transducers measure light attenuation. The absorbance of solutions reflects the absorption and scattering of suspended particle for incident light which could determine the turbidity of solutions. The response of this kind of turbidimeters varied uniformly when the turbidity became very large. So turbidimeters using transmission method were suitable for solutions with large turbidity.

The other kind of nephelometric turbidity transducers measure scattered light at a certain angle to the incident beam [4]. The standardized method makes use of the Tyndall effect, measuring scattered light at an angle 90 deg relative to the incident light. However, there are variations regarding the used lamps and measurement angles in the turbidimeters [5]. Until recently most instruments for water turbidity measurement have been constructed to meet the requirements in the U.S. standard. Such turbidimeters measure the particle scattering of light from a tungsten lamp and the wavelength of the spectral peak response of the detection system should be between 400 and 600 nm. A revised procedure has been introduced by The International Organization for Standardization. This standard (ISO 7027) prescribes that particle light scattering should be measured in the near infrared (λ= 860 nm) and it sets stringent conditions to parallelism for the incident light. The use of near infrared light minimized interferences caused by the presence of dissolved light-absorbing substances. Since colored organic substances are often present in surface water and drinking water from surface sources, it should be assumed that instruments using the longer wavelength will produce better turbidity measurements.

Apart from the two basic methods mentioned above, there are many other improved ways for the turbidity measurement. Kontturi et al. [5] developed a new sensor which makes use of the total internal reflection of a laser beam at the liquid–prism interface, and the turbidity is assessed using the concept of laser speckle pattern. Good correlation between the standard deviation of dynamic speckle and turbidity value. Shenoy [6] developed Optical fibre probes in the measurement of scattered light, and describe the application of turbidity sensing. Wang et al. [2] developed a instrument for real-time measurement of low turbidity by using time-correlated single photon counting technique.

In this paper, upon on the transmission method, we have developed a new system and algorithm realizing turbidity measurement based on chromaticity coordinates measurement. The mechanism underlying the turbidity measurement will be discussed, and supported by the experimental results. The proposed realtime turbidimeter in this paper can be used to continuously monitor the turbidity of water.

2. Experimental

2.1. Experimental setup

The visible spectral acquisition system consisted of a light source, a sample cell, and a spectrometer (Fig. 1). The pulsed xenon light source (PX-2, Ocean Optics, USA) produced visible light, which was collimated by a lens and then passed through the sample cell; the emergent light was collected into the spectrometer (USB2000+, Ocean Optics, USA) by the converging lens. The spectral data were shown in the computer.

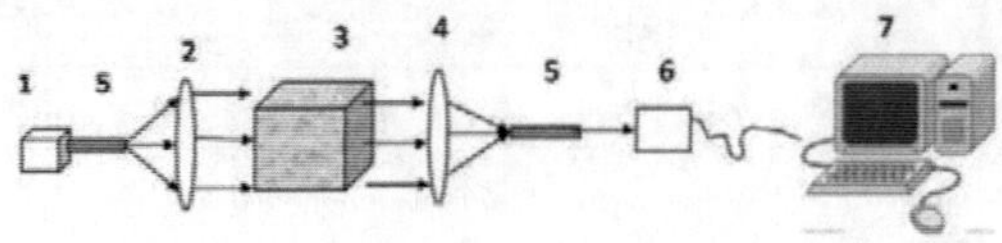

Fig. 1 Experimental setup

2.2. Measurement

In this study, turbidity solutions of 2, 5, 10, 20, 30, 40, and 50 NTU were prepared by diluting a 400 NTU formazine turbidity solution. The water used for all solutions consisted of deionized water supplied by a MilliQ water purification system (Millipore, Billerica, MA, USA). A series of experiments were carried out with the experimental setup shown in Fig. 1, and a 10 mm optical path was used during measurements.

3. Results and Discussions

3.1 The absorbance in the visible spectrum for different turbidity solutions

In theory, the absorbance was satisfied with Lambert Beer's law, the equation was shown as below:

$$I = I_0 \times 10^{-kCL} \tag{1}$$

where k is the absorption coefficient, C is the concentration of substance, L is the optical path, I_0 and I are the original and emergent light intensity.

The absorbance of solutions reflects the absorption and scattering of suspended particle for incident light which could determine the turbidity of solutions. The light absorption experiments for different turbidity solutions were carried out to explore the absorption and scattering of suspended particle. As shown in the Fig. 2, the loss of light intensity became larger with the increase of turbidity in the visible region. Moreover, the change of absorbance is larger when the wavelength is shorter. Therefore, chromaticity coordinate of solutions could reflect the turbidity.

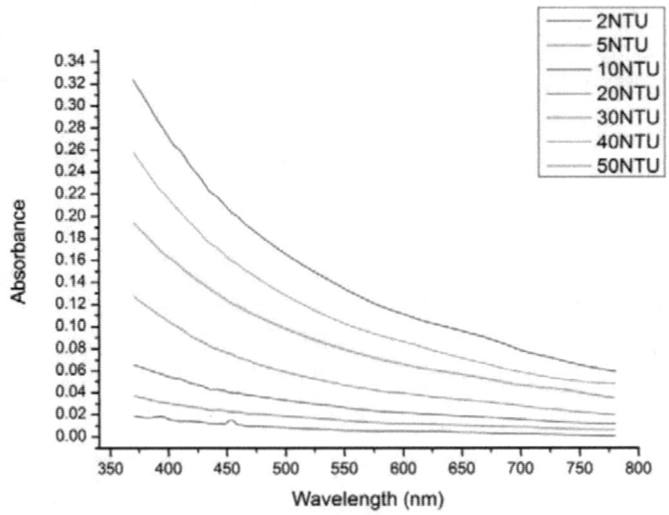

Fig. 2 The absorbance curves of different turbidity solutions in the visible range

3.2 The color coordinates of different turbidity solutions

Color tristimulus values are used to indicate the visual feeling for water color, the computational formula are listed below:

$$X = k\sum S(\lambda)10^{-A(\lambda)}x(\lambda)\Delta\lambda \tag{2}$$

$$Y = k\sum S(\lambda)10^{-A(\lambda)}y(\lambda)\Delta\lambda \tag{3}$$

$$Z = k\sum S(\lambda)10^{-A(\lambda)}z(\lambda)\Delta\lambda \tag{4}$$

where x(λ), y(λ), z(λ) are visual matching function, S(λ) is spectral power distribution of lighting luminous, the D65 light source is used to represent S(λ), A(λ) is absorbance of testing solutions.

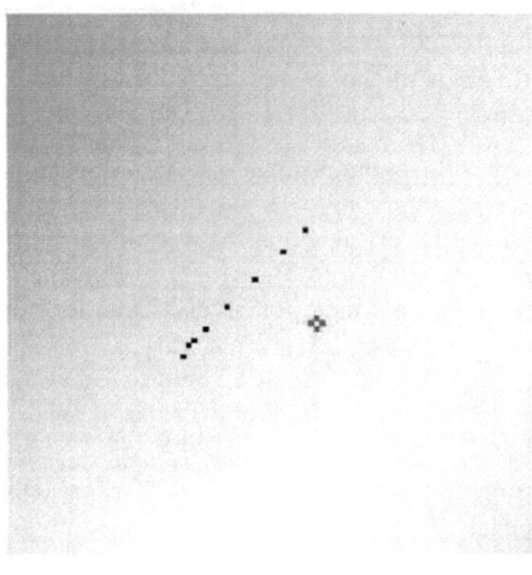

Fig. 3 The positions of different turbidity solutions in chromaticity diagram

The color coordinates xyz are normalization of XYZ values, the XYZ values of different turbidity solutions are calculated using equation (2), (3) and (4). The black squares shown in the Fig. 3 represented the positions in the chromaticity diagram of 0, 2, 5, 10, 20, 30, 40, 50 NTU turbidity solutions from the bottom up. These points can be distinguished clearly, and the relationships between xyz values in chromaticity diagram and turbidity are shown in Fig. 4.

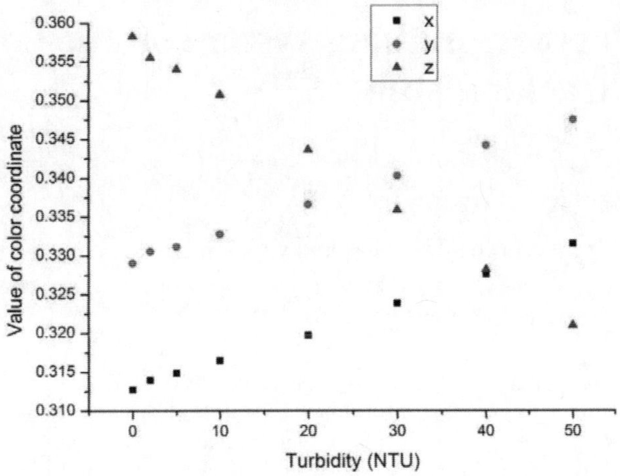

Fig .4 Relationships between xyz values in color diagram and turbidity

As seen in the Fig. 4, there are good linear relationships between xyz values and turbidity. The z value is more sensitive with the increase of turbidity. In the meantime, the method can be applied up to 50 NTU, and it is reliable up to that range.

4. Conclusion

In this paper, a new color measurement-based turbidimeter had been developed for accurate measurement of the turbidity. The feasibility of this instrument was demonstrated by series of turbidity measurement experiments. The results showed the turbidity could be stably measured in the range of 0-50 NTU.

Acknowledgment

The study was supported by National High Technology Research and Development Program of China (863 Program) (Project No.2014AA06A504)

Reference

[1] Octavian A. P *et al* 2007 Multibeam optical system and neural processing for turbidity measurement *IEEE Sens J* **7** 677-684
[2] Huanqin W *et al* 2015 Instrument for real-Time measurement of low turbidity by using time-correlated single photon counting technique *IEEE T Instrum Meas* **64** 1075-1083
[3] Haijiang T *et al* 2012 Design and characterization of a smart turbidity transducer for distributed measurement system *Sensors Actuat A-Phys* **175** 1-8
[4] D. H and G. A 1998 Comparison of nephelometric turbidity measurements using wavelengths 400-600 and 860 nm *Water Res* **32** 3143-3145
[5] Ville K *et al* 2009 Robust sensor for turbidity measurement from light scattering and absorbing liquids *Opt Lett* **32** 3743-3745
[6] M. R. S 2014 Optical fibre probes in the measurement of scattered light: Application for sensing turbidity *Pramana-J Phys* **82** 39-48

A fast assistant decision-making system on the emergent maneuver of the tracking ship

Qiong Huang[1], G H Xue[1] and X Q Ni[1]

[1]China Satellite Maritime Tracking and Control Department, Jiangyin

E-mail: huangqiong_nudt@163.com

Abstract. This paper studies a fast assistant decision-making system on the emergent maneuver of the tracking ship, adopting the design method of the emergent working state of the tracking ship based on the meteorological prediction , the virtual display technology based on the multi-stage mapping, and the 2-dimension area algorithm based on the line-scanning. It solves problems that the tracking ship met during working, such as the long TT&C time, the dense crucial observation arc, the complicated working flow, and the changeful scheme. It established the hard basement for the fast design of the emergency working state when the tracking ship in the awful sea conditions.

1. Introduction

In all kinds of spacecraft launching, the tracking ship undertake the key measuring assignment of the flight segment of spacecraft at sea. In some important tasks , such as the manned space flight, the tracking ship must work long and continuously, the working waters is in the higher latitude, the weather is often changeful, and the awful sea weather occur frequently. Thus it is usual for the tracking ship to adjust the work scheme to meet an emergent TT&C in preparation of tasks. It is difficult to complete a fast, accurate and visualized design of the emergent working state of the tracking ship by the traditional techniques.

For this, based on analyzing the influence of sea conditions on the fast design of the emergent working state of the tracking ship, This paper brings forwards a fast assistant decision-making system on the emergent maneuver of the tracking ship.

2. Key technology

2.1. *Design method of the emergent working state of the tracking ship based on the meteorological prediction*

The meteorological prediction of the waters in which the tracking ship working is imported into the working state design software with diagram, which is taken as the foundation to analyze spacecraft track, signal coverage, and task segment, to meet an emergent work condition design etc., and the analysis result can be demonstrated visually.

The total system adopts the three-stage structure, carrying out the separation that the data, logic operation and chart display. The total structure is divided into three modules, task analysis, graphics drive, and user interface. Figure 1 show the total system structure.

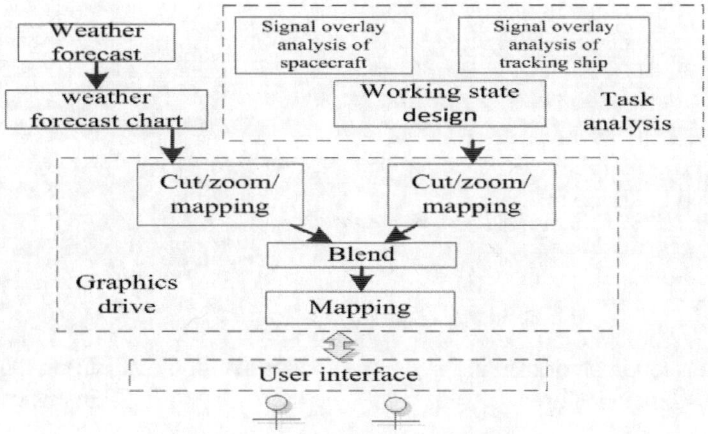

Figure 1. Total system structure.

2.2. Virtual display technology based on the multi-stage mapping

In order to make show district zoom and move freely, avoid the sketch flicker phenomenon when projecting directly that is easy to appear, and make the chart display more real-time, this system adopts a kind of virtual monitor technology, that is, before mapping the ground track of spacecrafts and the diagram of weather forecast to the show district, firstly complete all picture-related processing(such as overlying, cut, move, zoom etc.) in the memory, then project the picture to the show district, which can improve greatly the system processing efficiency and the final show effect.

Several graphic processing is needed in the human-computer interaction process, including a series of multi-level operation such as mapping, coordinate transition and area cutting. Therefore, the system adopts the multi-stage mapping technology which deal with the different level graphic operations respectively. The application of such technology can reduce the complexity and raise the flexibility of the system. Figure 2 shows the processing flow chart of the virtual display technology based on the multi-level mapping.

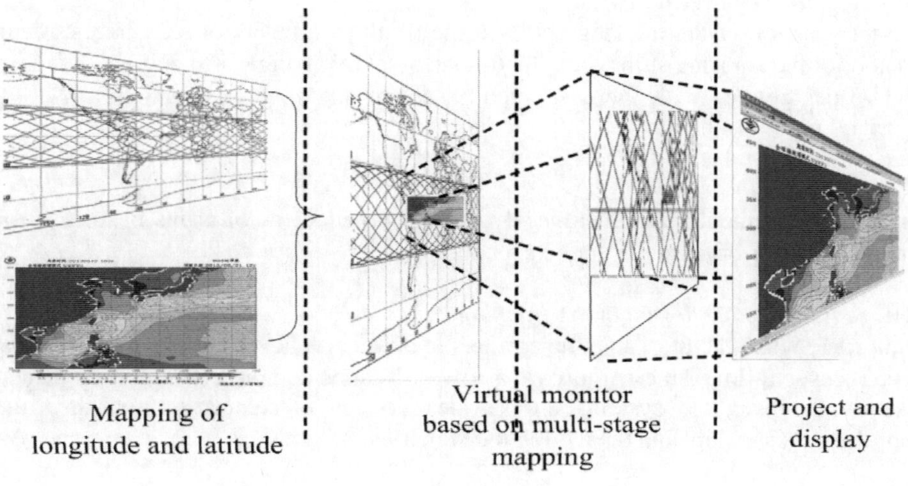

Figure 2. Processing flow of virtual display technology based on multi-stage mapping.

2.3. 2-dimensiona area algorithm based on the line-scanning

Journal of Physics: Conference Series **679** (2016) 012029 doi:10.1088/1742-6596/679/1/012029

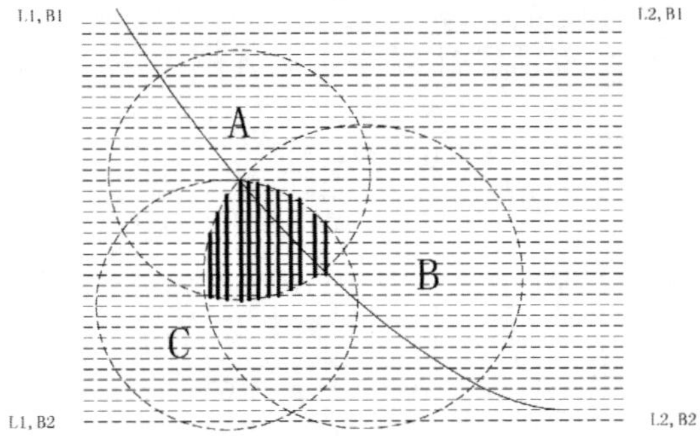

Figure 3. Intersection algorithm in 2-dimension area.

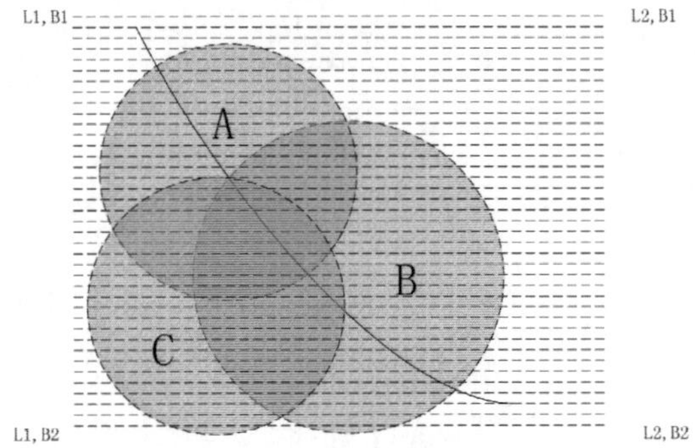

Figure 4. Combination algorithm in 2-dimension area.

Since the computer only can be used to deal with the discrete data, appropriate mechanism is needed for the continuous data processing of the 2-dimensiona area, like the intersection, combination and complement calculation. In this mechanism, the 2-dimensiona area should be dispersed firstly. Then a dispersed method is applied to actualize the logic calculation. The logic calculation sketch of the 2-dimensiona area is shown in figure 3 and figure 4.

3. Main functionality

The system has the following main functionality:

3.1. The weather forecast diagram can be imported as the base map of the work condition design.

According to the bound of longitude and latitude of sea area, the system can import and show the corresponding weather forecast diagram with the different resolution, size, and format that the weather system provides. It's convenient to design the emergent working state of the tracking ship based on the specific meteorological condition.

3.2. The tracking ship coverage analysis.

According to the location of the tracking ship and the track parameters of aerospaces, it can draw the coverage contour of the tracking ship when the tracking elevation angle is 0 degree, 3 degree, 5 degree, 7 degree and 75 degree etc, freely increase, modify, delete any coverage contour, and set their show style(type, weight, color of line etc.).

3.3. Signal coverage analysis.

Calculates and analyzes the signal coverage of aircrafts on all kinds of orbits by means of the flight track, the attitude parameters, and the antenna location.

3.4. The emergent work state design of the tracking ship.

Shows the sail route when the tracking ship carries out an emergency TT&C task. According to user's demands, calculates real-time the emergency maneuver distance and direction. To any measuring point users choose on the map, the system can provide services of calculating real-time the working parameters of the tracking ship, and displaying the statistics.

3.5. Show the trail of spacecrafts.

According to a variety of spacecrafts orbit elements, such as circular orbit, elliptical orbit etc., and flight arc, the system can show their ground-track and path number, and set the show style(type, weight, color of line etc.).It can also display the key TT&C segment and the common .TT&C segment differently.

3.6. Overlay area operation.
According to the cover district of spacecrafts on the sea, the projection on the sea of the cover district of the tracking ship to spacecrafts, and the districts users custom, this system can make the assemblage operations, such as combine, overlap, cut, move etc.

3.7. The user interface.
According to users' operations, the analysis results can be displayed visually, zoomed in and out in any multiple, and moved parallel. To any measuring point users choose on the map, the system can realize the real-time interactive functions, including of calculating the working parameters of the tracking ship, giving the tracking feature points, drawing the coverage diagrammatic sketch of the tracking ship to spacecrafts.

4. Conclusion
This system has been applied successfully in the fast and visualized adjustment of the emergent maneuver scheme of the tracking ship in the Shenzhou-10 mission, ensuring the design of the maritime TT&C implement scheme efficiently. The automated combination of the weather forecast result and the maneuver area of the tracking ship makes the design of the emergent maneuver scheme efficiently, the virtual display technology based on the multi-stage mapping improves the visualization of the system greatly, and the application of 2-dimensiona area algorithm based on the line-scanning raises the degree of automation of the design.

The system is of applied value in many fields like follow-up rendezvous and docking(RVD) task, as well as the technique support for the emergent maneuver in the subsequent maritime TT&C task. The future work is focused on the system flexibility optimization.

References
[1] Jian SL 2009 *The Conspectus on the TT&C technology based on the TT&C ship*（Beijing: National Defense Industry Press） pp 9-11,270-78

[2] Zhao WH 2013 *Terms in ocean-based space tracking technology*(Beijing: National Defense Industry Press)p 8

[3] Ni XQ, Huang K and Mao YX et al 2009 *Telecommunication Engineering*. **49(7)**pp13-7

[4] Liu B, Fu MH and Xue GH et al 2014 *Analysis and Design of the TT&C task based on the TT&C ship*(Beijing: National Defense Industry Press)pp168-84

A Combined Radio and Underwater Wireless Optical Communication System based on Buoys

Yuhang Song[1], Zheng Tong[2], Bo Cong[3], Xiangyu Yu[1,*], Meiwei Kong[1], Aobo Lin[1]
[1] Department of Ocean Engineering, Ocean College, Zhejiang University, Yuhangtang Road 866, Hangzhou, Zhejiang, 310058 P.R. China,
[2] Department of Petroleum Equipment, RIPED, PetroChina, No.20, Xueyuan Avenue, Haidian District, Beijing, 100083 P.R.China,
[3] China satellite maritime tracking and control department, Jiangyin, 214431 P.R. China,

Email: 21434093@zju.edu.cn

Abstract. We propose a system of combining radio and underwater wireless optical communication based on buoys for real-time image and video transmission between underwater vehicles and the base station on the shore. We analysis how the BER performance is affected by the link distance and the deflection angle of the light source using Monte Carlo simulation.

1. Introduction

With the rapid development of modern information technology and communication technology, communication systems are increasingly improved in many industries. However, the communication technology in underwater environments is still in a lower level. People are always confused about the underwater detection, control and communication. For example, if fishermen cannot get the information of fish resources and other underwater situation after overfishing, natural disasters or environmental pollution timely, there is no doubt that it will have a strong impact on the trading market of fisheries.

So how to monitor the underwater environment, collect underwater information, and transmit information to the shore is a serious problem to solve. As we all know that underwater vehicles transmit data to boats mainly by means of underwater acoustic communication which has low data rate and large time delay. And the collected data is saved to the memory and then analyzed on the shore. This method seriously restricts the development of large-scale, high-efficiency underwater operation.

In order to make up for the shortage of the method as mentioned above, we propose the scheme of combining radio and underwater wireless optical communication system based on buoys for real-time image, and video transmission between underwater vehicles and the base station on the shore.

Compared with acoustic technology, optical technology has the advantages of high bandwidth, cost-effectiveness and low-energy consumption [1-2]. Furthermore, long lifetime, low power consumption and high efficiency are the prominent advantages of light emitting diodes (LEDs) as light sources of underwater wireless optical communication systems [3-4]. Blue and green light with the

Journal of Physics: Conference Series **679** (2016) 012030 doi:10.1088/1742-6596/679/1/012030

outstanding advantages of low absorption loss and strong anti-interference ability is as preferred option.

2. System Description

The schematic diagram of our proposed scheme employing combined radio and underwater wireless optical communication system between underwater vehicles and base stations on the shore is shown in figure.1. The whole system mainly consists of a base station on the shore, a buoy on the sea and an underwater vehicle.

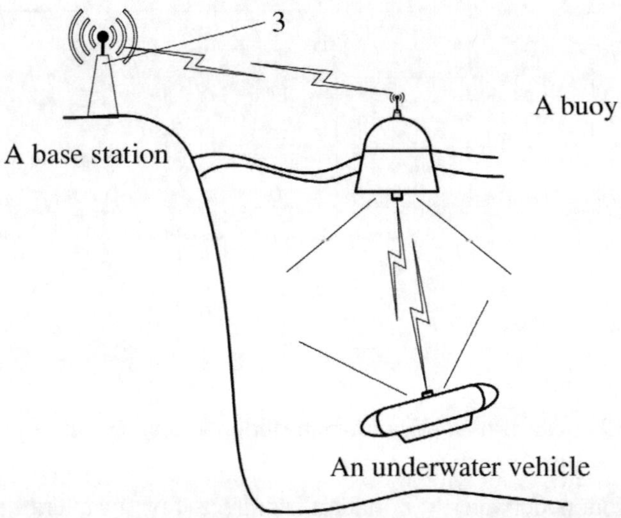

Figure 1. The schematic diagram of a combined radio and underwater wireless optical communication system based on buoys.

Underwater wireless optical communication is used to complete the transmission of image or video data and control signals between underwater vehicles and buoys. Radio communication is used for the transmission of signals between buoys and base stations. A main control unit equipped at the bottom of a buoy is utilized to process the light signal of visible light communication and radio communication.

Figure.2 (a), (b) and (c) show the principle diagram of the optical transmitter and receiver on the underwater vehicle, the buoy and the base station respectively.

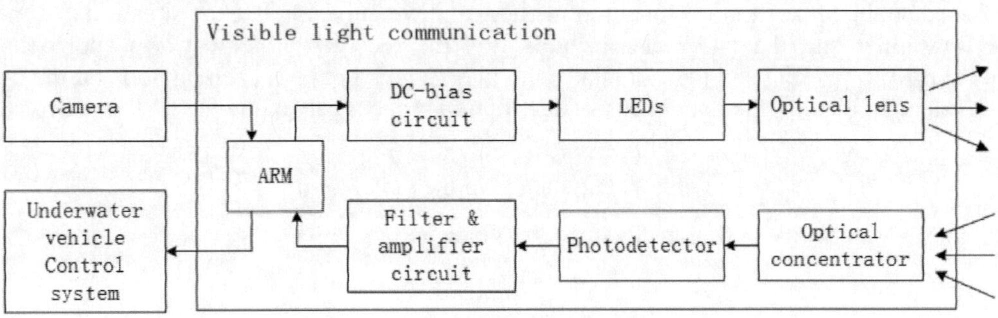

(a)

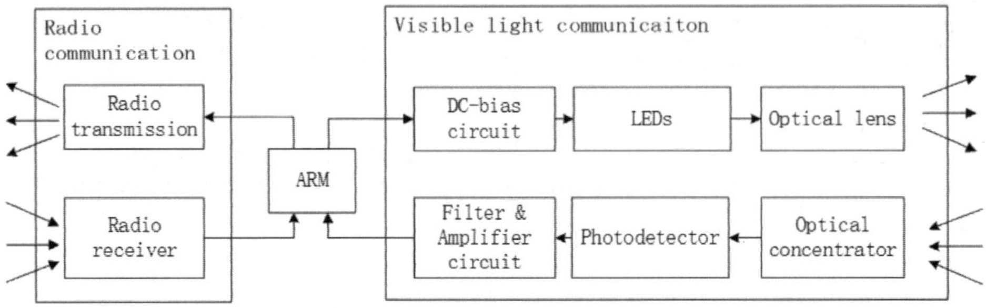

(b)

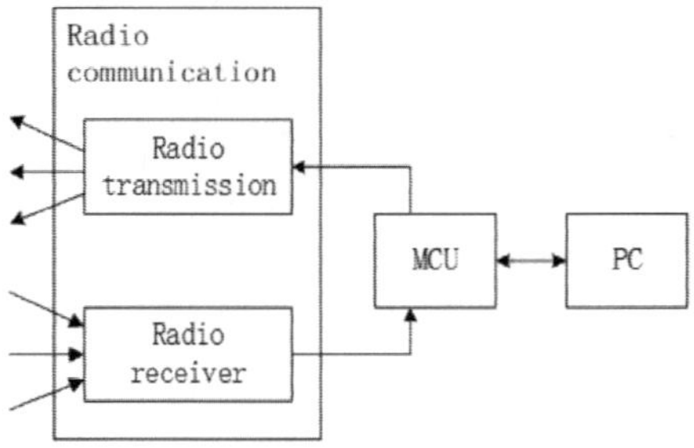

(c)

Figure 2. The principle diagram of the optical transmitter and receiver on: (a) the underwater vehicle, (b) the buoy, (c) the base station.

For uplink transmission, underwater information is collected by a camera carried on an underwater vehicle and captured by the ARM. The signal is modulated to LEDs by utilizing a bias-tee (BT). After emitted from LEDs, the light is diverged, launched into optical lens, transmitted in seawater, fed into the optical concentrator of a buoy, and focused on the high bandwidth photodetector. Through filtering and amplifying, the signal is sent to the ARM. Then the signal is transmitted to the base station on the shore by means of radio, demodulated and transmitted to the PC.

For downlink transmission, the base station on the shore sends the control signal through the reverse link to monitor the position of buoys and control the operation of the underwater vehicle.

3. Simulation and Results Discussion

In this simulation, we consider comprehensive modeling of the underwater optical communication channel based on the Monte Carlo simulation method, and we only simulate the situation that the buoy was the transmitter and the underwater vehicle was the receiver. Monte Carlo approach firstly generates numerous photon, and then simulates the interactions between each photon and the medium in the seawater [5]. Finally, the light intensity distribution at the receiver and the BER is calculated.

Table 1. Parameters in the simulation

Parameters	Value
λ	532nm
P	1W
ω	20°
a(λ)	0.069
b(λ)	0.08
c(λ)	0.15
θ	180°
d	20cm

The main parameters which are taken into consideration in our Monte Carlo simulation are displayed in Table 1. The characteristic of the transmitter includes the beam wavelength λ, the optical power P, and the divergence angle of the light source ω.

Absorption and scattering are the two main factors affecting light propagation in the seawater. These two effects are characterized by absorption coefficient $a(\lambda)$ and scattering coefficient $b(\lambda)$, and the summation of them is defined as extinction coefficient $c(\lambda)$.

The characteristic of the receiver contains field-of-view angle θ and aperture size d.

BER can be calculated by SNR considering OOK modulation. In the simulation, we consider that how the BER performance is affected by the link distance and the deflection angle of the light source.

To begin with, the underwater vehicle is right under the buoy, and the vertical length of the communication link is 45m and the light source has no deflection angle. Figure.3 (a) is the light intensity distribution at the receiver, and figure.3 (b) is the BER distribution of different distance from the center of the light spot. It shows that if the horizontal distance between the center of the light spot and receiver is less than 7.5m, the BER will be below the FEC limit ($2x10^{-3}$).

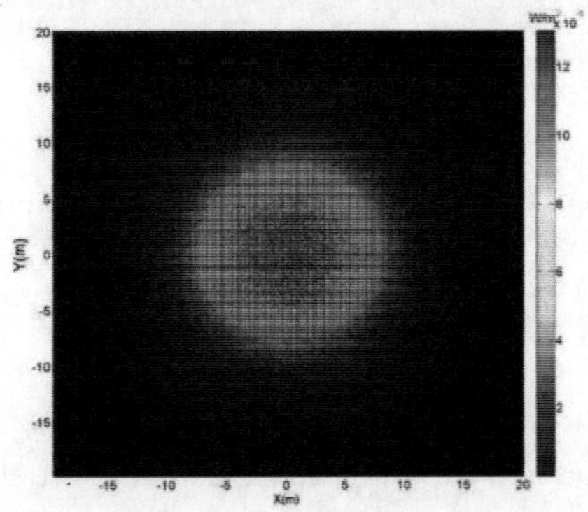

(a)

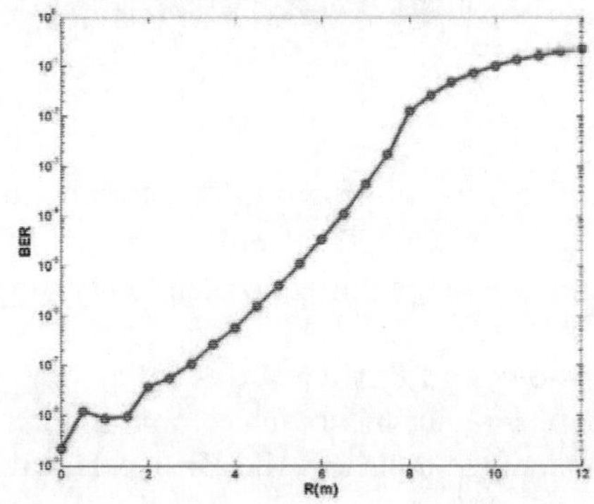

(b)

Figure 3. The vertical length of the communication link is 45m and the light source has no deflection angle. (a) The light intensity distribution at the receiver. (b) The BER distribution of different distance from the center of the light spot.

Then we only change the vertical distance between the buoy and the underwater vehicle from 46m to 58m. The result is displayed in figure.4, which shows that the maximum effective distance of communication can be 49.8m with the FEC coding.

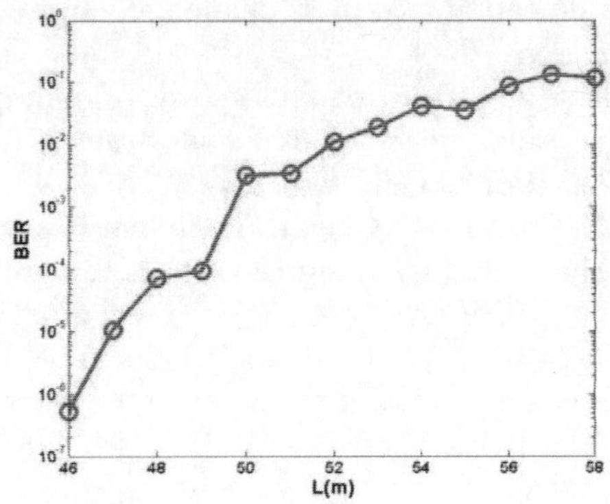

Figure 4. The BER distribution of the light spot center when the vertical length of the communication link is changed and light source has no deflection angle

In the end, we keep the vertical distance at 45m and change the deflection angle of the light source. Figure 5(a) is the light intensity distribution at the receiver when the deflection angle is 10°. Figure 5(b) illustrates that the maximum deflection angle of the light source can be 24° with the FEC coding. And the communication system can overcome the impact of wave by adjusting the position of the underwater vehicle.

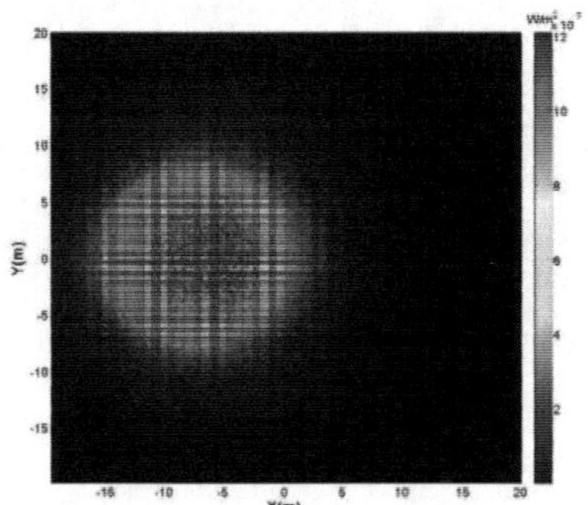

(a)

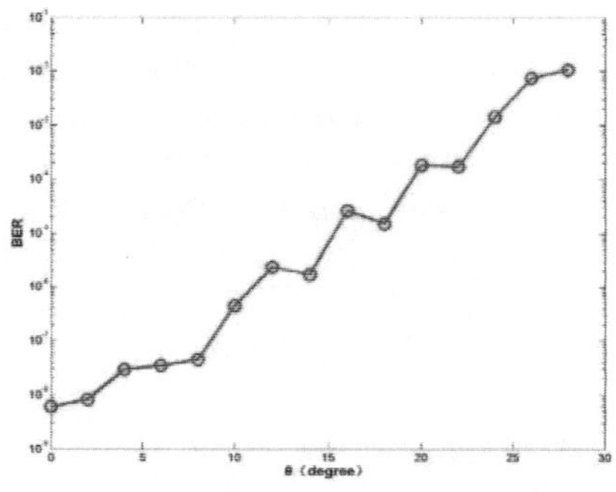

(b)

Figure 5. The vertical length of the communication link is 45m (a) the light intensity distribution at the receiver when deflection angle of the light source is 10° (b) the BER distribution of the center of the light spot when the deflection angle is changed.

4. Conclusions and Future Work

The combined radio and optical communication system based on buoys has been proposed in this paper. We analyze that how the BER performance is affected by the link distance and the deflection angle of the light source, and verify the feasibility of this communication system using Monte Carlo simulation method.

Building a mathematical mode, manufacturing a product prototype, and testing in an underwater environment are the next steps that we have to complete.

References

[1] Arnon S 2010 Underwater optical wireless communication network *Opt. Eng.* vol **49** no. 1 pp 1–6

[2] Anguita D, Brizzolara D, and Parodi G 2009 Building an underwater wireless sensor network based on optical: communication: research challenges and current results *2009 Third Int. Conf. Sens. Technol. Appl.* pp 476–9

[3] Tian B, Zhang F, and Tan X 2013 Design and development of an LED-based optical communication system for autonomous underwater robots *Advanced Intelligent Mechatronics (AIM), 2013 IEEE/ASME Int. Conf*

[4] Wu F M, Lin C T, Wei C C, Chen C W, Chen Z Y, Huang H T, and Sien C 2013 Performance comparison of OFDM signal and CAP signal over high capacity RGB-LED-Based WDM visible light communication *IEEE Photonics Journal* vol **5** no. 4 p 7901507

[5] Gabriel C, Khalighi M, Bourennane S, Leon P, and Rigaud V 2013 Monte-Carlo-based channel characterization for underwater optical communication systems *Opt. Commun. Networking, IEEE/OSA J.* vol **5** no. 1 pp 1–12

Acknowledgments

This work was supported in part by the National Natural Science Foundation of China (61301141) and Qianjiang Talent Program of Zhejiang Province (QJD1402014).

Journal of Physics: Conference Series **679** (2016) 012031 doi:10.1088/1742-6596/679/1/012031

Airborne Wireless Optical Communication System in Low Altitude Using an Unmanned Aerial Vehicle and LEDs

Meiwei Kong[1], Zheng Tong[2], Xiangyu Yu[1], Yuhang Song[1], Aobo Lin[1], Jing Xu[1,*]
[1] Department of Ocean Engineering, Ocean College, Zhejiang University, Yuhangtang Road 866, Hangzhou, Zhejiang, 310058 P.R. China
[2] Department of Petroleum Equipment, RIPED, PetroChina, No.20, Xueyuan Avenue, Haidian District, Beijing, 100083 P.R.China

E-mail: jxu-optics@zju.edu.cn

Abstract. In this paper, we demonstrate the feasibility of airborne wireless optical communication system using an unmanned aerial vehicle and LEDs. Monte Carlo simulation method is used to evaluate the performance of the communication channel. Considering OOK modulation, we illustrate how the BER performance is affected by the link distance, the divergence angel and the deflection angel of the light source.

1. Introduction

Ships, planes and satellites are three traditional platforms that can be used to implement data transmission with underwater equipment, such as unmanned underwater vehicles, submarines, underwater wireless sensor networks. However, shipborne and airborne communications spend a lot of manpower, material and financial resources and are inconvenience in bad weather. And satellite borne communication needs a relatively long period to obtain information, which seriously affects the efficiency of communication especially in the case of an emergency. With the advantages of convenience, cost-efficiency, and good motoring performance, an unmanned aerial vehicle system, including platform system, information collection system and ground control system, has been a new developing momentum. At present, unmanned aerial vehicles are mainly utilized to implement the collection of high-resolution images. There are almost no reports about unmanned aerial vehicles used for optical wireless communications with an underwater target.

Acoustic communication is the dominant method for underwater communication. However, it has some obvious disadvantages, such as large time delay, limited bandwidth, and high energy consumption. In order to make up for the deficiency as mentioned above, underwater wireless optical communication is becoming an increasingly popular alternative.

Light emitting diodes (LEDs), taking advantages of low cost, high efficiency[1], low attenuation in the 400nm-550nm range [2], and high data rates (up to 1 Gbps over a few meters) in seawater, are used as light sources for the proposed wireless optical communication system.

Based on lots of theoretical study, we proposed the airborne wireless optical communication system in low altitude using an unmanned aerial vehicle and LEDs which allows the transmission of real-time data, images and videos.

2. System Description

Figure.1 shows the schematic diagram of airborne wireless optical communication system in low altitude using an unmanned aerial vehicle and LEDs. Unmanned aerial vehicles and underwater vehicles are two important components of the system and both of them are equipped with wireless optical communication system.

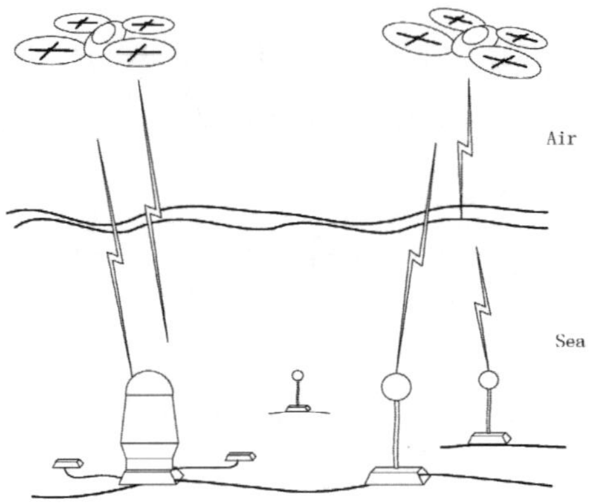

Figure.1. The schematic diagram of airborne wireless optical communication system in low altitude using an unmanned aerial vehicle and LEDs.

The principle diagram of the optical transmitter and receiver on the unmanned aerial vehicle and the underwater vehicles are presented in Figure.2 (a) and (b), respectively.

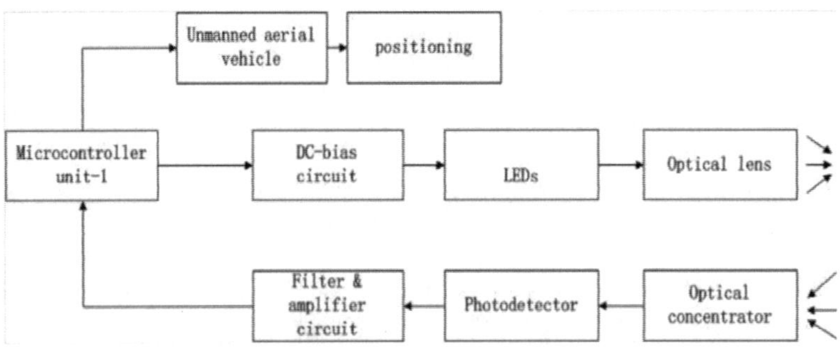

(a)

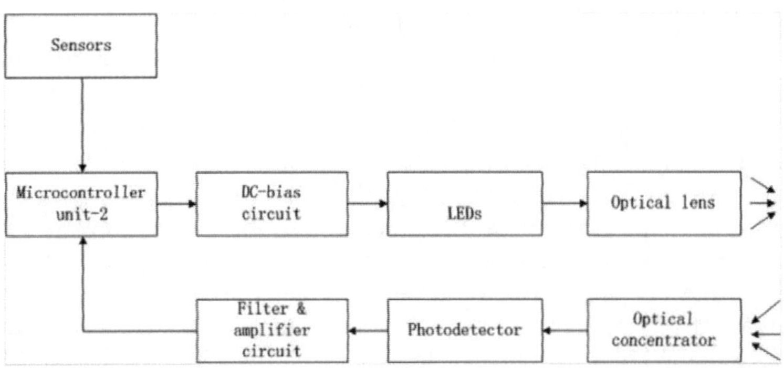

(b)

Figure.2. The principle diagram of the optical transmitter and receiver on: (a) the unmanned aerial vehicle (b) the underwater vehicles.

The aircraft through the positioning and navigation technologies detects underwater vehicles mounted at the fixed position. Light beam of LEDs, with bigger divergence angle and higher peak power, is launched into the first convex lens. When scanning certain area of the sea surface, a request signal is first sent to underwater vehicles.

Along with the background light of sky, light pulse beam is received by the optical detector, suffering from the attenuation of different depth and mass of sea water. In order to improve the sensitivity of the detector, we apply the diversity reception technology of multiple detectors. Meanwhile a filter is adopted to suppress the fluorescence in the sea and the residual background light of sky. Amplifier circuit is used to amplify the signal and further improve the signal noise ratio (SNR) at the receiver.

When receiving the request signal, underwater vehicles begin to send data to implement the function of uplink transmission.

3. Simulation and discussions

In this work, we use the Monte Carlo approach to simulate the wireless optical communication channel considering on–off-keying (OOK) modulation. We take into account different system parameters such as the water type (T) , absorption coefficient (a), scattering coefficient (b), extinction coefficient (c), beam wavelength (λ), optical power (P), field-of-view (FOV) angel (ψ) of the detector, bandwidth (B) of the detector and aperture size (d) of the detector, the vertical distance between the unmanned aerial vehicle and the sea level(H_1) , the vertical distance between the underwater vehicles and the sea level(H_2) , which are shown in Table1. What's more, we use the Henyey–Greenstein (HG) model for photon scattering, and bit-error-rate (BER) is obtained from the SNR.

Table1. The parameters in the simulation

Parameters	Value
T	Coastal water
a	0.069
b	0.08
c	0.15
λ	532nm
P	1W
ψ	180°
d	20cm
B	250MHz
H1	20m
H2	50m

Figure.3 (a) and (b) show the light intensity distribution at the receiver when the underwater vehicle can align with the unmanned aerial vehicle and the divergence angle of the light source are 10°and 20° respectively. With the increase of the divergence angle of the light source, the light intensity at the receiver becomes weak. Furthermore, we use the BER to demonstrate how the divergence angle of the light source affects the communication quality. Figure.4 displays the BER distribution of the center of the light spot when the light has different divergence angles. We can see if the divergence angle is less than 15.2°, the BER will be lower than the FEC limit ($2x10^{-3}$). Figure.5 is the BER distribution of different distance from the center of the light spot when the underwater vehicle cannot align with the plane and the divergence angle is 11°. If the horizontal distance between the center of the light spot and the receiver is less than 3.25m, the BER will be lower than the FEC limit.

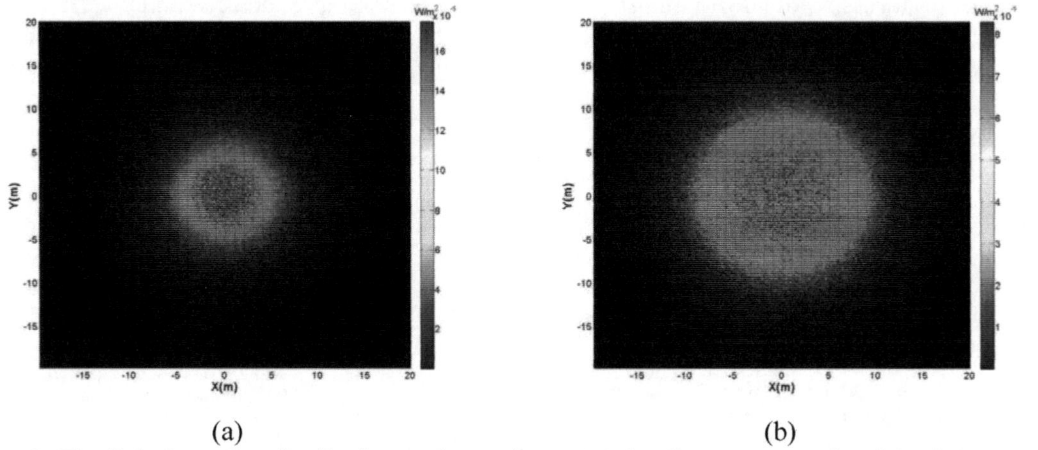

Figure.3. The light intensity distribution at the receiver: (a) the divergence angle of the light source is 10° (b) the divergence angle of the light source is 20°.

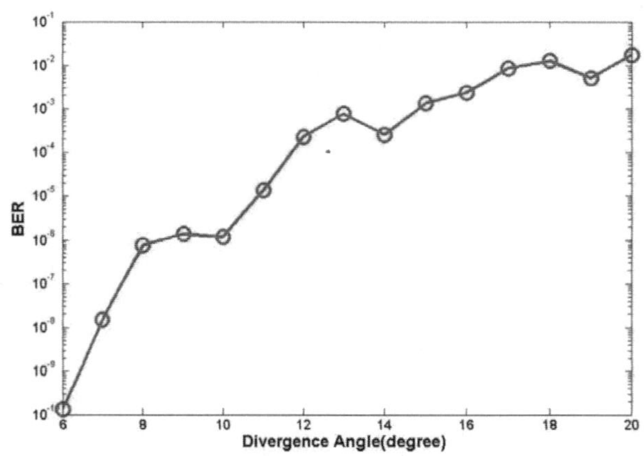

Figure.4. The BER distribution of the center of the light spot when the light has different divergence angles.

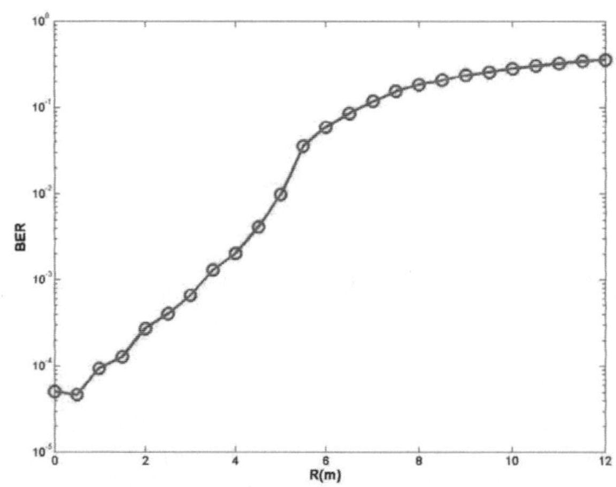

Figure.5. The BER distribution of different distance from the center of the light spot when the underwater vehicle cannot align with the unmanned aerial vehicle.

Figure.6 illustrates the BER distribution of the center of the light spot when the underwater vehicle cannot align with the unmanned aerial vehicle, the divergence angle is 11° and the light source has different deflection angle. If the deflection angel is less than 18°, the BER will be lower than the FEC limit.

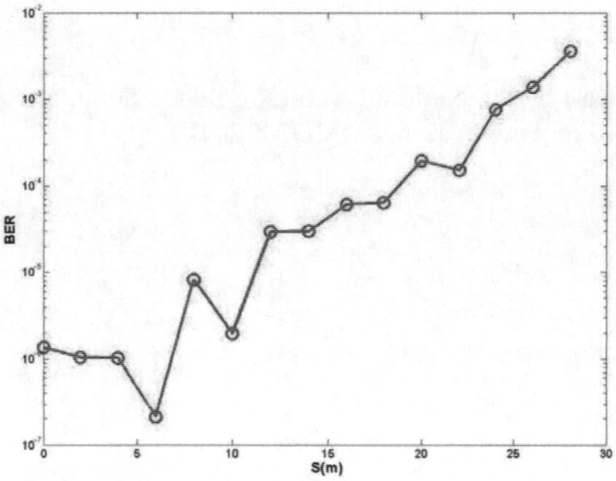

Figure.6. The BER distribution of the center of the light spot when the underwater vehicle cannot align with the unmanned aerial vehicle and the light source has different deflection angle.

Figure.7 is the BER distribution of the center of the light spot when the unmanned aerial vehicle is at different height and the divergence angle of the light source is 11°. If the height is less than 31m, the BER will be lower than the FEC limit.

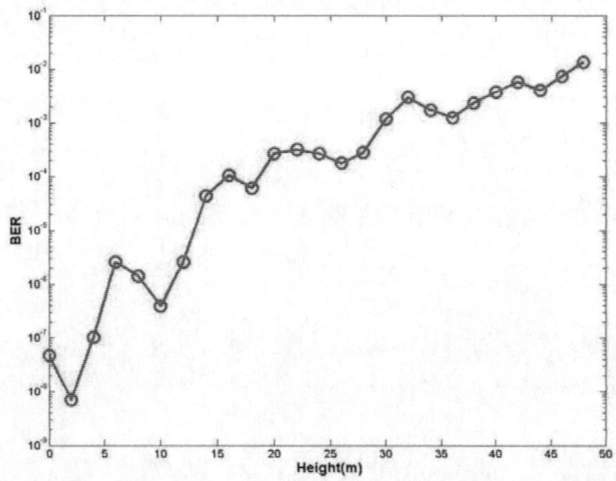

Figure.7. The BER distribution of the center of the light spot when the unmanned aerial vehicle is at different height and the divergence angle of the light source is 11°.

4. Conclusions and Future Work

In this paper, we have theoretically investigated the feasibility of airborne wireless optical communication system in low altitude using an unmanned aerial vehicle and LEDs and proposed the experimental scheme. Using Monte Carlo simulation method, we illustrate how the BER performance is affected by the link distance, the divergence angel and the deflection angel of the light source.

Building a mathematical mode, manufacturing product prototype, and testing in an underwater environment is the following work in the near future.

References

[1] Simpson J A 2008 A 1 Mbps underwater communications system using LEDs and photodiodes with signal processing capability (North Carolina State University Press)
[2] Chancey M A 2005 Short range underwater optical communication links (North Carolina State University Press)

Acknowledgments

This work was supported in part by the National Natural Science Foundation of China (61301141) and Qianjiang Talent Program of Zhejiang Province (QJD1402014).

Underwater Wireless Optical Communication System Using Blue LEDs

Aobo Lin[1], Zheng Tong[2], Yuhang Song[1], Meiwei Kong[1], and Jing Xu[1,*]

[1] Department of Ocean Engineering, Ocean College, Zhejiang University,Yuhangtang Road 866, Hangzhou, Zhejiang, 310058 P.R. China
[2] Department of Petroleum Equipment, RIPED, PetroChina, No.20, Xueyuan Avenue, Haidian District, Beijing, 100083 P.R.China

E-mail: 21434093@zju.edu.cn

Abstract. We demonstrate a self-designed underwater wireless optical communication system using blue LEDs. The performance of the transmitter and receiver was experimentally investigated. Four different square wave signals (10 KHz, 100 KHz, 500 KHz and 1 MHz) were successfully transmitted via a short water channel at the first phase.

1. Introduction
Nowadays underwater wireless sensor networks (UWSNs) have played a more and more important role in underwater surveillance field. As the traditionally used underwater acoustic communication is limited in bandwidth and celerity, there is a growing need for real-time high-rate data, image, and even video transmission between the nodes of a UWSN. Underwater wireless optical communication turns out to be an appropriate solution thanks to its high-bandwidth, cost-effectiveness and low-energy consumption [1-2].

Light emitting diodes (LEDs), with low power consumption, high efficiency and long lifetime, are good light sources for indoor visible light communication systems [3]. But their great potential to be as the light source for underwater wireless optical communications has not been fully investigated. So in this paper we propose the underwater wireless optical communication system based on blue LEDs.

2. System Description
The experimental block diagram of an underwater wireless optical communication system based on blue LEDs is shown in figure 1. The signal in a sensor was uploaded into Memory-1 in a serial way. When Receiver-1 had received a sending request from Emitter-1, ARM-1 sends the signal in Memory-1 to Emitter-2, where the signal directly drives the blue LEDs via a bias-tee (BT).

After transmitting in an underwater optical communication channel, light was focused on Receiver-2 which consists of photoelectric detecting circuit, filter circuit, and amplifying circuit. After clock recovery and level decision, the signal was captured by ARM-2, uploaded into Memory-2, and further processed offline via MATLAB to calculate the BER.

Figure 2 shows the transmitter and the receiver of the underwater wireless optical communication system. Emitter-2, Receiver-2 and convex lens are placed in the front. The main control circuit is placed in the middle. Serial communication and power interface are placed at the end. The whole system can realize high-speed data collection, storage and processing.

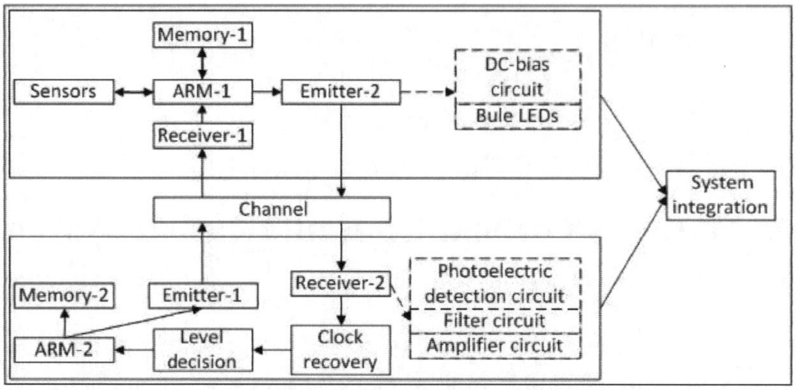

Figure 1. Experimental block diagram of an underwater wireless optical communication system based on blue LEDs.

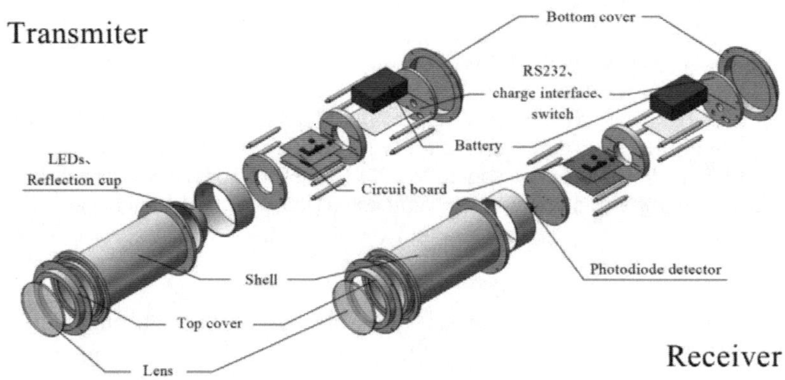

Figure 2. Structure of underwater wireless communication system.

The transmitter and the receiver have a similar structure, but with different circuits. As shown in figure 3 and figure 4, the transmitter includes reverse amplification circuit, driving circuit and bias circuit. The receiver consists of transimpedance amplifier circuit, follower, amplifier circuit, comparing circuit and limiting circuit.

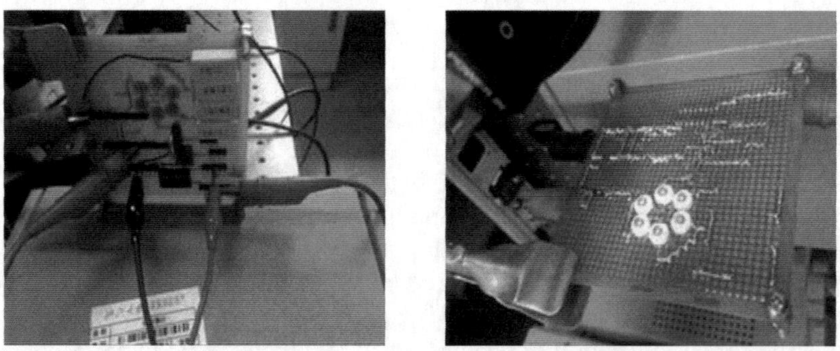

(a) (b)

Figure 3. Circuits of transmitter.

In figure 3, reverse amplification circuit is to transform the voltage level from low voltage level to high voltage level, which can improve load capacity, and its input-signal value is limited from 0V to

5V. The driving circuit is used to drive LEDs for optical signal generation. The bias circuit makes LEDs operate in the linear region. Supply voltages of these three circuits were 5V, 12V and 16V, respectively.

In figure 4, a silicon PIN photodiode with a detection diameter of 10mm is used to detect the optical signal, and its response spectrum ranges from 300 to 1100 nm. The transimpedance amplifier circuit transforms the current signal to voltage signal. Follower is used to decrease output impedance and signal loss. Between the follower circuit and amplifier circuit, we used a blocking capacitor to remove the direct-current part of signal. In order to satisfy the processor's requirement of input signal, a comparing circuit was designed to transform analog signal to transistor-transistor logic (TTL) signal and a limiting circuit was designed to restrict the peak-to-peak voltage below 3.3 V_{pp}. Supply voltages of these circuits were 4.5V, positive 6V, negative 6V and 2.9V, respectively.

Figure 4. Circuits of receiver.

3. Experiment and Discussions

Figure 5 illustrates the experimental setup using an arbitrary waveform generator (AWG) to generate electrical signal at the transmitter side and a digital storage oscilloscope (DSO) to display the received signal waveforms at the receiver side.

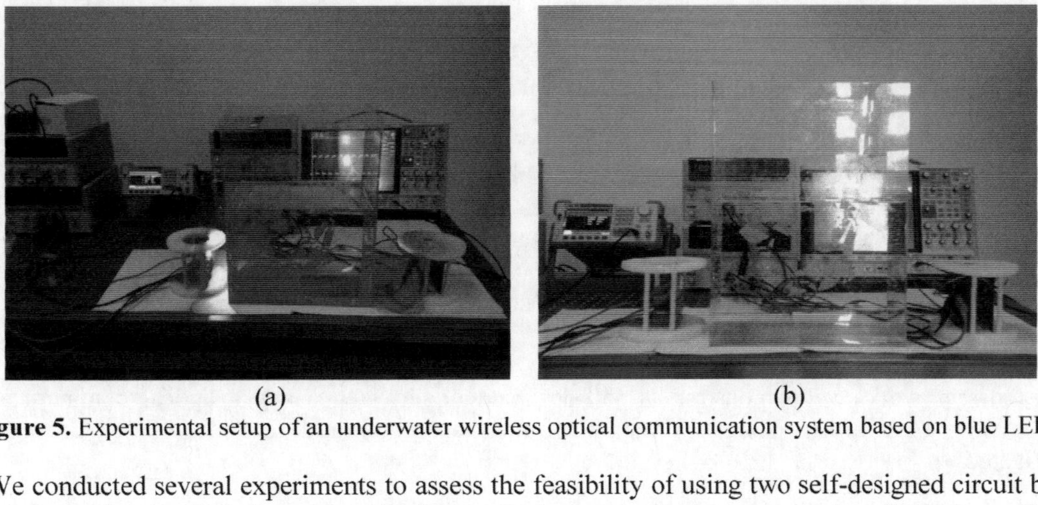

(a) (b)

Figure 5. Experimental setup of an underwater wireless optical communication system based on blue LEDs.

We conducted several experiments to assess the feasibility of using two self-designed circuit board as the transmitter and receiver in a wireless optical communication system described above. In the experiments, we used four different square wave signals of 10 KHz, 100 KHz, 500 KHz and 1 MHz as the transmitted radio frequency (RF) signals, which was used to drive the transmitter. The amplitude of the driving signal is 6 V_{pp}. After being emitted from Emitter-2, the light was transmitted in the free space and focused on Receiver-2. Then the signal waveforms were processed by the receiver circuits, and ultimately displayed on the DSO. Figure 6 shows the waveforms of the received signals with four different frequencies. The clock recovery and data storage will be done in the subsequent experiments.

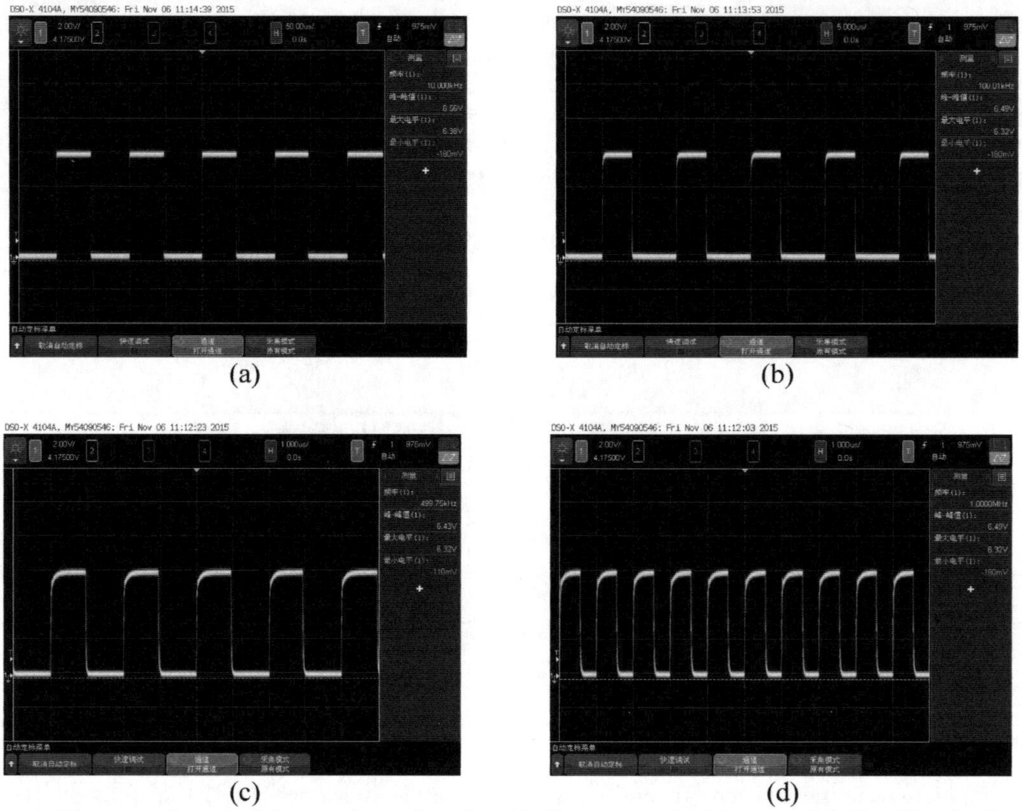

Figure 6. The waveforms of the received signals with five transmission rates (a) 10 KHz, (b) 100 KHz, (c) 500 KHz, (d) 1 MHz.

From the figure 6, we can see that when the signal frequencies were 10 KHz and 100 KHz, the signals can be well recovered. However, when the signal frequency is increased to 1 MHz, the signal was deformed. It is due to that the detection area of Photodiode is too large, which results in a too large input capacitance. At high frequencies, the total input capacitance can severely limit the available bandwidth from the circuit or cause unstable dynamic response. There are two main methods to solve the problem. One is to decrease the detection area of the photodiode, but the received optical energy will decrease. The other is to employ an additional buffer amplifier to actively charge and discharge the input capacitance as required. Compared to the first method, the latter one not only can receive more energy, but also the overall bandwidth of the circuit will be increased.

4. Conclusions and Future Work

In the current work, we demonstrate a self-designed underwater wireless optical communication system using blue LEDs. The performance of the transmitter and receiver was experimentally investigated.

Further work on circuit optimization, adding data storage system, manufacturing product prototype, testing in an underwater environment as well as improving the transmission distance and bit rate would be carried out.

References

[1] Arnon. S 2010 Underwater optical wireless communication network *Opt. Eng.* vol **49** no. 1 pp 1–6

[2] Anguita. D, Brizzolara. D, and Parodi. G 2009 Building an underwater wireless sensor network based on optical communication: research challenges and current results *Int. Conf. on Sensor Technologies and Applications (SENSORCOMM)* pp 476-9

[3] Wu F M, Lin C T, Wei C C, Chen C W, Chen Z Y, Huang H T, and Chi S 2013 Performance comparison of OFDM signal and CAP signal over high capacity RGB-LED-Based WDM visible light communication *IEEE Photonics Journal* vol **5** no. 4 p 7901507

Acknowledgments

This work was supported in part by the National Natural Science Foundation of China (61301141) and Qianjiang Talent Program of Zhejiang Province (QJD1402014).

A Novel Routing Strategy on Space Communication Network

Yujian LI[1], **Xu-Liang WANG**[1], **Zhen-dong XI**[1] and **Bing-hong WANG**[2]

[1]JLOMAC, China Satellite Maritime Tracking and Control Department, Jiangyin, China
[2]Department of Modern physics, University of Science and Technology of China, Hefei, China

E-mail: jinzhili@mail.ustc.edu.cn

Abstract. In this paper, we propose an effective routing strategy on the basis of the so-called nearest neighbor search strategy by introducing a preferential time delay exponent β. Traffic dynamics both near and far away from the critical generating rate R_c are discussed. Simulation results demonstrate that the optimal performance of the system corresponds to β=-0.5. Due to the low cost of acquiring nearest-neighbor information and the strongly improved network capacity, our strategy may be useful and reasonable for the protocol designing of modern communication networks and space communication networks.

1. Introduction

A variety of systems in nature can be described by complex networks and the most important statistical features of complex networks are the small-world effect and scale-free property.[1,2,3] It may serve as a very useful tool for understanding nature and our society. Since the discovery of some common interesting features of many real networks such as small-world phenomena by Strogatz and Watts[1] and Scale-free phenomena by Albert and Barabási, [2] processes of dynamics conducting on the network structure such as traffic congestion of information now have drew more and more attention from engineering and physical field, due to the importance of large communication networks such as **WWW**[4] and **Internet**[5] in modern society. Evolution of networks structure and interplay of traffic dynamics also play an important role in the research of the traffic system including Internet, highway network and so on , they want to understand and explain the process of dynamics on the underlying structure.

Many previous excellent works focus on the evolution of structure driven by the increment of traffic[6,7,8] and some explore how different topologies impact the traffic dynamics[12,13]. Some works[10,11] gave several models to mimic the traffic routing on complex networks by introducing randomly selected source as well as particles (packets) generating rate and destination of each particles[18-22]. Those models define the capacity of networks described by critical generating rate, too. At this critical rate, a continuous phase transition from free flow state to congested state occurs. In the free state, the numbers of created and delivered particles are balanced, leading to a steady state. While on the jammed state, the number of accumulated particles increases with time is due to the limited delivering capacity or finite queue length of each vertex.

We believe that the study on the network search is very important for traffic systems, for the existence of particles routing from origin to destination and communication cost is very meaningful. A few previous studies incorporate the search strategies and the traffic processes on networks[9]. In this paper, we present a traffic model in which particles are routed only based on local topological information with a single tunable parameter β. In order to maximize the nodes handling and delivering capacity of the networks which can be measured by an introduced order parameter η, the optimal β is found out. We also check the dynamic properties in the steady state for different α including average number of particles versus vertex's degree, particles distribution and particles traveling time distribution. The dynamics right after the critical generating rate Rc exhibits some interesting properties independent of β, which indicates that although the system enters the jammed state, it possesses partial capacity for forwarding particles. Our model can be considered as a preferential walk among neighbor vertexes. We arrange the paper as follows: in the following section we describe the model in detail, then simulation results of traffic dynamics are provided in both the steady and congested states, A conclusion and discussion are given in the last section.

2. Traffic model

The first paragraph after a heading is not indented (Bodytext style). Our traffic model is described as follows: at each time step, there are R particles generated in the system, with randomly chosen sources and destinations, and all vertexes can deliver at most C particles toward their destinations, which is one of the most interesting properties of the whole traffic network. The capacity of each vertex is set to be C. As a remark, there is difference between the capacity of network and vertexes. The capacity of the whole network is measured by the critical generating rate R_c at which a continuous phase transition will occur from free state to congestion. The free state refers to the balance between created particles and removed particles at the same time. When the system enters the jam state, it means particles continuously accumulate in the system and at last few particles can reach their destinations. In order to describe the critical point accurately, we use the order parameter:

$$\eta(R) = \lim_{t\to\infty} \frac{C}{R} \frac{<\Delta N_p>}{\Delta t} \tag{1}$$

$\Delta Np = N(t+\Delta t)-N(t)$ with <···> indicates average over time windows of width Δt and Np(t) represents the number of data particles within the networks at time t, here. For $R < R_c$, $<\Delta N>= 0$ and $\eta = 0$, indicating that the system is in the free state with no traffic congestion. Otherwise for $R > R_c$, $\eta \to r$, where r is a constant larger than zero, the system will collapse at last. To navigate particles, each vertex performs a local search among its neighbors. If a particle's destination is found in the searched area, it will be delivered directly to its target, otherwise, it will be forwarded to a neighbor j of vertex i according to the probability:

$$\Pi_{i\to j} = \frac{\exp \beta * l_j/C}{\sum_i \exp \beta * l_j/C} \tag{2}$$

Here, l_i is the queue length in vertex i, so $\beta * l_j/C$ represents the waiting (delay) time of one new arrival particle. The sum runs over the neighbors of vertex i on the searched area and β related to waiting time is an adjustable parameter studied by us in the next context. Once a particle reaches its destination, it will be canceled from the system.

Here, we choose m = 5 and network size N = 1000 fixed for simulations. We should also note that the queue length of each vertex is assumed to be unlimited and the **FIFO** (First in First out) discipline is applied at each queue[9]. Another important rule called path iteration avoidance (PIA) is that a link between any pair of vertexes is not allowed to be visited more than twice by the same particle.[9] Without this rule, the capacity of the network is quite low due to many times' unnecessary visiting to the same links by the same particles, which does not exist in the real traffic systems. We note that this PIA routing algorithm does not damage the advantage of local routing strategy. If each particle records the links it has visited, the PIA can be easily performed. One can find that this rule does not need the global topological information. Therefore, we think this rule is rational and can considerably improve the network capacity.

3. Results and discussion

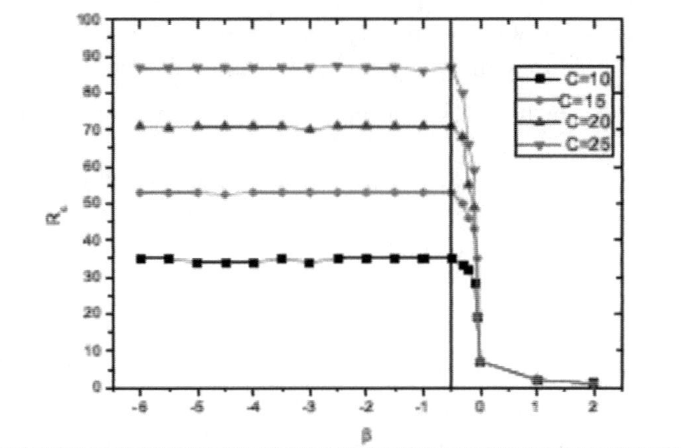

Figure 1 The critical R_c versus beta with network size N=1000, m=4. The critical beta =-0.5 marked by a black solid line.

As shown in Figure 1, It is easy to find that the capacity of the system is not the same for different beta. We can preliminarily determine that it is the critical situation. when the beta increases , the capacity of BA network measured by R_c considerably finishes the optimal performance at β =-0.5. we study every situations under different vertex capacity, and found that the critical system β is -0.5.

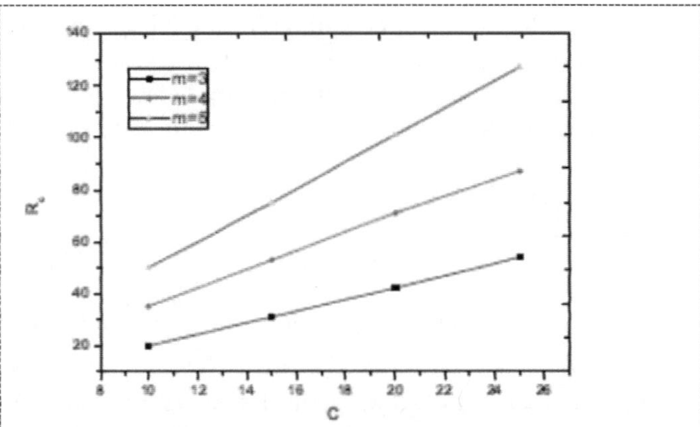

Figure 2(color online). The variance of R_c versus vertex capacity C with the increasing of m with network size N = 1000 and different vertex capacity.

As shown in Figure 2, the increment of vertex capacity C and link density m considerably enhances the capacity of BA network measured by R_c due to the fact that with higher link density, particles can more easily find their target vertexes.

4. Conclusion

We have introduced a new routing strategy based on local information, trying to give a solution to the problem of traffic congestion in modern communication networks. The optimal parameter β =-0.5 is obtained with maximizing the whole system's capacity. In addition, the property that scale-free network with occurrence of congestion still possesses partial delivering ability suggests that only improving processing ability of the minority of heavily congested vertexes can obviously enhance the capacity of the system. The variance of critical value R_c with the increment of C is also discussed. Our study may be useful for designing communication protocols for large communication networks due to the local information the strategy only based on and the simplicity for application. The results of current work may also shed some light on alleviating the congestion of modern technological networks.

Acknowledgment
This work is funded by the National Natural Science Foundation of China (Grant Nos. 61403421), the JLOMAC Open Grant (FOM2014OF001), and the Young Talents Research Grant (ZZ2014).

References

[1] Watts D. J. and Strogatz S. H., 1998 *Nature* **393** 440.
[2] Barabási A.-L. and Albert R., 1999 *Science* **286** 509.
[3] Albert R. and Barabási A. -L., 2002 *Rev. Mod. Phys.***74** 47.
[4] Pastor-Satorras R., Vázquez A., and Vespignani A., 2001 *Phys. Rev. Lett.* **87** 258701.
[5] Albert R., Jeong H., and Barabási A.-L., 1999 *Nature* **401** 103.
[6] Barrat A., Barthélemy M., and Vespignani A., 2004 *Phys. Rev.Lett.* **92** 228701.
[7] Barrat A., Barthélemy M., and Vespignani A., 2004 *Proc. Natl. Acad. Sci. U.S.S.* **101** 3747.
[8] Wang W.-X., Wang B.-H., Hu B., Yan G., and Ou Q., 2005 *Phys. Rev. Lett.* **94** 188702.
[9] Wang W.-X., Wang B.-H., Yin C.-Y., Xie Y.-B., and Zhou T., 2006 *Phys. Rev.* E **74** 016101.
[10] Yin C.-Y., Wang B.-H., Wang W.-X., Yan G., and Yang H.-J., 2006 *Eur. Phys. Jour.* B **49** 205.
[11] Wang W.-X., Wang B.-H., Yin C.-Y., Zhou and T., 2006 *Physical Review* E **73** 026111.
[12] Singh B. K., and Gupte N., 2005 *Phys. Rev.* E **71** 055103.
[13] Ashton D. J., Jarrett T. C., and Johnson N. F., 2005 *Phys. Rev. Lett.* **94** 058701.
[14] Zhao L., Park K., and Lai Y. C., 2004 *Phys. Rev.* E **70** 035101.
[15] Lee E. J., Goh K.-I., Kahng B., and Kim D., 2005 *Phys. Rev.Lett.* **71** 056108.
[16] Petter Holme, and Beom Jun Kim, 2002 *Phys. Rev.* E **65** 066109.
[17] Goh K.-I., Lee D.S., Kahng B. and Kim D., 2003 *Phys. Rev. Lett.* **91** 148701.
[18] Kim B. J., Yoon C. N., Han S. K. and Jeong H., 2002 *Phys. Rev.* E **65** 027103.
[19] Zhao L., Lai,Y.-C. K. Park, and Ye N., 2005 *Phys. Rev. E* **71** 026125.
[20] Zhang Huan, Liu Zonghua, Tang Ming, and Hui P. M., 2007 *Phys. Lett.* A **364** 177.
[21] M. A. de Menezes, and Barabási A.-L., 2004 *Phys. Rev. Lett.***92** 028701.
[22] Liu Zonghua, Ma Weichuan, Zhang Huan, Sun Yin, and Hui P. M., 2006 *Physica* A **370** 843.

A Novel Ship-rocking Forecasting Method based on Hilbert-Huang Transform

De-yong Kang[1] Yu-jian Li[1,†] Xu-liang Wang[1] and Chen Zhi[2]

1JLOMAC, China Satellite Maritime Tracking and Control Department, Jiangyin, China 214431

2 Department of Modern physics, University of Science and Technology of China, Hefei, China 230026.

jinzhili@mail.ustc.edu.cn

Abstract. The ship-rocking is a crucial factor which affects the accuracy of the ocean-based aerospace vehicle measurement. Here we have analysed groups of ship-rocking time series in horizontal and vertical directions utilizing a Hilbert based method from statistical physics. Based on these results we could predict certain amount of future values of the ship-rocking time series based on the current and the previous values. Our predictions are as accurate as the conventional methods from stochastic processes and provide a much wider prediction time range.

1. Introduction

In the ocean-based flight vehicle measurement, a very important task is to characterize special properties of the objective from massive quantity of data. Based on this one may have a way to increase the accuracy of the measurement. To increase the accuracy of the measurement, much effort has been done to study two crucial time series: the ship rocking and the ship deformation. The current strategy is to make appropriate corrections based on these two time series, and thus to reduce or even eliminate the periodicity in the signals of interest. However, after these corrections the measurement results still show considerable difference from those basement standard results obtained through other ways, such as GPS, laser, etc[1].

In this paper we consider the ship-rocking time series based on the approach from statistical physics. To those time series which show strong periodicity, an effective method is to investigate the instantaneous analytic signal of a time series[2,3]. The instantaneous analytic signal is obtained based on a Hilbert transform. The original time series as well as its Hilbert transform construct an analytic signal at each measurement time. On the complex plane one can calculate the instantaneous amplitude and the instantaneous phase of the analytic signal. Such instantaneous amplitude and phase may reveal certain intrinsic properties of systems which are not seen from the conventional methods. For example, previous research shows that, the change of the correlation in human auto-regulation may not happen in the original signal, however it does show in the correlation of its instantaneous phase increments [2]. Here, similarly we investigate the characteristics of the analytic signal. We could predict certain amount of future values of the ship-rocking time series based on the current and the previous values.

1.1. DATA

We have obtained four groups of ship-rocking time series: GX1129JL839RW839RW840. The data starting with "GX" are those obtained when the ship navigates freely, while others are those obtained when the ship navigates while maintaining a fixed orientation. All other information is unknown to us. For each group we have the time series in three directions of the Euler angles: the navigation direction, the horizontal direction and the vertical direction. In the following we mark the navigation direction as "KC", the horizontal direction as "OC", and the vertical direction as "PC". If not specially indicated, the unit of all angles is in radian. All time series are measured every 50 milliseconds.

1.2. BASIC TECHNIQUE

Our results in this paper are based on the Hilbert transform (HT) to a stationary time series. To any stationary time series s(t), its HT is defined as:

$$\tilde{s}(t) \equiv \frac{1}{\pi} P \int_{-\infty}^{\infty} \frac{s(\tau)}{t-\tau} d\tau$$

Where P denotes the Cauchy principal value. In the frequency domain, the Fourier component at any frequency f for the Fourier transform of ˜s(t) can be obtained very easily from the Fourier transform of s(t), i.e., that of s(t) at the same frequency f rotates 90° clockwise (for f > 0) or anticlockwise (for f < 0) in the complex plane. As a simple example, if s(t) = sin(at), then one would obtain ˜s(t) = cos(at). Based on this fact, one could define an "analytic signal" for this time series:

$$S(t) \equiv s(t) + i\tilde{s}(t) = A(t)e^{i\varphi(t)}$$

Where A(t) and phy(t) are the instantaneous amplitude and phase of s(t), respectively.

2. Results and discussion

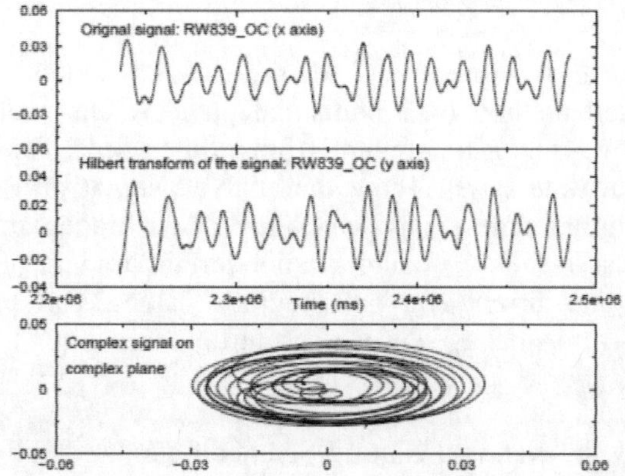

Figure 1. The original ship-rocking time series and its Hilbert transform, as well as the corresponding analytic signal for the group RW839 in the horizontal (OC) direction.

As shown in Figure 1, we show the patterns of the ship-rocking time series for the group RW839 in the OC direction. Similar behaviors are observed in signals of other groups and in other directions. In the analyses, we always remove the zero frequency part in the Fourier transform of s(t) (which is related to the average of s(t)). From the lowest panel of Figure 2 one could find that the evolution of the analytic signal S(t) is very slow when seen in the complex plane. It usually follows the similar behavior when rotating around the origin.

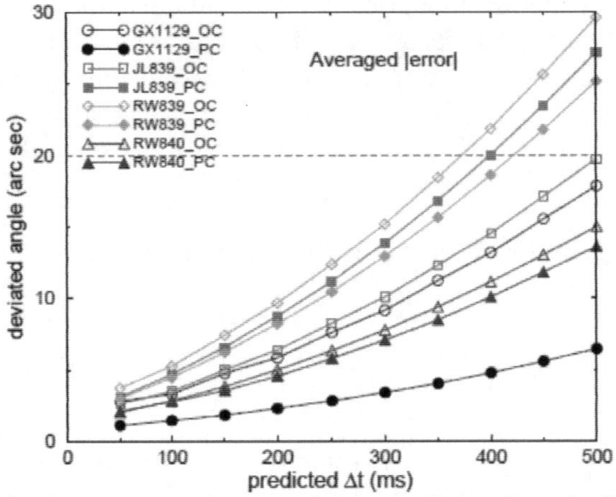

Figure 2. (color online). The absolute errors of the prediction from the Hilbert transform based method as a function of the prediction time interval.

From these known values we first construct its corresponding analytic signal. We next extend the track of the analytic signal on the complex plane from the helix model. The ten predicted values are then compared to the actual values in the original time series. The results of comparison are shown in Figure 2. From the results of both the averaged absolute error we find that our predicted values are very close to that of the original signals within a much wider range of the prediction time window than that of current methods. The results seem better in the averaged absolute error than that in the standard deviation. We attribute this to certain outliers in the instantaneous phase increments. Some tunning at positions near the outliers should be able to further improve the current accuracy.

3. Conclusion

In summary, we have analyzed four groups of ship-rocking signals in horizontal and vertical directions utilizing a Hilbert based method from statistical physics. Our method gives a way to construct an analytic signal on the two-dimensional plane from a time series. The analytic signal share the complete property of the original time series. From the analytic signal of a time series, we have found some information of the original time series which are often hidden from the view of the conventional methods. Based on these results we could predict certain amount of future values of the ship-rocking time series based on the current and the previous values. Our predictions are as accurate as the conventional methods from stochastic processes and provide a much wider prediction time range.

Acknowledgment

This work is funded by the National Natural Science Foundation of China (Grant Nos. 61403421), the JLOMAC Open Grant (FOM2014OF001, FOM2014OF002), and the Young Talents Research Grant (ZZ2014).

References

[1] Li Xiaoyong, Du Zhijun and Chen Guiming. 2005 *Manned Spaceflight* **3** 38. (in Chinese).
[2] Chen Z., et al. 2006 *Phys. Rev.* E **73**, 031915, 2005 *Phys. Rev. E* **71** 011104, 2002 *Phys. Rev.* E **65** 041107.
[3] Peng C.-K., et al. 1994 *Phys. Rev.* E **49** 1685.

8th International Conference on Advanced Infocomm Technology (ICAIT 2015) IOP Publishing
Journal of Physics: Conference Series **679** (2016) 012035 doi:10.1088/1742-6596/679/1/012035

Research of Gear Fault Detection in Morphological Wavelet Domain

Shi Hong[1], Shan Fang-jian[1], Cong Bo[1] and Qiu Wei[1]
[1] China Satellite Maritime Tracking and Controlling Department, Jiangyin Jiangsu 214431, China

E-mail: woai0528@163.com

Abstract. For extracting mutation information from gear fault signal and achieving a valid fault diagnosis, a gear fault diagnosis method based on morphological mean wavelet transform was designed. Morphological mean wavelet transform is a linear wavelet in the framework of morphological wavelet. Decomposing gear fault signal by this morphological mean wavelet transform could produce signal synthesis operators and detailed synthesis operators. For signal synthesis operators, it was just close to orginal signal, and for detailed synthesis operators, it contained fault impact signal or interference signal and could be catched. The simulation experiment result indicates that, compared with Fourier transform, the morphological mean wavelet transform method can do time-frequency analysis for original signal, effectively catch impact signal appears position; and compared with traditional linear wavelet transform, it has simple structure, easy realization, signal local extremum sensitivity and high denoising ability, so it is more adapted to gear fault real-time detection.

1. Introduction

Running gear, which is the usual transmission component for speed and transmission power, plays an important role on the whole machine operation, but the manufacturing process, work environment, machine wear and other factors existed led to the gear fault easily, and even the subtle fault also can cause the whole machine fault, result in great economic loss. Therefore, the analysis of the gear fault detection has very important significance. Fourier transform and wavelet transform are commonly used analysis methods in time-frequency domain[1], but the traditional Fourier transform, reflecting the signal overall characteristics only, has significant limitations for non-stationary nonlinear signal because of the lack of refinement function for the details of the signal and especially in the time and frequency mutations; The wavelet analysis with its multi-scale and multi-resolution advantage is more superior to the Fourier analysis, because that the wavelet transform can do signal decomposition in different directions, different scales, and different resolution， and to some extent, it can give more meticulous description on the details of non-stationary signals for the time-frequency information can be effectively extracted from the signal, it has extraordinary significance in the field of signal processing and becoming more and more widely used in gear fault diagnosis[2-3]. And for the traditional wavelet, based on linear decomposition of frequency, determine gear faults and types by analysis the interval of impact signal in time domain but frequency information of the gear faults feature such as multi-step meshing frequency, side frequency cannot be directly extracted[4]. In recent years, a new wavelet analysis has become the focus of attention, Morphological Wavelet transform

Journal of Physics: Conference Series **679** (2016) 012035 doi:10.1088/1742-6596/679/1/012035

were more popular, however, the theoretical analysis of new wavelet was still in the early stages of research, lots of advantages in the application has not been approved yet[5-6].

This paper seeks to take into account a morphological wavelet suite for fault diagnosis and analysis, used in vibration signal analysis of one-dimensional, verify the feasibility and effectiveness of morphological multi-resolution analysis, and inspect the stability in noise interference.

2. Program design

When gears running if there is local damage fault, the fault point and another gear engage each other will produce a shock response, with the continuous transmission, each of the rotation produce a periodic impulse response, i.e., the vibration signal which is a typical non-stationary nonlinear signal, it needs to extracted some information of effective representational gear fault status from signal to analyses and detect the fault of gears. Morphological Wavelet transform is an effective method undoubtedly, and the DTC contains signals analysis operator as low-pass filter and details analysis operator as high-pass filter. The fault signals passe through low-pass filter produce approximation of the original signal retained the basic information on normal running gears. While the detail information by high-pass filter includes vibration and noise, we can extract relevant profiles of gear fault.

Mean filter is one of the more effective means of signal denoising[7], we consider using morphological mean wavelet transform to detect the fault signal of gear[8]. Make the mean filter as the low-pass filter of morphological wavelet, so we can get the signal analysis operator and details analysis as follows:

$$\psi^{\uparrow}(x)(n) = mean(\mathrm{x}(2n), \mathrm{x}(2n+1))$$
$$\omega^{\uparrow}(x)(n) = \mathrm{x}(2n) - \mathrm{x}(2n+1) \quad (1)$$

According to the perfect reconstruction condition, we can get synthesis operators as follows:

$$\psi^{\downarrow}(x)(2n) = \psi^{\downarrow}(x)(2n+1) = x(n)$$
$$\omega^{\downarrow}(x)(2n) = 0.5 * \omega(n)$$
$$\omega^{\downarrow}(x)(2n+1) = -0.5 * \omega(n) \quad (2)$$

Then the morphological midpoint wavelet which achieves perfectly reconstruction without redundancy is constructed. Specific steps in gear fault detection as follow:

(1) Morphological mean wavelet transform of the original signal, the gear fault signal is decomposed into signal coefficients and detail coefficients for each layer;

(2) Basic information signal stored in the signal coefficient, but the details of coefficients contains vibrations' and most of noise generated by the fault. By thresholding detail coefficients of morphological mean wavelet are processed to eliminate noise;

(3) We observe every detail de-noised layer, comprehensive analysis to determine the fault feature to guide troubleshooting.

3. Experiment

The simulation is completed in two parts, vibration signal without noise and with noise, construction simulation signal generated by the superposition of two parts. Suppose the time-domain waveform of original vibration signals shown in Figure 1(a), the impact signals generate by gear fault signal shown in Figure 1(b), so, the total simulated signal is . As shown in Figure 1(c), the sampling frequencies of 15KHz.

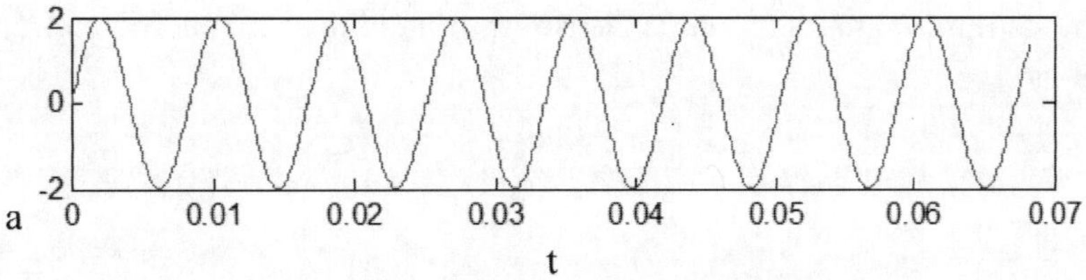

(a) The original gear vibration signals

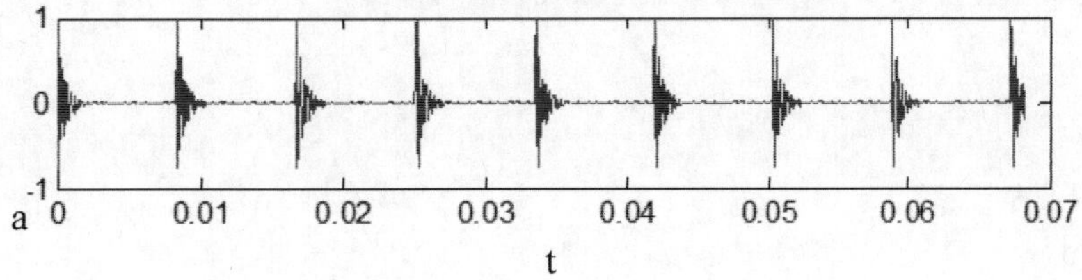

(b) Gear Fault repeated impulse response

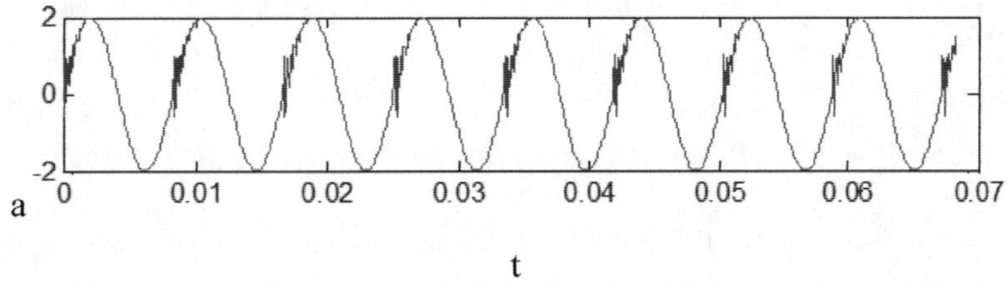

(c) Synthetic gear fault signals

Figure 1. Simulation signals.

A spectrum obtained by Fourier transform of the gear fault simulation signals as shown in Figure 1(c) is shown in Figure 2. It can be seen that the frequency of normal gear vibration signal is captured at 120Hz, however, the fault repeated impact response was not clearly portrayed. Therefore, there has limitations of Fourier transform analysis for non-stationary signals such as gear fault.

Fourier transform of the gear fault simulation signals as shown in Figure 1(c), obtain a spectrum as shown in Figure 2. As can be seen, the frequency of normal gear vibration signals is captured at 120Hz, however, the repeated impact response of fault was not clearly portrayed. Therefore, Fourier transform analysis of non-stationary signals such as gear fault has limitations.

Morphological mean wavelet transform has the characteristics of traditional wavelet transform. It can effectively solve the contradictions in time and frequency domain localization. Here, use morphological mean wavelet transform to decompose the fault signals for 5 layers, then the coefficient magnitude plot obtained is shown in Figure 3. As can been seen from the figure, the morphological mean wavelet transform is a multi-resolution analysis, according with the rule which can observe the high-frequency on small-scale and the low-frequency on big-scale when analyzing signals by the traditional wavelet. Thus it can illustrate the effectiveness of morphological mean wavelet transform for signal analysis.

Figure 4 is a schematic of signal and high-frequency details after morphological mean wavelet transform for 5 layers. Figure 4(a) is the part of signal which can extract the characteristics of original signal. Figure 4(b), 4(c) and 4(d) are the part of three layers details before decomposition. It can observe or extract relevant fault impact signal's characteristics by other means to guide the fault determination. It can be seen that signal portion are consistent with the characteristics of original signal, but the details portion especially the part of details after decompose 1 layer has more accurate portrayal. And after signal reconstructing by synthetic operator, it has no different with Figure 1(c),

Journal of Physics: Conference Series **679** (2016) 012035 doi:10.1088/1742-6596/679/1/012035

therefore, from the capture signal singularity sense, morphological mean wavelet analysis is better than Fourier analysis.

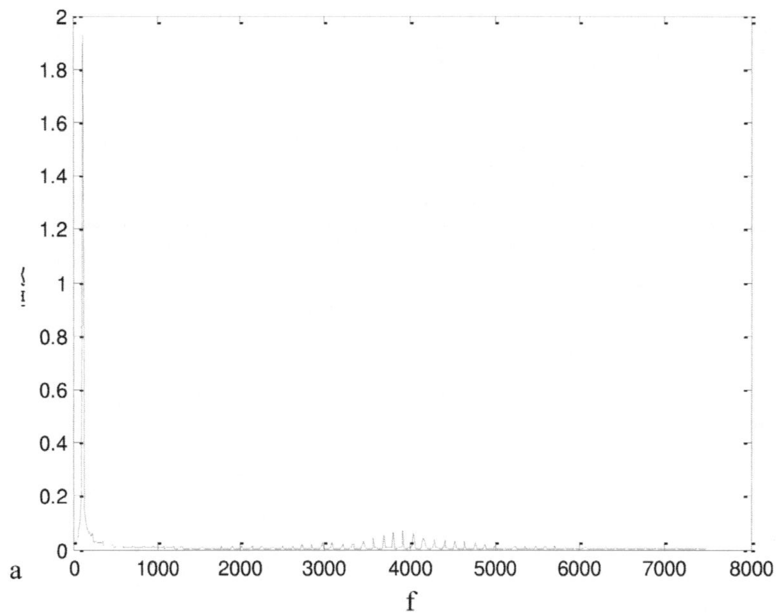

Figure 2. Spectrum of Fourier transform for fault signal.

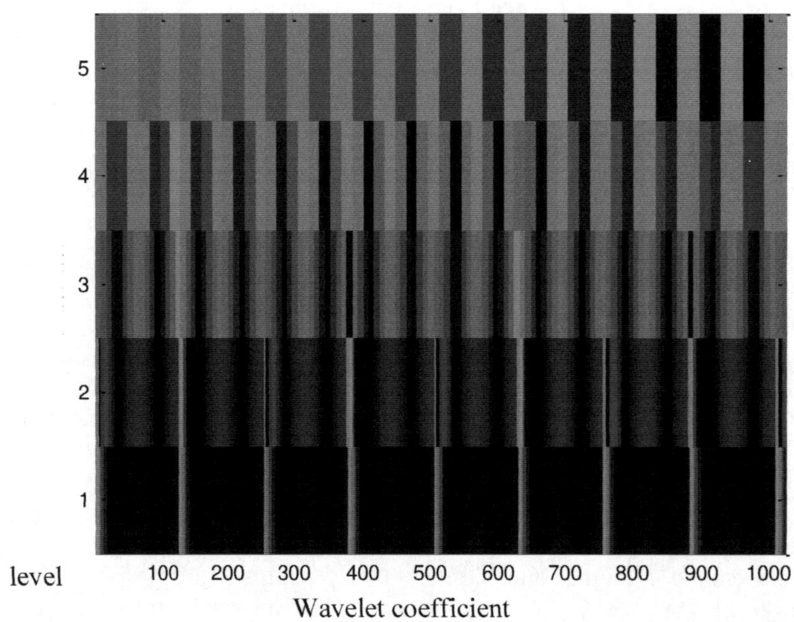

Figure 3. The morphological mean wavelet coefficient for fault signal.

Journal of Physics: Conference Series **679** (2016) 012035 doi:10.1088/1742-6596/679/1/012035

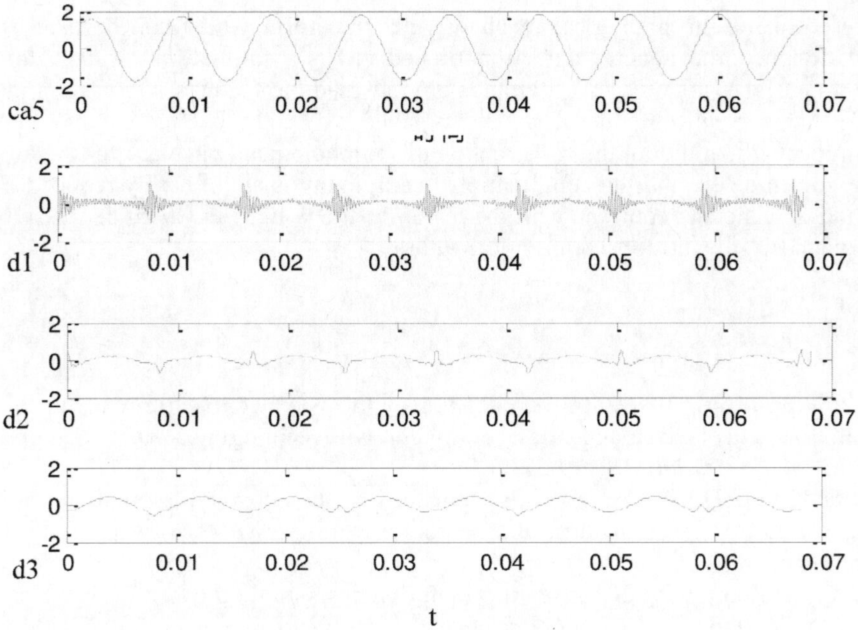

Figure 4. Signal and details after morphological mean wavelet transform.

The results obtained when noise as shown in Figure 5. Compare de-noising results, the morphological mean wavelet transform has a stronger signal singularity positioning capability, therefore, morphological mean wavelet transform has good prospects in gear fault detection. In addition, morphological mean wavelet has simple structure, easy to achieve, the building is conducive gear fault line real-time system.

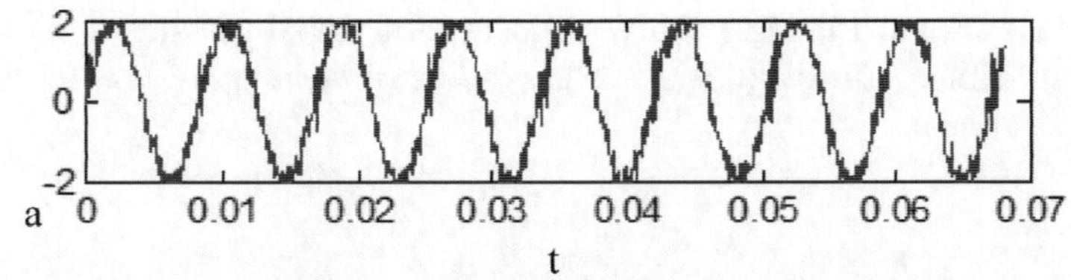

(a) Gear fault signal affected by noise

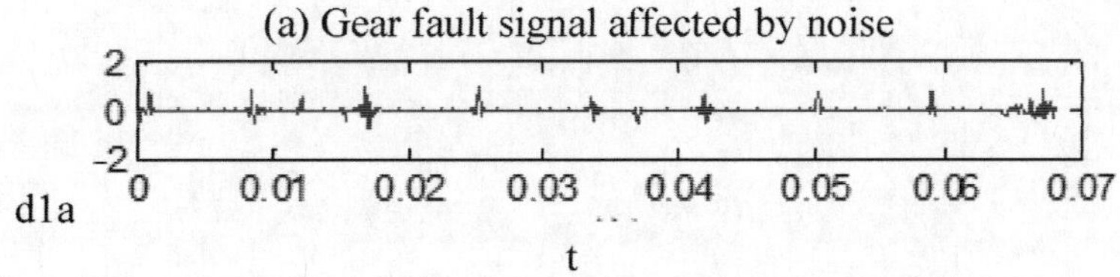

(b) Db4 wavelet analysis for noisy fault signal

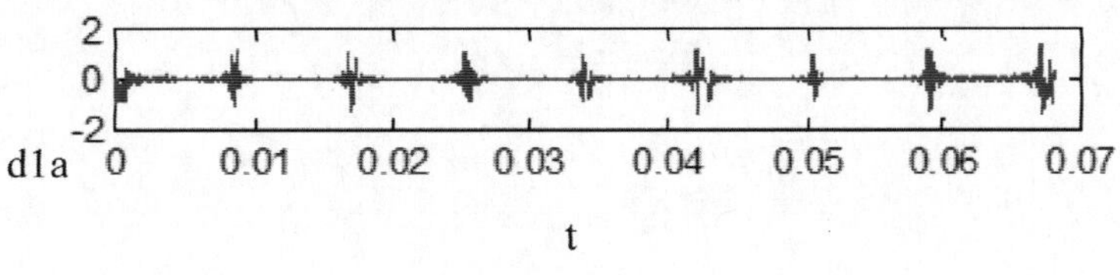

(c) Morphological mean wavelet analysis for noisy fault signal

Figure 5. The result in this paper.

4. Conclusion

This paper presented the morphological mean wavelet transform which can be used for gear fault detection, and designed the general detection procedure. The simulation results showed that the denoising effect on details after signal separate is obvious and the features are more prominent in this paper. The non-stationary information of signal has saved well and the detection and analysis of fault signal is more reliable. In addition, the construction of morphological wavelet is based on simple math and easy to be implemented, and the computational complexity is significantly reduced. Further work such as a method of this paper combine with the feature extraction means to guide the actual gear fault detection and diagnosis may be considered in the future.

Reference

[1] PENG Z K and CHU F L 2009 Application of the wavelet transform in machine condition monitoring and fault diagnostics: a review with bibliography *J. Mechanical. Systems. and.Signal. Processing.* **18** 199-221

[2] Wu J D and Liu C H 2009 An expert system for fault diagnosis in internal combustion engines using wavelet packet transform and neural network *J. Expert. Systems. with. Applications. Part1.* **36** 4278-4286

[3] Zhang J C and Zhang J M 2012 Bearing fault diagnosis method based on wavelet analysis and nonlinear prediction *J. Chinese. Journal. Of. Scientific. Instrument.* **33** 127-131

[4] Wang X D and Zi X Y 2011 Multiwavelet denoising with improved neighboring coefficients for application on rolling bearing *J. Mechanical. Systems. And. Signal. Processing.* **25** 285-304

[5] Heijmans H J A M and Goutsias J 2000 Nonlinear multiresolution signal decomposition schemes— Part Ⅱ: Morphological wavelets *J. IEEE. Tans. Image. Processing.* **9** 1897-1913

[6] Peilin Z and Bingyi L 2011 Max-lifting morphological wavelet transform based gear fault feature extraction *J. Chinese Journal of Scientific Instrument* **31** 2736-2741

[7] Gonzalez C and Woods E 2007 *Digital. Image. Processing. Second edition* Beijing Publishing house of electronics industry pp88-91

[8] Zhu Z K, Ruqiang Y and Liheng L 2009 Detection of signal transients based on wavelet and statistic for machine fault diagnosis *J. Mechanical Systems and Signal Processing* **23** 1076-1097

Journal of Physics: Conference Series **679** (2016) 012036 doi:10.1088/1742-6596/679/1/012036

An underwater ship fault detection method based on Sonar image processing

Shi Hong[1], Shan Fang-jian[1], Cong Bo[1] and Qiu Wei[1]
[1] China Satellite Maritime Tracking and Controlling Department, Jiangyin Jiangsu 214431, China

E-mail: woai0528@163.com

Abstract. For the research of underwater ship fault detection method in conditions of sailing on the ocean especially in poor visibility muddy sea, a fault detection method under the assist of sonar image processing was proposed. Firstly, did sonar image denoising using the algorithm of pulse coupled neural network (PCNN); secondly, edge feature extraction for the image after denoising was carried out by morphological wavelet transform; Finally, interested regions Using relevant tracking method were taken, namely fault area mapping. The simulation results presented here proved the feasibility and effectiveness of the sonar image processing in underwater fault detection system.

1. Introduction

Ships may encounter unpredictable situations while its sailing on the ocean, and there may be bad weather, threat from the other ships, also the damage caused by long time operation of various equipment of the ship itself. Currently, ships fault detection and troubleshooting can be solved on the water basically, but for underwater fault, there has not yet provided a feasible solution, even the type of specific underwater failure can be determined from remote security system. It is a week line for the realization of fault real-time maintenance underwater particularly turbid waters. If this problem has not solved, the entire ship manoeuvre maintenance system will be flawed on the sea. In the turbid and low visibility waters, underwater video equipment can be seen in the visual range of 1-2 meters. Therefore, the ship's underwater fault detection and immediate repair is necessary by means of high-resolution imaging sonar to obtain high-resolution image data. High-resolution imaging sonar has important applications in the muddy waters of proximity detection. It can provide essential technical support for underwater ship maintenance program by processing and analyzing the image data obtained through high-resolution imaging sonar, and greatly improve the efficiency and reliability of the ship maintenance implementation.

2. Program design

The system flow chart in this paper is shown in Fig. 1. Taking the characteristics of sonar image itself into account: noise pollution, blurred or discontinuity edges, low contrast, non obvious target area characteristic features and so on, it is necessary to take sonar image denoising at first, which can remove the interference to acoustic imaging generated by complex underwater environments to make the image clearer and more obvious target feature. Pulse coupled neural network (PCNN) is taken in this paper to complete image denoising by adjusting the brightness of pixels, according to whether

Journal of Physics: Conference Series **679** (2016) 012036 doi:10.1088/1742-6596/679/1/012036

each neuron and its neighboring neurons can inspire output pulse train or not, it can be determined and distinguished between noise and pixel gray values, in order to take appropriate measures[1]. Pulse coupled neural network is a feedback network consists of several interconnected PCNN neurons. In the practical application of image processing, PCNN network is a single two-dimensional network of local connections. The number of neurons is equal to the number of pixels in the image input, and they are correspondence. The output of each neuron has only two states, i.e. excitation (ignition) or inhibition(no ignition). Compared with the conventional classical neural network, training process is not required for PCNN to implement image processing, segmentation and recognition, which is different from the traditional multi-layer network, making it ideal for real-time image processing environment.

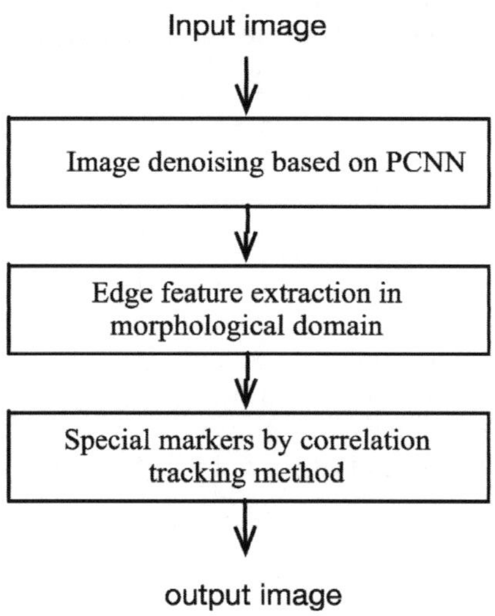

Figure 1. The system flow chart in this paper.

Secondly, feature extraction by extracting the relevant characteristics of the target area is taken to the image after de-noising for further special tag localization and identification of workers, which can largely reduce system complexity and running time. This paper does domain transform processing for image by morphological wavelet transform to extract image edge feature[2].

Morphological wavelet, which is produced by making mathematical morphological combine with wavelet analysis, can take the characteristics of non-linear filter of morphological and the characteristics of multi-resolution of wavelet transform into account, make the various types of wavelet analysis brought in its framework. Mathematical morphological solves a variety of image processing problems from expansion and corrosion, and the program of morphological wavelet signal analysis can be get by using morphological filter instead of the analysis and synthesis filter of linear wavelet transform. The sketch map of two-dimension morphological wavelet decomposing is showed in Fig. 2, where the detailed analysis operators are consist of a vertical direction operator $y_v(n)$, a horizontal direction operator $y_h(n)$ and a diagonal direction operator $y_d(n)$.

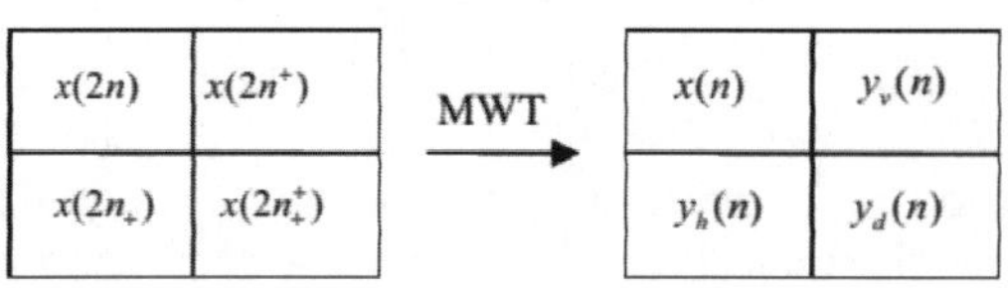

Figure 2. 2-D morphological wavelet transform(2-D MWT).

Studies have shown that it has different information sensitivity characteristic of human eyes to different frequencies and in different directions. It has high sensitivity for intermediate frequency information, leading the information sensitivity is weaker above a certain frequency and the diagonal direction. Since morphological wavelet is nonlinear multi-resolution analysis, the image reconstructed from different scales of space can highlight the signal characteristics of different frequencies after appropriate treatment.

For the nonlinear multi-resolution analysis of morphological wavelet, reconstruct image from different scales of space can highlight the signal characteristics of different frequencies after appropriate treatment[3]. Combined with the characteristics of morphological wavelet, multi-channel adaptive image enhancement structure can be constructed, based on this, image edge detection can get a better result conducive to solving practical problem.

Finally, to improve efficiency of fault detection and failure exclusion for the workers, certain parts of the ship which require special mark is tagged on the image, which is called special tag localization[4]. In this paper, correlation tracking method by matching between the target template feature and various sub-regions sequences of the actual image is carried out, to identify the sub-image position most similar to target template as the current target position[5-7]. The correlation tracking system is insensitive for all other scene dissimilar with the selected tracking target image. When the resolution is good and the noise and interference are not serious, it has better tracking ability and anti-background interference, which is conducive to tag localization for interest area such as fault region.

3. Experiment

In our experiment, we selected a representative sonar image which is a 256×256 underwater hull image showed in Fig. 3(a), then noisy image can be get by adding Gaussian noise whose noise standard deviation is 0.1 showed in Fig. 3(b). It is obviously can be seen that this image has serious noisy pollution phenomenon, and the hull edge is ambiguous even some places are incomplete.

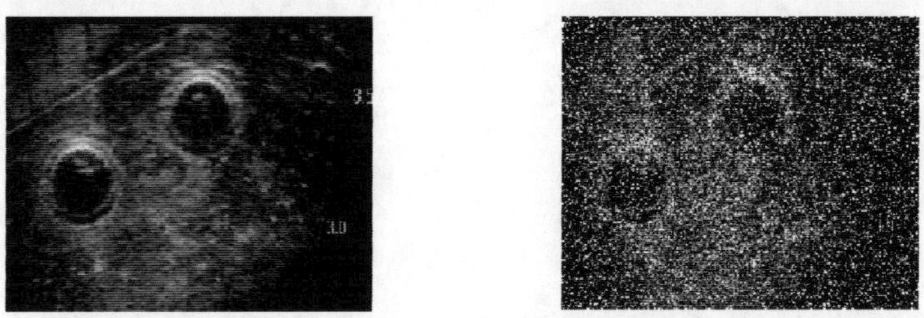

(a) Hull image (b) Noisy image

Figure 3. The underwater hull image.

The sonar image denosing result by PCNN method and some classical methods is showed in Fig. 4 and Fig. 5. Comparing with other denoising method, our method has better denoising effect, which is clearly to be seen that in the comparison curve of sonar image denoising under the different SNR in Fig. 5. In addition, it also can be seen from Fig.3 that the PCNN method not only have better performance in denoising, but also reserve the detail of the image well. To sum up, PCNN method is more suitable for sonar image denoising.

Journal of Physics: Conference Series **679** (2016) 012036 doi:10.1088/1742-6596/679/1/012036

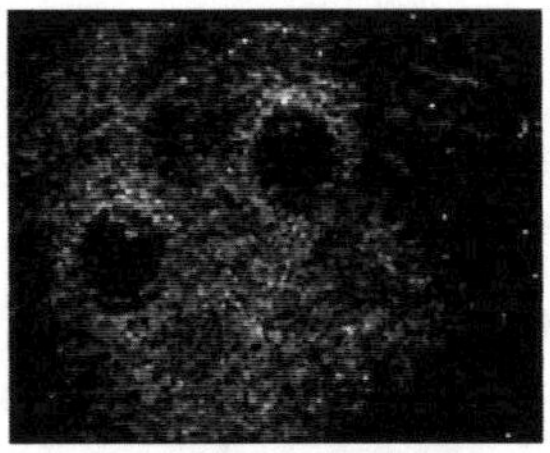

(a) Mean filter

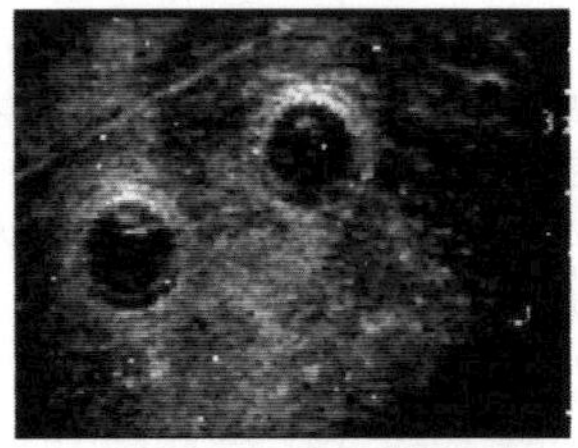

(b) Med filter

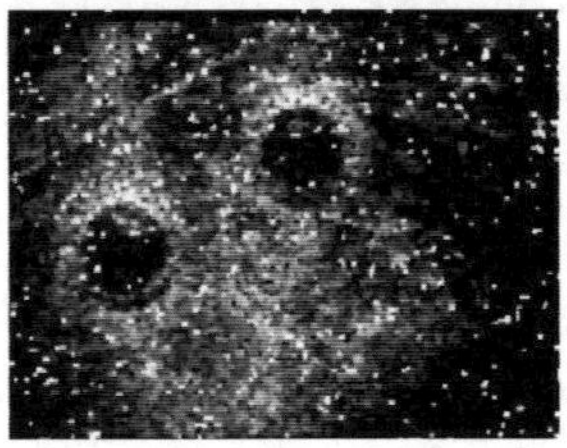

(c) Morphological filter

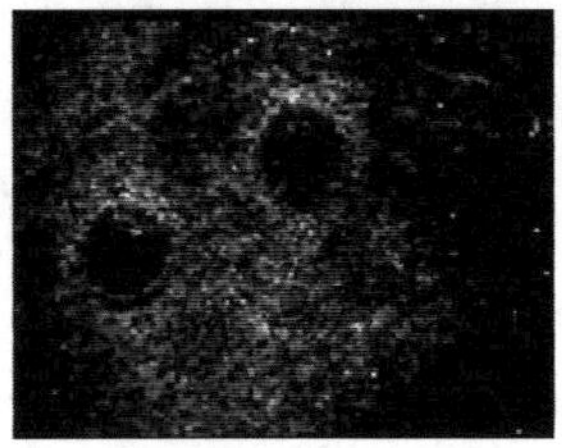

(d) PCNN filter

Figure 4. Denoising image.

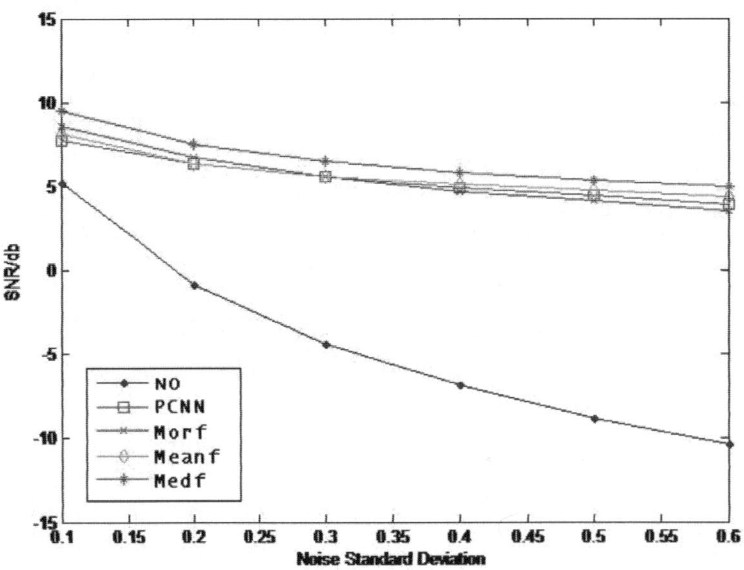

Figure 5. The comparison curve of sonar image denoising under the different SNR.

The image feature extraction results of the hull edge is showed in Fig. 6.

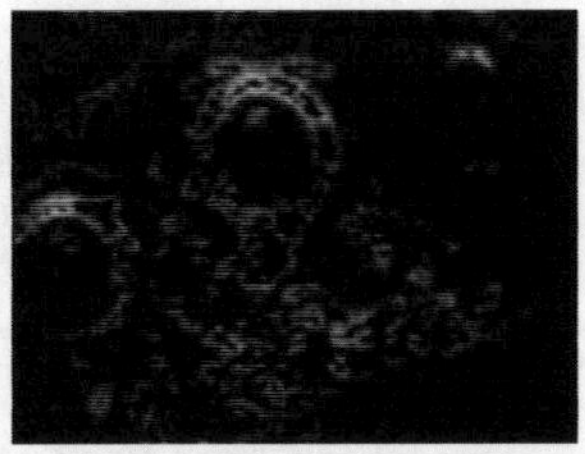

Figure 6. Edge feature extraction.

The special markers result is showed in Fig. 7. When the resolution is good and the noise and interference are not serious, it has better tracking ability and anti-background interference, which is conducive to tag localization for interest area such as fault region.

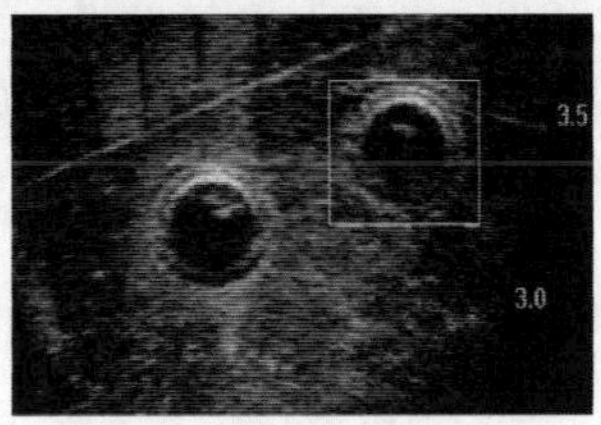

Figure 7. Special makes.

Finally, these simulation results prove the feasibility of this paper.

4. Conclusion

For the application of sonar image processing on ship underwater fault detection, this paper has realized location markers of a particular hull by de-noising, feature extraction and special markers of hull sonar image. Appling to the actual ship underwater inspection and maintenance, setting particular location to be the fault area, it is possible to achieve the auxiliary for ship underwater maintenance and repair work on a certain extent by acquiring high-resolution sonar images. In summary, this paper provides technical support for inspection and maintenance.

Reference

[1] Wei Z and Xue X 2010 AUV sonar image processing based on improved Pulse Coupled Neural Network model and Morphology *8th. IEEE. International. Conference. On. Control. And. Automation.* **1**

[2] Qingyun D and Yinglin Y 2003 A line Feature Extraction Method for Online Palmprint Images Based on Morphological Median Wavelet *J. Chinese. Journal. Of. Computers.* **26** 234–239

[3] Zhang J and Wu Q 2006 Morphological undecimated wavelet decomposition for fault location on power transmission lines *IEEE. Transactions. On. Circuits. And. Systems: Regular. Papers.* **53** 1395–1402

[4] Zhang X and Jia F 2014 A marker-based watershed method for X-ray image segmentation *J. Computer. Methods. And. Programs. In. Biomedicine.* **113**

[5] Habermehl D and Henkner K 2013 Evaluation of different fiducial markers for image-guided radiotherapy and particle therapy *J. Official. Organ. Of. The. Japan. Radiation. Research. Society.* **54**

[6] Lane D M and Chantler M 1998 Tracking and classification of multiple objects in multibeam sector scan sonar image sequences *J. Underwater. Technology.*

Journal of Physics: Conference Series **679** (2016) 012036 doi:10.1088/1742-6596/679/1/012036

[7] Trucco A and Garofalo M 2009 Processing and Analysis of Underwater Acoustic Images Generated by Mechanically Scanned Sonar Systems *IEEE Transactions on Instrumentation and Measurement.* **58**

Journal of Physics: Conference Series **679** (2016) 012037 doi:10.1088/1742-6596/679/1/012037

Intelligent Optimization of Modulation Indexes in Unified Tracking and Communication System

Wei-wei Yang, Bo Cong, Qiong Huang and Li-wei Zhu

China Satellite Maritime Tracking and Control Department, China

E-mail: nudtyww@163.com

Abstract. In the unified tracking and communication system, the ranging signal and the telemetry, communication signals are used in the same channel. In the link budget, it is necessary to allocate the power reasonably, so as to ensure the performance of system and reduce the cost. In this paper, the nonlinear optimization problem is studied using intelligent optimization method. Simulation analysis results show that the proposed method is effective.

1. Introduction

Unified tracking systems, such as USB(Unified S-Band), UCB(Unified C-band) are widely deployed in ground tracking stations to support space flight missions including spaceship, satellite, deep space explorer, etc. In unified tracking system, subcarriers for ranging, telemetry, command are modulated to one carrier by PM(Phase Modulation) or FM(Frequency Modulation) and the power allocation between them are determined by modulation indexes. Appropriate modulation indexes will lower the transmitted power, thus lower the cost while guaranteeing the specified system performance[1], so an optimization procedure is needed in the mission design. However, in this scenario, the optimization problem is nonlinear and quite complicated.

In this paper, we propose an approach based on Multimethod Collaborative Optimization (MCO) algorithm and verifies its effectiveness, efficiency by simulation.

2. Power Allocation Model of Downlink Signal

In the downlink of the communication between the satellite and the ground station, the link equation is usually expressed as [2]:

$$\left.\frac{S}{\Phi}\right|_{\Sigma} = EIRP + G/T + 2\lambda - 2R - 20\lg(4\pi) - k - L_{\Sigma}^{'} - M \tag{1}$$

Where S is the power of the signal, Φ is the power spectral density of the noise, G/T is the quality factor of the ground receiving system, $EIRP$ is the equivalent isotropic radiated power of the satellite transponder, λ is downlink carrier wavelength, R is the operation range, k is Boltzmann constant, $L_{\Sigma}^{'}$ is other losses except feed line loss, and M is system design margin.

Angle modulation is adopted for each signal in the unified tracking and communication system. In downlink channel, phase modulation is used for all signals while the residual carrier is used as a tracking lock signal. In this paper, S-band system is taken for example, there are three sub-carriers in downlink, which are major tone, minor tone, and the telemetry sub-carrier. Signal noise power spectral density ratios of residual carrier and each sub-carrier are as follows:

$$\begin{cases} \left.\dfrac{S}{\Phi}\right|_c = \left.\dfrac{S}{\Phi}\right|_\Sigma \cdot \prod_{i=1}^{3} J_0^2(m_i) \\ \left.\dfrac{S}{\Phi}\right|_j = \left.\dfrac{S}{\Phi}\right|_\Sigma \cdot 2J_1^2(m_j) \cdot \prod_{\substack{i=1 \\ i \neq j}}^{3} J_0^2(m_i) \quad (j = 1,2,3) \end{cases} \tag{2}$$

where m_1, m_2, m_3 are modulation indexes of each sub-carrier, J_0, J_1 are zero-order and first-order Bessel function individually.

In the link budget, it is necessary to allocate the power of the various signals reasonably, so as to ensure the performance of the work, and to make the lowest emission power, in order to reduce the cost.

Taking into account the main constraints of power allocation, the optimization model is established as follows:

$$\begin{cases} Find \quad m_1, m_2, m_3 \\ Min \quad EIRP \\ S.t. \\ \qquad \left.\dfrac{S}{\Phi}\right|_c \geq \left.\dfrac{S}{\Phi}\right|_{c,th} \\ \qquad \left.\dfrac{S}{\Phi}\right|_j \geq \left.\dfrac{S}{\Phi}\right|_{j,th} \quad (j = 1,2,3) \\ \qquad m_\Sigma = \sqrt{\sum_{j=1}^{3} m_j^2} < \pi/2 \end{cases} \tag{3}$$

where $\left.\dfrac{S}{\Phi}\right|_{c,th}$, $\left.\dfrac{S}{\Phi}\right|_{j,th}$ are thresholds of carrier and each sub-carrier. m_Σ is the total modulation index.

3. Multimethod collaborative optimization algorithm

Multimethod Collaborative Optimization (MCO) algorithm is proposed based on comprehensive analysis of the optimization problem and a variety of optimization methods, which combines multiple effective and collaborative methods with a certain strategy to obtain better optimization performance.

The optimization information of one collaborative step is obtained by a certain number of iterations of each chosen methods, which is processed and used in next collaborative step. The algorithm is terminated when the collaborative optimization termination criterion is met. Using MCO algorithm to solve the optimization problem can get better results than the single optimization method.

In this paper, genetic algorithm, pattern search method and Powell method are chosen for collaborative optimization using parallel collaborative strategy. The genetic algorithm has better global optimality while the latter two are local search methods. Collaborative optimization with these three methods, can make use of their advantages to obtain good results.

4. Simulation Analysis

Assuming that the signal noise power spectrum density ratio of the residual carrier, the error rate of the telemetry signal and ranging accuracy are known, the threshold for carrier and each sub-carrier are calculated as follows:

(1) the residual carrier. $\left.\dfrac{S}{\Phi}\right|_{c,th} = 43\text{dBHz}$;

(2) the ranging tone. major tone (100kHz) ranging accuracy (R_0) is required for 10m, the signal noise power spectrum density ratio required by ground terminal is :

$$\left.\frac{S}{\Phi}\right|_{1,th} = 20\log\frac{c}{18\times R_0\times 10^5} + 10\log 5 = 32\text{dBHz}\,, \qquad \left.\frac{S}{\Phi}\right|_{2,th} = 27\text{dBHz}\,;$$

(3) telemetry signal. the error rate $P_e = 1\times 10^{-5}$ request signal to noise ratio S/N =7.4 dB, taking into account the demodulation loss 2.5dB, bit rate is 1024 bit / s, so ground telemetry terminal signal to noise power spectrum density ratio is :

$$\left.\frac{S}{\Phi}\right|_{3,th} = 7.4 + 10\log(1024) + 2.5 = 40\,\text{dBHz};$$

The optimal power allocation is based on the detection threshold of the modulation signal, and make the ratio of the useful power to reach the maximum, so that the power component of the modulated signal reaches the maximum.
Total index modulation should be no more than 1.5, i.e. $m_\Sigma \le 1.5$.

Table 1. System parameter setting

Variables	unit	value	Variables	unit	value
R	km	1000	G/T	dB/K	22.5
f	MHz	2250	M	dB	6
$\left.\frac{S}{\Phi}\right\|_{c,th}$	dBHz	43	$\left.\frac{S}{\Phi}\right\|_{1,th}$	dBHz	32
$\left.\frac{S}{\Phi}\right\|_{2,th}$	dBHz	27	$\left.\frac{S}{\Phi}\right\|_{3,th}$	dBHz	40

Table 2. Results of optimization

Variables	carrier	Main side tone	The secondary tone	telemetry
m_i /rad		0.391	0.223	0.898
m_Σ /rad	1.0041			
L_i /dB	-2.29	-13.29	-18.29	-5.28
η_i (%)	59.02	4.69	1.48	29.64
η (%)	94.83			
$\left.\frac{S}{\Phi}\right\|_i$ /dBHz	43	32	27	40
$\left.\frac{S}{\Phi}\right\|_\Sigma$ /dBHz	45.29			
$EIRP$ /dBW	-40.32			

5. Conclusion

In this paper, the power allocation optimization model of downlink signal is established for the unified tracking and communication system. Multimethod collaborative optimization is proposed to solve this nonlinear optimization problem in order to obtain optimal modulation indexes, which can reduce the cost while guaranteeing the specified system performance. Simulation results show that the proposed method is effective and all the constraints are satisfied.

References

[1] L.Zhang, G. Li 2007 Optimized selection of modulation index of TTC system *Spacecraft Engineering* **16** 38
[2] Yefu Zhao *et al* 2013 *Radio Tracking System*(Beijing, Defense Industry Press) p 391

Journal of Physics: Conference Series **679** (2016) 012038 doi:10.1088/1742-6596/679/1/012038

Failure Analysis and Quality Control of Microwave Devices

Zhimao Ming, Bo Zong, and Xiaoshu Bai

China Satellite Maritime Tracking and Control Department, Jiangsu China 214431,

E-mail: mzmnudt2008@126.com

Abstract. Microwave devices have been widely used in the communication and navigation and navigation fields. The reliability level of microwave devices is an important factor to affect the reliability of electronic equipment. The statistical analysis for the failure of sixty microwave devices was presented and the main causes of failure were given. The failure is mainly analyzed from three aspects below, the surface failure mechanism, the interior failure mechanism and the failure mechanism of electrode system and encapsulation. The surface failure mechanism is analyzed from four aspects, ionic contamination on the surface, silicon dioxide layer defect, the influence of interface states between silicon and silicon dioxide, radiation ionization trap. The interior failure mechanism is analyzed from two aspects, failure caused by thermal breakdown and failure due to latch-up effect of integrated circuit. The failure mechanism of electrode system and encapsulation is analyzed from two aspects, failure mechanism of metallization system and failure mechanism of metallization system of bonding. Meanwhile, the results showed that operation, process, adjustment, components and using problems were the main causes of failure. The valuable statistical data and analysis results were provided for the quality control of microwave devices.

1. Introduction

Communication and navigation products play an important role in civilian life and military equipment nowadays. Microwave device is the critical component of electronic system in communication and navigation products [1]. A variety of microwave devices develop rapidly. Integrated circuit is representative in microwave devices, and its design and manufacture technology develops rapidly to the direction of larger scale, faster rates, and higher level of integration [2]. To improve the reliability of communication and navigation products, we must improve the reliability of microwave devices firstly. Failure of microwave devices may occur due to various factors in its production process or service environment, so understanding about the failure mechanism of microwave devices is the premise to improve the reliability in communication and navigation products application.

From the statistical analysis for the failure of sixty microwave devices, we can summarize that failure modes of microwave devices are as follows [3], the failure which is obtained by appearance observation and electric characteristics test such as flaking of oil paint, looseness of outer lead, inverse current leakage and short circuit, the failure which is due to the defect in interior of the device such as serious injury of inner lead, extraneous matter on the chip and crack of chip. Meanwhile, the failure mechanism of microwave devices is summarized from the follow three respects, and it can illustrate the reason of why the failure occurs.

2. The surface failure mechanism

2.1. Ionic contamination on the surface
With the stained moisture or conductive material on the surface of chip and ionizing radiation or electrostatic charge accumulation, positive ions and negative ions appear on the surface of oxide layer. These ions can move along the surface under the effect of bias voltage, inducing that positive ions accumulate around the negative electrode and negative ions gather around the positive electrode. When the contamination is serious, silicon surface barrier alters considerably. The accumulation of moving electric charge on the surface reduces the surface conductance, and induces surface leakage and breakdown creep. Ionic contamination on the surface can also induces corrosion of metal, and further induces rust and fracture of the electrode and packaging system of the device [4].

2.2. Silicon dioxide layer defect
Surface passivation technology defect which induces surface degradation is an important failure mode. According to its degradation parameters and mechanism, electric conduction of Si-SiO_2 boundary layer results in reverse voltage degradation of p + n junction and change of threshold voltage VTH of MOS device. Recombination of surface carrier induces degradation of current amplifier facto of transistor or deterioration of the low frequency noise. Reasons of surface degradation are as follows.

1) Swimming ion of encapsulation materials invades the passivation membrane on Si surface.
2) Interface charge on interface between Silicon and passivation membrane.
3) The pinhole, island caused by defects of passivation membrane on Silicon surface.
4) Leakage of the surface passivation membrane caused by electric field.

2.3. The influence of interface states between silicon and silicon dioxide on device performance
Interface states between silicon and silicon dioxide is different from electronic state in internal silicon wafer. It can give free electrons to internal silicon wafer and the positive charge at the interface form on the interface. Meanwhile, it can also give electron-hole to internal silicon slice, and the negative charge at the interface form on the interface. Due to the existence of interface states charge, space charge region with opposite charge forms on silicon wafer near the interface, and it induces that electron-hole concentration near silicon wafer surface is different from the interior. Exhausted and inversion layer forms on silicon wafer surface owing to electrostatic induction effect, and it will effect reverse characteristic of the device, and it will also make h_{FE} decline.

2.4. Radiation ionization trap
When Si-SiO_2 system is irradiated by gamma rays, X rays, the trap density in Silicon dioxide layer increases, meanwhile, electron-hole pairs form. These surplus electron-hole pairs will induce ionization damage. With transition of electron from valence band to conduction band, electrical properties, chemical properties and physical and mechanical properties of microwave devices will be influenced seriously, and finally the reliability and stability of the device will be influenced [5].

3. The interior failure mechanism
The interior failure mechanisms mainly include failure caused by thermal breakdown and failure due to latch-up effect of integrated circuit[6].

3.1. Radiation ionization trap
Thermal breakdown is a typical interior failure, and it seriously threats secure use of power transistor and high inverse voltage. After the p-n junction reaching first breakdown point, and when the voltage increases continually, the electric current will increase to a critical value. After that, the voltage between the two-terminal of device decreases suddenly and the electric current increases sharply, which shows the negative resistance phenomenon. The phenomenon is called secondary breakdown. When operating point moves to low voltage and large current meanwhile, the device will be burned without protect. Secondary breakdown is a thermal breakdown, and its destructive effect is irreversible. The first breakdown and secondary breakdown is shown in figure 1.

Journal of Physics: Conference Series **679** (2016) 012038 doi:10.1088/1742-6596/679/1/012038

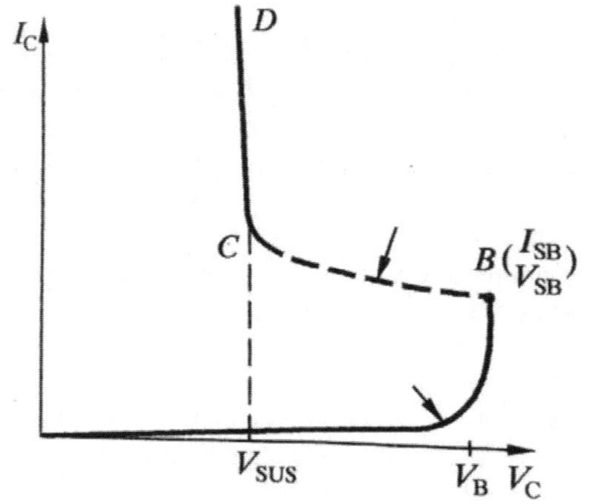

Figure 1. Current-voltage characteristic curve.

There is intrinsic difference between first breakdown and secondary breakdown. The first breakdown is an electric breakdown, and device can still be back to normality when reverse voltage decreases, so the first breakdown is nondestructive and reversible. But secondary breakdown is a thermal breakdown, and it is destructive and irreversible. When the secondary breakdown occurs, device will be burned due to the excessive current through the p-n junction. There is no exact theory to perfectly explain the secondary breakdown now, and suppositional theories include thermal model and current mode [7].

3.2. Failure due to latch-up effect of integrated circuit

CMOS integrated circuit has the characteristics of low power dissipation, fast speed, big noise tolerable limitation, high output impedance and wide voltage range, so it plays an important role on the large scale and very large scale integrated circuit. But input impedance of CMOS integrated circuit is high, so there is a special failure mechanism, latch-up effect except Electrostatic discharge damage (ESD). Owing to existence of silicon controlled rectifier in circuit construction, when the triggered conduction condition forms, low resistance high current between the power and ground will occur. When integration density is higher and the size of the device is smaller, latch-up effect will be more sensitive. Failure due to latch-up effect seriously influences reliability of CMOS integrated circuit [8].

4. Failure mechanism of electrode system and encapsulation

Electrode system of microwave devices includes metallized layer, wire bonding, ohmic contact between the chip and the base, and they play the role of transmitting power. Electrode system on the device reliability is very important. According to pertinent data, Failure due to metallized layer is the most common in electrode system, and it accounts for 26% of device failure [9]. And the second factor is the wire bonding and chip bonding.

4.1. Failure mechanism of metallization system

1) Mechanical injury

The metallized aluminum layer of device consists of strips of aluminum, and it combines the active region of the device with soldered joint. The aluminum is easy to scratch owing to its softy and poor mechanical strength. For example, slide of measuring probe, scratch of metal tweezers and scribing can make aluminum film of the chip scratch, and they seriously affect the yield of product. Scratch should be eliminated before encapsulation through strict microscopy.

2) Electro migration

The metallized aluminum layer of device combines the active region of the device with soldered joint, or metallized aluminum layer connect several active regions. When there is enough high current passing through the aluminum bar, the aluminum ions in energized aluminum strip transfer along the

Journal of Physics: Conference Series **679** (2016) 012038 doi:10.1088/1742-6596/679/1/012038

electron source direction. The transmission is obvious under the condition of high temperature and large current. The aluminum ions accumulate topically in metallic membrane and finally induce short circuit. Cavity will occur in aluminum strip after a long time, which will induce open circuit. The phenomenon is called electro migration. Electro migration is sensitive to current density and temperature, and it can lead to open circuit, short circuit and parameter degeneration of the device. Rational design, increasing the cooling, decrease of the current density, reduction of temperature and strict control on the process are measures to improve migration resistance of the aluminum layer system.

3) Discontinuity of metallic membrane on the oxidation layer stair

Aluminum membrane on the silicon dioxide stair is usually very thin for planar device. So the aluminum membrane on the oxidation layer stair especially around ohmic contact is very likely to be fractured.

Discontinuity of aluminum membrane on the oxidation layer stair is due to the masking effect of the aluminum membrane in evaporation process. Incident particles can't reach the stair surface, and it induces the blind angle, and then aluminum membrane gap occurs. Meanwhile, the evaporation aluminum grains are relatively coarse, and they grow up slowly in the flank inducing that low atom density area forms around the corner. Existence of low atom density area decreases strength of aluminum membrane, and it is the essential reason for the discontinuity of metallic membrane.

In order to reduce the failure phenomenon, zero angle of θ in sputtering process, multisource evaporation, rotation of substrate when evaporating and increasing the temperature of the substrate can be adopted [10].

4.2. Failure mechanism of bonding

About a quarter to a third of failure on microwave devices is caused by the bonding, so it has a great influence on the quality of the device.

1) Failure caused by error in process

In the failure of bonding, a part is caused by error in process. For example, if the lead is too long, it is easy to touch the bare chip and burn device. Overweight pressure welding can damage lead, and finally induce breaking. Too light pressure welding or too dirty aluminum surface make the pressing point easy to fall off. Oblique pressure welding is likely to cause a short circuit.

2) Cracked silicon filament caused by glue in encapsulation process

Low temperature performance of glue layer is poor, and it will crack seriously after long-term low temperature test. With thicker glue layer, lower temperature and longer time, the damage will be greater, and finally silicon filament may be snapped. If the glue is excessive in encapsulation process, it will be piled into the interior and jointed with the lead, and finally silicon filament is snapped.

3) Failure of encapsulation system

As for the production process of microwave devices, lead quality and encapsulation process are also very important on reliability of microwave devices. For any device, if the chip is fine but without good encapsulation to protect the chip, tube core will be affected by external high and low temperature, moisture, salt fog, low air pressure, vibration, impact and centrifugal effect, so the quality of the device cannot be assured.

Due to the simple process, low cost of plastic encapsulation plus device and significantly improvement of purification technology, reliability of plastic encapsulation device is improved, so a large number of civilian devices generally use plastic package. However, because of poor moisture resistance, big effect by harmful impurities in the resin and poor thermal shock résistance, performance of plastic encapsulation device decreases, meanwhile, corrosion occurs in the electrodes and wiring. Due to above defects in plastic encapsulation device, reliability of plastic encapsulation device is lower than device of hermetically metal sealing, plastic encapsulation device is generally used in civil products which are in good environmental conditions. Epoxy resin adhesive piece is also usually used in integrated circuit encapsulation. Failure caused by epoxy resin adhesive piece is as follow. Due to its poor heat dispersion, catalysis on epoxy resin occurs above 200 ℃, and it induces that the chip is easy to fall off. For using silver paste or epoxy resin as die bonding material, thermal fatigue occurs in thermal cycle of the device caused by thermal change and different expansion coefficient between chip and tube socket, and it results in failure because of increase of contact resistance.

4) Failure caused by rust and fracture of the base pin

Corrosion of base pin in microwave devices includes chemical corrosion and stress corrosion. The most general material of base pin is Kovar alloy which is sensitive to corrosion, so it seriously influences the reliability of the device. Reasons of corrosion and fracture on base pin include two aspects, internal cause and external cause [11]. Internal causes are as follows.

a. Kovar alloy lead is sintered badly resulting in boundary oxidation.
b. Kovar alloy grains are coarse.
c. Electrodeposited coating is unsound or it has pinhole defect.
d. Excessive sulfur content in the plating brightener.
e. There is residual corrosion substance before and after the plating.

External causes are as follows.

a. Soldering paste and flux remain on base pin
b. Cartons for filling device contain acidic material.
c. In the process of shipping process, base pin suffers damage from mechanical stress.

Excessive humidity or corrosive gas exists in storage or work environment of the device, and it induces corrosion failure of base pin.

5. The main failure causes and quality control methods of microwave devices

Statistical analysis for the failure of sixty microwave devices showed that operation problems, process problems, adjustment problems, components problems and using problems were the main causes of failure. Based on the failure causes and statistical analysis, we can summarize that main failure causes in using aspect are unsuitable operation and electrical-over-stress. Main failure causes in the quality of its components and module were lead bonding, ruggedized protection glue, chip defect, chip bonding, coil separation and so on. The respective percentage of the above causes was 27%, 13%, 13%, 10%, 10%. Ruggedized protection glue and coil separation are specific failure cause. Statistical data of components and modules is relatively consistent with Rome Air Development Center, but the failure rate of domestic components caused by chip bonding is much higher than the foreign components and modules. It is worth of attention for the manufacturer. Process improvement which is based on the main failure causes of microwave devices can improve the quality and reliability for the manufacturer. Meanwhile, quality evaluation and then eliminating defective devices through specific inspection and analysis methods before the actual service can reduce the failure rate in the microwave components' service and improve the system reliability.

6. Conclusion

Failure of microwave devices affects the work reliability of communication and navigation products. From the statistical analysis for the failure of sixty microwave devices, the failure mechanism was summarized. Failure mechanism of microwave devices includes the surface failure mechanism, the interior failure mechanism and the failure mechanism of electrode system and encapsulation. The surface failure mechanism includes ionic contamination on the surface, silicon dioxide layer defect, the influence of interface states between silicon and silicon dioxide, radiation ionization trap. The interior failure mechanism includes failure caused by thermal breakdown and failure due to latch-up effect of integrated circuit. The failure mechanism of electrode system and encapsulation includes failure mechanism of metallization system and failure mechanism of metallization system of bonding. Being familiar with failure mechanism of microwave devices is beneficial to radically reduce failure factor and improve the application reliability. Meanwhile, operation problems, process problems, adjustment problems, components problems and using problems are the main causes of failure. Quality control methods are mentioned briefly at last.

Acknowledgments
I want to give my sincere appreciation to XinXin Zhang who gives me much help in experimental research, and Xiaoshu Bai who helps me a lot in article writing.

References
[1] Friedrich Beck and Stephen Wilson 1997 *Wiley Ser. Qual. Reliab. Eng.* **14** 5
[2] Kuroki and Oota K 1989 *Proc. Electron. Compon.* **11** 88
[3] Ciappa M, Carbognani F and Cova P 2002 *Microelectron. Reliab.* **42** 1653
[4] Jon MC and Sturm GW 1982 *NDT Int.* **15** 185
[5] Zhao Y H and Fu G C 2015 *Eng. Fail. Anal.* **47** 229
[6] Jin M and Wang X H 2005 *Equip. Environ. Eng.* **2** 34
[7] Cao X P, Fang B and Yin Z T 2008 *Electron Prod. Relativ. Environ.Test.* **26** 46
[8] Yang C C 2009 *Sci. Technol. Innov. Herald* **10** 4
[9] Liu Y H, Jian W T, Zhang R Z and Dong W C 2009 *Chin. Integr. Circuits* **12** 51
[10] En Y L, Luo H W and Lai P 2006 *Fail. Anal. Prev.* **1** 40
[11] Friedrich Beck and Stephen Wilson 2012 *Electron. Encapsulation* **12** 34

The Bayesian Reliability Assessment and Prediction for Radar System Based on New Dirichlet Prior Distribution

Zhimao Ming, Xiaodong Ling, Xiaoshu Bai, and Bo Zong

China Satellite Maritime Tracking and Control Department, Jiangsu China 214431,

E-mail: mzmnudt2008@126.com

Abstract. This article studies on Bayesian reliability growth models of complex system based on new Dirichlet prior distribution when the sample of system is small. The model briefly describes expert experience as uniform distribution, then equivalent general Beta distribution of uniform distribution can be solved by optimization method when prior parameters are variables, mean is constraint condition, and variance is regarding as the optimization objective. The optimization method solves the problem of how to determine values of hyper-parameters of new Dirichlet distribution when these parameters have no specific physical meaning. Because the multidimensional numerical integration of posterior distribution is very difficult to calculate, WinBUGS software is employed to establish Bayesian reliability growth model based on a new Dirichlet prior distribution, and two practical cases are studied under this model in order to prove validity of model. The analysis results show that the model can improve the precision of calculation, and it is easy to use in engineering.

1. Introduction

With lots of new developing products more and more complex, the procedure of new product developing usually have following characteristics, such as smaller sample, non-homogeneous, heritage and requiring higher reliability. In general, development of a new product is the process of design-test-redesign-retest, the last test is continuing and evolution of early one, there is therefore lot of history information and early test data. At the same time, many experts have prior knowledge of a new product development. Therefore, Bayesian reliability growth model is usually used to the reliability assessment of new product when having prior information, history experience and data of similar product.

In recently, the main Bayesian reliability growth models are Smith model [1], Barlow Scheuer model [2], Logistic model [3], discount factor method [4], and so on. These models usually base on non-prior information hypothesis as well as testing information discount principle, so they can only give the reliability assessment of current testing product, rather than reliability estimation of final product. In 1991, Mazzuchi T. A. and Soyer R. [5], [6], [7] presented the reliability growth model of product based on Dirichlet prior distribution. The marginal distribution and joint distribution of Dirichlet distribution were used to combining opinion and experience of experts as well as test information of similar products. Meanwhile these distributions can predict product reliability and achieve management of reliability growth, such as how to foresee cost and periods of product before test and/or during test, how to trace the process of reliability growth of product, how to determine the

Journal of Physics: Conference Series **679** (2016) 012039 doi:10.1088/1742-6596/679/1/012039

stop time, etc. Erkanli, Mazzuchi and Soyer [8] discussed the calculation issue of reliability growth model based on Sequential Dirichlet prior distribution, and introduced the application of MCMC in computation for Bayesian posterior distribution. The Sequential Ditichlet family has (m+1) independent parameters, namely $\beta,\alpha_1,...,\alpha_m$, by choosing { α_i } the Sequential Dirichlet prior describes well knowledge of engineer, relevant decision methods were discussed in van Dorp etal [9]. Moreover, Somerville, Dietrich and Mazzuchi [10] gave the analysis method base on Sequential Dirichlet prior distribution as well as its application on the accelerated life test (ALT), considering step stress ALT the inverted sequence Dirichlet prior distribution had used because of product reliability decreasing during whole test period. Based on Dirichlet prior distribution Zhang and Li [11] [12] discussed the issue of Bayesian reliability of multinomial distribution aiming at robustness of likelihood function. Yu [13] and Liu [14] studied on the issue of reliability growth based on Sequential Dirichlet distribution. Furthermore, Li and Wu [15] pointed out some disadvantages of the Sequential Dirichlet distribution, such as reliability prior information of difference stages is described by only one parameter β, which is not consistent with various present practices. In fact the prior density of various phases generally is single peak, therefore in order to accord with this characteristic, the value of β must be lager value, i.e. the variance of various phases is all small. So sequential prior Dirichlet distribution has become a dominant prior distribution in this situation, thus the prior information will has a main effect on calculation of posterior information based on a small sample test, i.e. test results will lightly influence on posterior inference. Furthermore, all of prior parameters of Dirichlet distribution must be determined before reliability growth test, but in engineering experts usually give accurate reliability assessment of next phase only under knowing test results and innovative approach of last one.

Therefore, Li and Wu [16] studied further on a new prior distribution class in their paper, and established Bayesian reliability growth model based on the new Dirichlet prior distribution. Moreover, the value of reliability assessment and its confidence are described by two parameters in each reliability growth phase. However, parameters of the new prior distribution class $\alpha_1,\alpha_2,\ldots,\alpha_m$ have not definitely physical meaning, it is difficult to determine their values by experts experience, and will be "respectable" when it will be able to solve the problem of practice engineering.

In order to solve the above problems, this paper focuses on Bayesian reliability growth model of new product developing based on a new Dirichlet prior distribution. The physical meaning of the hyper-parameters of the new Dirichlet prior distribution has been defined and determined by optimization method. Moreover, discussing the problem of how to calculate multidimensional numerical integration of posterior distribution, WinBUG software package is employed to describe Bayesian reliability growth inference model of a new Dirichlet prior distribution. Finally, the paper presents an application of the model to two cases, the results show that the model is brief and effective and can be used easily to the reliability growth assessment.

2. Bayesian reliability growth model and a new prior distribution

2.1. Model hypothesis

I. Binomial distribution model is usually used to reliability analysis of complex large system. Assuming products have been undergone m test period. In each test stage, there are n_k products undergoing a time-terminated test, s_k products survived. Therefore, we can collect data of success and failure of product.

II. After each test stage, correcting products and removing failure. Therefore, reliability of products is an evolution process during the whole test phases. Assuming the reliability of product is Rk in the kth period, thus,

$$0 \le R_1 \le R_2 \le \ldots \le R_{m-1} \le R_m \le 1 \tag{1}$$

Based on above hypothesizes and combining the information of similar products as well as expert experience, this paper studies on how to assess reliability growth of product by Bayesian theory and technology.

2.2. Likelihood function
According to the model assumption, success data of the kth time-terminated test stage s_k follows a Binomial distribution, and the parameters are n_k and R_k, i.e.

$$s_k \sim Binomial(n_k, R_k) \tag{2}$$

Likelihood function as following.

$$L(R_k; n_k, s_k) = \binom{n_k}{s_k} R_k (1-R_k)^{n_k - s_k} \tag{3}$$

2.3. Prior distribution family
I. Order Dirichlet distribution
The expression of order Dirithlet distribution is following,

$$\pi(\tilde{R} \mid \tilde{\alpha}, \beta) = \frac{\Gamma(\beta)}{\prod_{i=1}^{m+1} \Gamma(\beta\alpha_i)} \prod_{i=1}^{m+1} (R_{i-1} - R_i)^{\beta\alpha_i - 1} \tag{4}$$

Where, $\tilde{R} = (R_1, R_2, ..., R_{m-1}, R_m)$, $\beta > 0$, $\alpha_i > 0$, $\sum_{i=1}^{m+1} \alpha_i = 1$, the β parameter is shape parameter, the α parameter is location parameter and $R_0 = 0$, $R_{m+1} = 1$. Moreover, the α parameter can be determined by prior information and well describes engineering experience of reliability assessment of various phases.

According to characteristics of Dirilchlet distribution marginal distributions of above several distributions are a Beta distribution, namely,

$$R_i \sim Beta(\beta\alpha_i^*, \beta(1-\alpha_i^*)) \tag{5}$$

Where, $\alpha_i^* = \sum_{k=1}^{i} \alpha_k$ .Furthermore, according to characteristics of Beta distribution the following expressions are given.

$$E[R_i] = \sum_{j=1}^{i} \alpha_k = \alpha_i^* \tag{6}$$

$$\alpha_i = E[R_i] - E[R_{i-1}] \tag{7}$$

Thus, the α parameter expresses reliability growth between each stage. In general product reliability increases quickly in the early phase of product development, corresponding value of α is bigger. Contrary in the late of product development α is usually small because it is difficult to have bigger improvement in such stage. The β parameter describes the degree of confidence of estimate value, assuming α is determined, the larger β the smaller standard deviation of prior, therefore confidence of reliability is higher, the reverse is true.

Inspection of above analysis of order Dirichlet distribution shows that in whole period of product development there is only a shape parameter β describing the variance of product reliability in various test phases, however experts usually need to know test results and correct approach of last phase so that they can give accurate reliability assessment of next one.

II. A new prior distribution family based on Beta distribution

A disadvantage of the order Dirichlet distribution is that it only has one shape parameter β to describe the variance of product reliability in various test periods, and is not consistent with present practice. Based on general Bate distribution of various test phases Reference [17] established a new prior distribution family by conditional distribution modality, which is fit for reliability growth model of new product. The distribution eliminates some of disadvantage of order Dirichlet distribution, and can express well the prior information.

Regarding the kth test interval, constructing general Beta distribution of $(R_{k-1},1)$ as reliability distribution of product in this interval, then,

$$g_k(R_k \mid R_{k-1}) = g_k(R_k \mid R_{k-1}; a_k, b_k) = \frac{(1-R_{k-1})^{1-a_k-b_k}}{B(a_k,b_k)}(R_k - R_{k-1})^{a_k-1}(1-R_k)^{b_k-1} I_{(0,1)}(R_k) \quad (8)$$

Where $a_k > 0, b_k > 0, R_0 = 0$, $B(\cdot,\cdot)$ is Beta function, And $B(a_k,b_k) = \dfrac{\Gamma(a_k)\Gamma(b_k)}{\Gamma(a_k+b_k)}$. Then, the density function of joint prior distribution of $\tilde{R} = (R_1, R_2, ..., R_{m-1}, R_m)$ described as following,

$$\pi(\tilde{R} \mid \tilde{\alpha}, \tilde{\beta}) = \prod_{k=1}^{m+1} g_k(R_k \mid R_{k-1}) \quad (9)$$

Equations (8) and (9) show that there are two parameters in each test period, which can accurately express experts' experience. Moreover, for each reliability growth stage the reliability estimation and its confidence can be described by both parameters. Therefore, the new prior distribution is better than the order dirichlet prior distribution.

2.4. Mathematics characteristic of prior distribution family

The condition distribution, expressed by equation (8), is general Beta distribution in $(R_{k-1},1)$, which characteristic is similar with standard Beta distribution in (0, 1). Several typical Beta density curves show in figure 1.

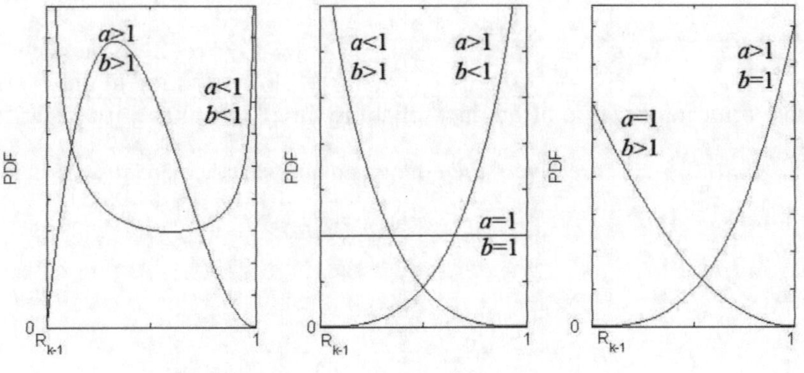

Figure 1. Several typical Beta density curves.

Considering general Beta distribution in $(R_{k-1},1)$, when $R_1, R_2, ..., R_{k-1}$ is known, we can solve the condition mean and condition variance of R_k by equation (8) as following,

$$\mu_k^* = \frac{a_k + b_k R_{k-1}}{a_k + b_k}$$
$$v_k^* = \frac{a_k b_k (1 - R_{k-1})^2}{(a_k + b_k)^2 (a_k + b_k + 1)} \tag{10}$$

Based on above analysis of characteristic, it stands to reason that the physical meaning of parameters of a new prior distribution family is not clear. Therefore, it is the first step that how to determine parameters of prior based on prior information, if we wish to use the new prior distribution for reliability assessment of product.

3. The method of determining prior parameters

The prior information of reliability is used to presenting by continuous interval modality. For instance, product reliability of kth test phase, $R_k \in (R_{k,L^*k,H})$, may be given according to information of similar product and expert experience. In this paper, firstly prior information of product is described by uniform distribution, then the general Beta distribution is conducted by optimization method, which mostly approximate to the uniform distribution, when prior parameters are variables as well as mean is constraint condition, and variance is regarding as the optimization objective. Then the special Beta distribution is used to describing prior distribution as depicted in figure 2.

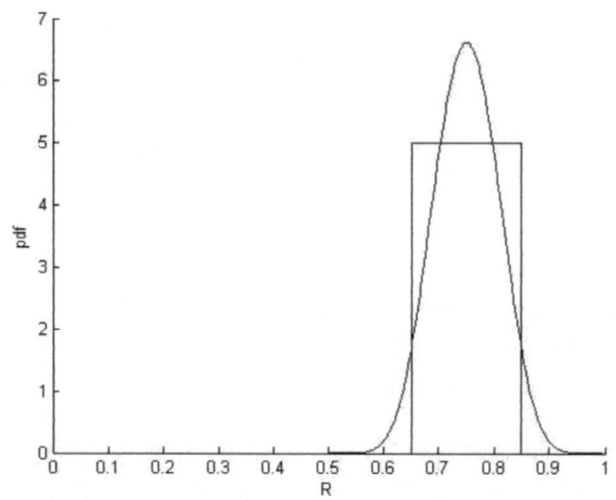

Figure 2. Equivalent general Beta distribution of uniform distribution.

If there are more prior information of product reliability in k^{th} test phase, for instance h continuous phases $(R^j_{k,L}, R^j_{k,H})$ j=1, 2, …, h, are given, then the mean and variance are shown as following.

$$ER_k = \frac{R_{k,L} + R_{k,H}}{2} \tag{11}$$

$$VR_k = \frac{(R_{k,H} - R_{k,L})^2}{12} \tag{12}$$

How to determine the prior parameters, one of natural ideas is that firstly making the mean and variance of general Beta distribution expressed by equation (10) to equal the mean and variance of uniform distribution $U(R_{k,L}, R_{k,H})$ expressed by equation (11) and (12), then we can determine parameters of prior distribution by solving equation (11). Unfortunately, the solutions of equation (11)

Journal of Physics: Conference Series **679** (2016) 012039 doi:10.1088/1742-6596/679/1/012039

are sometime negative, which is not according with non-negative characteristic of prior parameters. Therefore, this paper solves prior parameters of kth test stage by the optimization model describing as following,

$$\begin{aligned} &\text{Min} \quad (v_k^* - VR_k)^2 \\ &s.t. \\ &\begin{cases} \mu_k^* = ER_k \\ a_k > 0, b_k > 0 \end{cases} \end{aligned} \tag{13}$$

After considering equation (10) (11) and (12), the optimization model, equation (13), can solve prior parameters which exactly describe prior information. Where R_{k-1} of equation (10) can been substituted by condition mean of k-1th test stage.

If there are more prior information depicting product reliability R_k of kth test phase, such as the number of continuous phases $(R_{k,L}^j, R_{k,H}^j)$ j=1, 2, …, h, is h, then the mean and variance as following,

$$ER_k = \frac{1}{h}\sum_{j=1}^{h}\frac{R_{k,L}^j + R_{k,H}^j}{2} \tag{14}$$

$$VR_k = \frac{1}{h}\sum_{j=1}^{h}\left[\frac{(R_{k,H}^j - R_{k,L}^j)^2}{12} + \frac{(R_{k,H}^j + R_{k,L}^j)^2}{4}\right] - \frac{1}{h^2}\left(\sum_{j=1}^{h}\frac{R_{k,H}^j + R_{k,L}^j}{2}\right)^2 \tag{15}$$

Of course, the prior parameters of product reliability of kth test phase can also been solved by optimization model expressed by equation (13).

4. The method of calculating value of reliability by WinBUGS

The Bayesian statistical theory is one main branch of statistical theories, since 1763 the first paper of Bayesian statistical theory had been published, until 80th in twenty century Bayesian statistical theory has been in theory research phase, integral calculation is a big barrier in his development and application. However, Markov Chain Monte Carlo (MCMC) has been used to Bayesian statistical inference in recently, a main characteristic of this method is Metropolis-Hastings updating and Gibbs sampling, it can solve well the problem of numerical integration and sampling in multidimensional distribution, which is convenient for posterior inference of parameters and accelerate the application of Bayesian theory.

Regarding the theorem and application of MCMC, many research works have been done and make great progress, for instance the software WinBUGS developed in 1991, which is special software, is mainly used for Bayesian analysis of complex statistical model by MCMC approach. Therefore, in order to calculate multidimensional numerical integration of multivariate distribution, such as marginal posterior distribution, mean of posterior and quantile of posterior depicted in section 3 of the paper, this paper employed the software WinBUGS to solve posterior inference of parameters and have enough calculation accuracy.

4.1. Introductions of WinBUGS

MRC (Medical Research Council, Cambridge UK) and Imperial College of Science, Technology and Medicine joined efforts in developing WinBUGS (Bayesian Inference Using Gibbs Sampling), the exclusive software solves Bayesian inference by MCMC method. Gibbs sampling of many common models and distribution can easily finished by WinBUGS, programmer need not to know accurate equation of likelihood function or prior density of parameters, as long as setting prior distribution of

Journal of Physics: Conference Series **679** (2016) 012039 doi:10.1088/1742-6596/679/1/012039

variants and describing the interested model generally, then can easily achieve Bayesian analysis of model rather than complex compilation.

Considering WinBUGS software, Directed Graphical Model can be used in describing the interested model directly, alternatively programming the model codes, and presenting the Dynamic Graph of Gibbs sampling of parameters, based on Smoothing method getting core density estimation graph of posterior distribution, auto correlation graph of sampling value, transfer graph of mean and confidence interval, and so on. Having finished all above works, the result of sampling would be direct and reliable. After Gibbs sampling converging, we can get mean of posterior distribution of parameters, as well as standard variance, 95% confidence interval, quantile etc.

4.2. Description of Bayesian reliability growth model by WinBUGS codes

According to equation (5) which is general Beta distribution in $(R_{k-1},1)$, by substitution of variable,

$$X_k = \frac{R_k - R_{k-1}}{1 - R_{k-1}} \tag{15}$$

Then

$$X_k \sim Beta(a_k, b_k) \tag{16}$$

Where $Beta(\bullet,\bullet)$ is a standard Beta distribution in (0, 1). For analysis convenient, equation (2), (15) and (16) are rewrite as following,

$$s_k \sim Bino(n_k, R_k) \tag{17}$$

$$X_k \sim Beta(a_k, b_k) \tag{18}$$

$$X_k = \frac{R_k - R_{k-1}}{1 - R_{k-1}} \tag{19}$$

Where equation (17) and (18) express the relationship of statistic, equation (19) expresses determine relationship.

According to above equations, Bayesian inference model of reliability growth can be established by WinBUGS codes. Depicting as following,

```
model {
        for (i in 1： N) {
                s[i] ~ dbin(R[i]， n[i])
                x[i] ~ dbeta(a[i]， b[i])
        }
        for (i in 2： N){
                R[i]<-R[i-1]+(1-R[i-1])*x[i]
        }
                R[1]<-0+(1-0)*x[1]
        }
```

5. The cases study

5.1. ***Case one***

Assuming the development of an aero engine is consisted of three experimental stages, in the first stage two success and one failure, in the second stage four success and one failure, the last stage five success and no failure. According to the experts' assessment and history information, the estimation value of reliability is separately $R=(0.70, \quad 0.90, \quad 0.95)$ in each test stage, and 0.975 is prior value of product reliability after reliability growth test of three stages.

The reference [18] analyzed the reliability growth of product by order Dirichlet distribution, and assuming value of parameter β is 44. Number of Reliability growth test n_i, success number s_i as well as point estimation value of R_i based on prior information have been shown in the second column, the third column and the fourth column of table 1.

Table 1. Reliability growth test data and prior estimate.

t_i	n_i	s_i	R_i	$(R_{i,L}, R_{i,H})$
1	3	2	0.70	(0.4, 1)
2	5	4	0.90	(0.8, 1)
3	5	5	0.975	(0.95, 1)

In fact interval estimation of reliability of various test stages can be presented according to experts' experience and history data, moreover interval estimation describes prior information more accurate than point estimation. Assuming based on experience of experts interval estimation of each test stage (R_{iL}, R_{iH}) has shown in the fifth column of table 1. Inspection of the table 1 shows that in the first test stage no test data and correcting information of product can be available, therefore estimation value of reliability is not accuracy, and interval estimation for instance given by experts is much wider. After reliability growth test and improvement, the interval estimation is more accuracy and wide decreasing.

The prior parameters of general Beta distribution of various test phases α_k, b_k have been solved by optimization model expressed equation (13), where general Beta distribution is equivalent to uniform distribution $U(R_{k,L}, R_{k,H})$. Mean ER_k and variance V_k of the two distributions are shown in the table 2.

Table 2. Statistic characteristic parameters of equivalent general Beta distribution and uniform distribution.

t_i	Uniform Distribution		Equivalent General Beta Distribution			
	ER_k	V_k	ER_k	V_k	a_k	b_k
1	0.7	0.03000	0.5	0.03000	4.200	1.800
2	0.9	0.00333	0.75	0.00333	3.333	1.667
3	0.975	0.00021	0.95	0.00021	6.001	2.000

The figure 3 is the curve of uniform distribution based on experience of experts as well as the equivalence general Beta distribution.

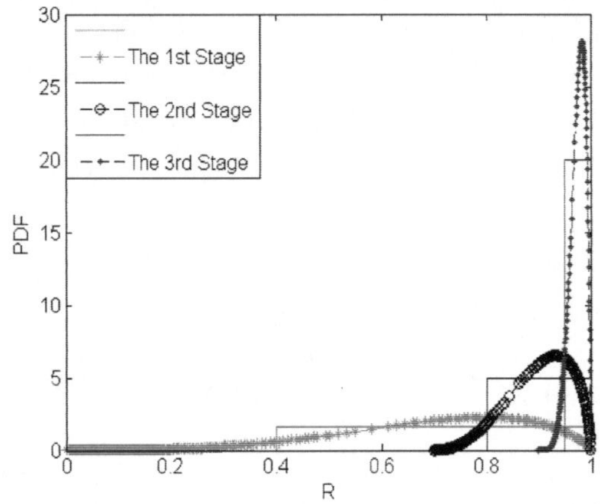

Figure 3. The curve of Equivalent general Beta distribution and uniform distribution in various test stages.

Inspection of this plot shows that prior distribution, which solved by optimization equation (13), has a good fitting with prior information.

After getting prior parameters based on prior information, posterior estimation values of reliability of various test phases can be solves by WinBUGS model expressed in section 3.2, and table 3 shows the results.

Table 3. Reliability posterior estimate values of various test stages

t_i	New Distribution(OPT) [a]		Order Dirichlet [b]		New Distribution(Li) [c]	
	$R_{kL}(0.9)$	V_k	$R_{kL}(0.9)$	V_k	$R_{kL}(0.9)$	V_k
1	0.4757	0.01924	0.6108	0.0047	\	0.0420
2	0.776	0.00555	0.8303	0.0020	\	0.0115
3	0.9395	0.00064	0.9497	0.0005	0.934	0.0015

[a] a optimization method presented by this paper.
[b] order distribution.
[c] method of reference [15].

Notice in table 3 that variance of posterior estimation value based on a new prior distribution is bigger than variance of order Dirichlet distribution, especially in the early two phases. Comparing the lower bound $R_{k,L}(0.90)$ of 90% confidence limit, the posterior estimation value based on a new prior distribution is smaller. Whereas the posterior estimation value based on order Dirichlet distribution is evidently effected by prior information, i.e. order Dirichlet distribution has a mainly influence on posterior estimation value.

5.2. *Case two*

Considering another situation in case one, assuming prior estimation value of product based on experience of expert is not various, the achieved reliability of product is lower because of test results of various stages becoming worse, and supposing test result of various stages (n_i, s_i) is (3, 1), (5, 3) and (5, 4) separately. By the same calculating procedure been used in table 3, posterior estimation values of various test stages are shown in table 4.

Inspection of the table 4 shows that if employing order Dirichlet distribution to depict the expert experience when the test sample is smaller, the prior distribution will be whip hand so that posterior estimation is mainly determined by prior information. It is not consistent with present practice.

According to the above cases, the new prior Dirihclet distribution can accurately describe expert experience, the Bayesian estimation value of product reliability during developing phase can be solved by optimization method presented by this paper. Moreover, the method has a good adaptability when test sample is small. Bayesian model based on a new Dirichlet distribution can take full advantage of history information, expert experience and test date of various phases.

Table 4. Reliability posterior estimate values based on worse test results of various test stages.

t_i	New Distribtion(OPT)		Order Dirichlet	
	$R_{kL}(0.90)$	V_k	$R_{k,L}(0.90)$	V_k
1	0.3049	0.01924	0.5886	0.00452
2	0.606	0.01059	0.8047	0.00230
3	0.8569	0.00233	0.9187	0.00081

6. Conclusion

This paper presents a new method for the application of Bayesian theory and technology to product reliability growth during product development phase. The research work mainly focuses on how to determent prior distribution parameters of a new Dirichlet distribution and presents the relevant optimization model and method, finally demonstrate the validity of Bayesian reliability growth model by WinBUGS software. The conclusions as following:

(1) The optimization method describes experience of expert as a general Beta distribution, which is fitted for Bayesian reliability growth model based on a new Dirichlet prior distribution, and solve the problem of how to determining hyper-parameters.

(2) Regarding the issue of multidimensional numerical integration of posterior, the paper employs WinBUGS software to establish the reliability growth model based on a new Dirichlet prior distribution, and solves posterior inference of parameters at the same time keeping enough calculation accuracy.

Acknowledgments

I want to give my sincere appreciation to XinXin Zhang who gives me much help in experimental research, and Haiying Yuan who helps me a lot in article writing.

References

[1] Smith A F M 1977 *IEEE Trans. Reliab.* **26** 346
[2] Barlow R E and Scheuer E M 1966 *Technometrics* **8** 53
[3] Sohn S Y 1997 *Comput. Ind. Eng.* **33** 741
[4] Tian G L 1992 *J. Astronautics* **15** 55
[5] Mazzuchi T A and Soyer R A 1991 *Reliab. Maintainability Symp. Proc.* **23** 322
[6] Mazzuchi T A and Soyer R 1992 *Reliab. Maintainability Symp. Proc.* **23** 468
[7] Mazzuchi T A and Soyer R 1993 *IEEE Trans. Reliab.* **42** 503
[8] Erkanli A, Mazzuchi T A and Soyer R 1998 *Am. Stat. Assoc. Am. Soc. Qual. Technometrics* **40** 14
[9] VanDorp J R, Mazzuchi T A and Soyer R 1997 *J. Stat. Planning Inference* **62** 207
[10] Somerville I F, Dietrich D L and Mazzuchi T A 1997 *Nonlinear Anal.-Theory Method & Appl.* **30** 4415
[11] Bilal M Ayyub and Richard H McCuen 1997 *Probability, Statistics & Reliability for Engineers* (New York: CRC Press)
[12] Zhang S F, L i R 1996 *Electron. Prod. Reliab. Environ. Test.* **6** 12
[13] Yu T X, Song B F, Feng YW 2006 *Syst. Eng. Theory Pract.* **11** 13
[14] Liu F, Dou Y F, Zhang W H 2006 *J. Solid Rocket Tech nol.* **29** 56
[15] Li G Y, Wu Q G, and Zhao Y H 2002 *Jpn Statist. Soc.* **32** 1
[16] David S, Andrew T and Nicky B 1996 *BUGS 0.5: Bayesian inference Using Gibbs Sampling*

Manual (version II) (U. K: MRC Biostatistics Unit)

[17] Mao S S, Wang J L and Pu X L 1998 *Advanced Mathematical Statistics* (Beijing: High Education Press and Springer Press)

[18] Joseph G Ibrahim, Chen M H and Debajyoti Sinha 2001 *Bayesian survival analysis* (New York: Springer)

A review of the technology and process on integrated circuits failure analysis applied in communications products

Zhimao Ming, Xiaodong Ling, Xiaoshu Bai, and Bo Zong

China Satellite Maritime Tracking and Control Department, Jiangsu China 214431,

E-mail: mzmnudt2008@126.com

Abstract. The failure analysis of integrated circuits plays a very important role in the improvement of the reliability in communications products. This paper intends to mainly introduce the failure analysis technology and process of integrated circuits applied in the communication products. There are many technologies for failure analysis, include optical microscopic analysis, infrared microscopic analysis, acoustic microscopy analysis, liquid crystal hot spot detection technology, optical microscopic analysis technology, micro analysis technology, electrical measurement, microprobe technology, chemical etching technology and ion etching technology. The integrated circuit failure analysis depends on the accurate confirmation and analysis of chip failure mode, the search of the root failure cause, the summary of failure mechanism and the implement of the improvement measures. Through the failure analysis, the reliability of integrated circuit and rate of good products can improve.

1. Introduction

With the progress of science and technology, communication products play an increasingly important role in the development of countries and improvement of daily life, and the integrated circuits are the core components of communication products. At present, the scale of integrated circuits enlarges gradually, functional integrations extend, and the requirement of reliability also improves. Therefore, the failure analysis of integrated circuits becomes more and more important. This paper intends to mainly introduce the failure analysis technology and process of integrated circuits applied in the communication products.

2. The failure analysis technology of integrated circuits

2.1. The optical microscopic analysis

Optical microscope has been the important tool for failure analysis applied in integrated circuit. The stereoscopic microscope and metallurgic microscope are the main microscopes. Combined with these two kinds of microscopes, the appearance damage and sticky dirt of chips, burnout and breakdown result from electrostatic discharge (ESD) and electrical over stress (EOS) [1], wire bonding, scratches and cracks of chips as well as other phenomena are observed.

2.2. The infrared microscopic analysis

Journal of Physics: Conference Series **679** (2016) 012040 doi:10.1088/1742-6596/679/1/012040

The infrared microscope is a technology which adopts near-infrared light source and converts tube imaging by infrared rays [2]. To near-infrared light, the semiconductor materials such as germanium and silicon as well as thin metal are transparent, as a consequence, the conditions inside the chip.

2.3. The acoustic microscopy analysis

Due to the transmission and reflection in different mediums of ultrasonic are different, the ultrasonic testing equipment based on this character is used to observe the fractures on and beneath the surface [3]. The fracture and the inhomogeneous structure of the chips are detected. This technology develops quickly due to no damage to products, because the internal defects of chip without destroying the chips are adopted in many cases.

2.4. The liquid crystal hot spot detection technology

The process defects, electrostatic discharge or electrical over stress of integrated circuits will result in leakage currents. The leakage currents are produced in the input terminal or output terminal, and heat is generated in the conductive path, so the heat is called by a hot spot [4]. Liquid crystal hot spot detection technology can detect the accurate position of the hot spots. With the natures of crystal and liquid at the same time, when the temperature of liquid crystal is lower than the critical temperature T_c, then liquid crystal presents the anisotropic crystal, and when the temperature of liquid crystal is higher than the critical temperature T_c, then liquid crystal presents the isotropic liquid. The liquid crystal is besmeared on the surface of chip, and the heat generated by the leakage current will make liquid crystal turn out the phase transformation. Therefore, the location of the hot spots can be determined by observing the phenomenon of phase transformation.

2.5. The optical microscopic analysis technology

Under the excitation of the electric stress, the semiconductor devices have transitions from the carriers to photons, thermionic analyzer instruments can detect these photons, they are also known as Light Emission Microscope. This technique is often used to find the unreasonable design, the defects produced by material and process such as the luminous point, pinhole and cutting-edge diffusion on the reverse PN junction as well as local heat in the aluminum membrane.

2.6. The micro analysis technology

Micro analysis technique is a modern technology which carries out analysis of micro morphology, material and chemical composition of electronic devices through the use of electron beam, ion beam, laser beam and X-ray [5]. As the materials of electronic technology become more and more complex, the structure of electronic components become more and more refined, the general analysis methods can't meet the need, so the micro analysis technique grows up quickly. The method commonly includes scanning electron microscope [6] (SEM), X-ray energy spectroscopy (EDX), auger electron spectroscopy (AES) and geometric X - ray imaging technology.

2.7. The electrical measurement

The electrical measurement is the first step when carries out the failure analysis of integrated circuits, the purpose is to detect the electric properties between chip pins, and the phenomenon of abnormal short circuit, open circuit and leakage between pins. Then the following analysis of the process can be determined. The commonly measuring tools are the millimeters, oscilloscope, curve trace and so on. Usually the results of the analysis can be obtained through the comparison of analysis curves or numerical values between the good products and bad products under the same condition.

2.8. The microprobe technology

Due to the high level of integration in the integrated circuit, it is necessary for internal circuit to directly measure by mechanical microprobe, because the general measurement devices such as pens or probe are hard to contact with the internal circuit [7]. The mechanical microprobe system is generally

Journal of Physics: Conference Series **679** (2016) 012040 doi:10.1088/1742-6596/679/1/012040

composed of microprobe and mechanical positioning device. The ability of the measuring system is determined by the diameter of probe tip and the accuracy of mechanical positioning. Recently, the diameter of probe in modern system of microprobes can reach to the nanometer. The positioning is performed by a computer system, and the isolation of the circuits is carried out by ultrasonic and laser cutting devices. However, this measure may scratch the metallization layer on the surface of chip, which brings the difficulty to the analysis in some extent.

2.9. The chemical etching technology

Chemical etching refers to the processing method of the sample with the effect of a variety of chemicals including various acids and antioxidants [8]. It is widely used in dealing with cleaning pollution on the surface of the chip, removing encapsulation to make chips exposure, eliminating the passivated layer, metallized layer and oxide layer, and so on. The most important thing of etching process is selecting the appropriate reagent and controlling reaction conditions (temperature, pressure and other factors) and reaction time.

2.10. The ion etching technology

Ion etching belongs to one of the etching technology based on ion processing technology, and ion etching technology is generally used in failure analysis. The reaction gases such as CF_4 and O_2 are activated by the plasma, and form the F^* free radical and O^* free radical, then the free radicals react with a variety of passivation layers such as FSG, Si_3N_4 and SiO_2, thus the plasma etching can be used to remove passivation layer without damaging the metallized layer below the passivation layer at the same time, and it also can be used for etching plastic package. Due to the isotropic plasma etching, ion etching can accurately control the degree of etching and avoid corrosion compared with the wet etching.

3. The process of failure analysis

The main processes [9] of failure analysis are shown in figure 1. And the detailed procedures are as follow.

(1) After getting the failed samples, the failed samples and the good samples used for comparison are labeled.

(2) The failure process and the failure circumstances of the sample are confirmed, at the same time, the state before failure is definite.

(3) To determine whether the appearance is soiled, the appearance of the sample is inspected.

(4) A preliminary test result of test chip is obtained by using automatic test equipment (ATE). And the kind of failure of the sample is determined and verified.

(5) According to the above analysis, the scheme of the sample is determined.

(6) Check whether the lead is normal, whether the encapsulation is layered or inane, and confirm the location and size of the wafer by scanning acoustic microscope or X – ray diffraction microscope.

(7) This step is mainly the electrical performance test. The phenomena that whether the electric property is normal, whether there are short circuit, open circuit and leakage are checked. The test results by automatic test equipment are verified by mini test machine or debugger, and the failure module and the failure reasons are analyzed.

(8) If the sample failed under some condition, then the failure in the corresponding conditions need to be verified. For example, some chips fail under room temperature or low temperature, but the other chips fail in a high pressure environment. Thus it is necessary for the chips to be tested before drying at 150 ℃ at the furnace for 24 hours in order to avoid the interference by the fouling failure of ions.

(9) Breaking seal

The process of breaking seal has a potential for the damage of the bond line, passivation and some of the top metals. The technology of breaking seal can be divided into chemical breaking seal and mechanical breaking seal. The process should not create the image for the signal and function of electrical property when the encapsulation is removed. The types of plastic package should be

Journal of Physics: Conference Series **679** (2016) 012040 doi:10.1088/1742-6596/679/1/012040

confirmed before breaking seal, and the best way of breaking seal is determined. At the end of breaking seal, the chips are tested, and the consistency with the chips before and after breaking seal are examined.

(10) Internal observation

The surface of chips, chip bonding, inner lead, the second bonding points and external pins are observed by low-magnification microscope to distinguish if there are some abnormal phenomena. Then the abnormal parts are observed by the high-powered microscope, and the contrasts the tested chips with the good chips are made. Record the abnormal parts by taking photos as the evidence and reference of the next step. Make the analysis of the failed parts and abnormal parts to see whether there are correlations and causal relationships between the failed parts and abnormal parts.

(11) The defects are captured and positioned through hot spot. The hot spot on the surface is caught through light emission or liquid crystal when the chip failure program runs. If there is a hot spot, the reference diagram and the schematic diagram are made contrast, the circuit where hot spot exists is inspected, and the relevance of hotspots and failure condition is analyzed.

(12) Remove the passivation layer on the surface of the chip. According to the chip technology, choose the reasonable ratio and etch the passivation layer by reactive ion etching.

(13) The microprobe analysis

After removing the passivation layer, the microprobe analysis is carried out to see if the failure changes. The lead wire of the suspicious circuit is detected as a probe point through the analysis of reference diagram and the schematic diagram as well as the combination with the location of hot spot. When the chip is tested, test program is load, the signal of the failed chip is observed and the comparison with the signals of failed chip and the good signals under the same situation is made. If necessary, the circuit can be cut and separated by laser for the purposes of measurement.

(14) The analysis of layer-stripping

The stripping plan is determined in accordance with the process of chip, every layer are stripped off and analyzed by high power microscope, scanning electron microscope (SEM) and transmission electron microscopy (TEM) so as to avoid excessive stripping and the damage of the failure location. Taking pictures and recording after finding the failure position. If foreign bodies lead to the failure, it is necessary to adopt the energy dispersive X-ray energy spectroscopy in order to confirm the composition. The electrical measurement of the chip is not carried out after stripping, and therefore the stripping plan starts until the prophase steps are verified.

(15) The failure analysis report is concluded through analyzing the data. All the failure analyses of above steps are not need to be undertaken. The ultimate aim is to find the failure position and confirm the failure mechanism, and therefore some unnecessary steps can be left out according to the different failure conditions.

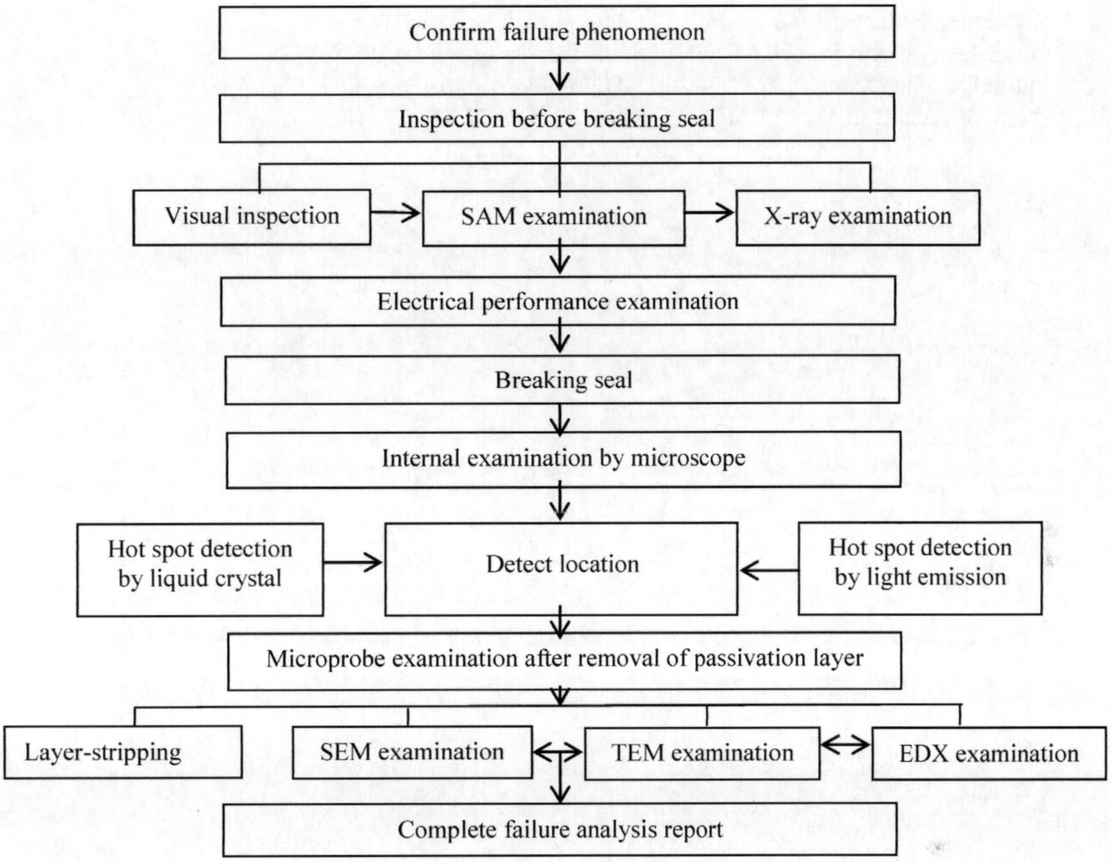

Figure 1. The processes of failure analysis.

4. Conclusion

The integrated circuits failure analysis applied in communications products plays a very important role in the improvement of the reliability in the process of production and use. There are many technologies for failure analysis, such as optical microscopy, infrared microscopy, acoustic microscopy, liquid crystal hot spot detection technology, optical microscopic analysis technology, micro analysis technology, microprobe, chemical etching, and ion etching. This paper also introduces the specific process of failure analysis, and the process can reflect the application of concrete failure analysis method. The integrated circuit failure analysis depends on the accurate confirmation and analysis of chip failure mode, the search of the root failure cause, the summary of failure mechanism and the implement of the improvement measures. Through the failure analysis, the reliability of integrated circuit and rate of good products can improve, and gain good economic benefits.

Acknowledgments

I want to give my sincere appreciation to XinXin Zhang who gives me much help in experimental research, and Haiying Yuan who helps me a lot in article writing.

References

[1] Wu C L, Yao S Y and Berges C 2015 *Microelectron. Reliab.* **55** 463
[2] Schmidt Ch, Altmann F and Breitenstein O 2012 *Mater. Sci. Eng.* **177** 1261
[3] Jon MC and Sturm GW 1982 *NDT Int.* **15** 185
[4] Chen Y 2010 *Electron. Prod. Reliab. Environ. Test.* **13** 99

[5] Robert Heiland and Alan Leslie 1994 *Microelectron. Eng.* **24** 51
[6] Gonzales A. J and Powell M. W 1978. *J. Vac. Sci. Technol.* **15** 837.
[7] Faure D. D and Waggoner C. A 2002 *Microelectron. Reliab.* **42** 1767
[8] Baerg W. and Rao V 1991 *Microelectron. Reliab.* **31** 105
[9] Zhao W, Wu F G, Yang S H, Wang Y and Xu Z L 2011 *Ind. Eng. J.* **14** 141

Journal of Physics: Conference Series **679** (2016) 012041 doi:10.1088/1742-6596/679/1/012041

The Present and Future of China's Sea-based Space Tracking and Communications

Bing Liu, Bo Cong, Yong Liu and Junshan Mu

China Satellite Maritime Tracking and Control Department, China

E-mail: gfkdly@aliyun.com

Abstract. Space tracking ship plays a crucial role in China's spaceflight tracking and data network. In this paper, the history of China's space tracking ship is reviewed, the main subsystesms of the tracking ship are introduced with the emphasis being put on its technical features. Special problems such as antenna stabilization, EMC, system calibration are elaborated in detail. Future possible missions and the technical challenges are also analyzed in the final part.

1. Introduction

In spaceflight missions, ground tracking support and telecommunication, usually being referred to as TT&C(Tracking, Telemetering and Command), play important roles. The TT&C system is able to track the spacecraft, obtain its trajectory and telemetry data, monitor the status of, and communicate with, the satellite or spaceship and control its behavior.

To meet the needs of space program, early in 1960s, China's space tracking network was planned and reliability, economy were its main concerns. Nowadays, the whole tracking network mainly consists of land-based, sea-based and space-based tracking stations and operation centers.

In 1965, China began to consider the tracking ship program. The first two space tracking ships were constructed in 1979 and since then the ground support of China's spacecraft was extended to the ocean. China became the forth country which owns the space tracking ship after the United States, the former Soviet Union and France.

Coming into the 21st century, as more space programs, such as space station and deep space exploration are being planned and more challenging requirements are proposed, new-generation tracking ships were constructed to replace the old ones. Unified S Band, unified C band, unified X band, tracking radar, etc. were installed and many advanced techniques, new materials were widely used which made the performance further and greatly improved.

The tracking ship can support manned or unmanned spaceship, lunar exploration satellite and other various flight missions. Compared with other approaches such as establishing or renting oversea ground stations, the construction and maintenance cost of the tracking ship are relatively higher but as mobile tracking station, the convenience and flexibility in deployment, and consequently the ability to support missions that can't be supported by land-based stations, is its unique advantage.

Journal of Physics: Conference Series **679** (2016) 012041 doi:10.1088/1742-6596/679/1/012041

Figure 1. China's new-generation space tracking ship.

2. Main systems of the space tracking ship

2.1. Radar Tracking System

The radar tracking system aboard the Yuanwang space tracking ship is capable of providing continuous accurate spherical-coordinate information of the target in real time for mission operational decisions and for future evaluation of performance. The radar can be used as a single system, or as one of a chain of radars along a flight path to ensure continuous tracking throughout the entire flight.

Figure 2. tracking radar on space tracking ships.

This tracking radar is high-accuracy, long-range, amplitude-comparison monopulse radar which is capable of manual or automatic acquisition and tracking of objects in flight or orbit. Provisions are incorporated for selecting either a single-pulse output for skin tracking or a coded-pulse group for beacon tracking.

The transmitter output pulse is channeled through the microwave components to a multimode feed and parabolic reflector radiating system. A pencil beam directivity pattern is formed by the combination of the feedhorn and the antenna reflector.

The RF return signal is directed by the multimode feed into a comparator where it is separated into a reference component, an elevation error component and an azimuth error component. These signals (azimuth, elevation, and reference) are feed to separate channels in the tracking section of the receiver. Error outputs from the azimuth and elevation error channels are fed to the angle tracking subsystem of the radar set, where they are amplified and used to position the antenna.

Range tracking subsystem determines the slant range of the target in space, and provides range data outputs which are fed through the digital data subsystem to the computer.

Journal of Physics: Conference Series **679** (2016) 012041 doi:10.1088/1742-6596/679/1/012041

To increase the probability of acquisition during the normal acquisition mode, the radar generates analog (joystick) or accepts digital scan command from the computer.

In addition to accepting stabilized designation data, the radar will accept feed-forward signals to reduce the effects of ship's motion during tracking. Further refinement of ship's motion compensation is achieved by the use of rate gyros which are part of the radar mount.

2.2. Unified Systems

Unified system integrates separate functions onto one carrier and can accomplish tracking, telemetry, command, communication etc. simultaneously.

The USB (Unified S-Band) operates in five modes, i.e., standard TT&C, coherent SS(Spread Spectrum) TT&C, noncoherent SS TT&C, DDT(Digital Data Transmission) and PCM/FM.

The UCB(Unified C-Band） operates in the mode of noncoherent standard TT&C which doesn't have the ability of range rate measuring. But compared with USB, UCB has the ability of synchronizing control of spinning satellite. In addition to coded telemetry, it also has the function of analog telemetry. Furthermore, on the new-generation tracking ship, UCB shares antenna and servo system with C-band pulse radar, shares BBE with USB.

2.2.1 Antenna Sharing As mentioned above, on the new-generation tracking ship, UCB and C-band pulse radar shares the same antenna and servo system. This is due to limited room for large antenna on the tracking ship and the requirement to increase the effect-cost ratio.

This composite system, called "C+C"， has three operation modes: pulse radar operating independently, UCB operating independently, pulse radar and telemetering of UCB. Several radio switches were installed in the feed network to realize transmitter switch between UCB and pulse radar, and polarization switch, and to provide radio channels for three operation modes

2.2.2 Spread Spectrum TT&C The technique of spread spectrum was adopted in the unified systems onboard the new-generation tracking ship to improve security and anti-jamming ability.

SS TT&C (Spread Spectrum TT&C) operates as following. In the BBE, command information is modulated with short PN code onto the carrier using BPSK, ranging PN code (long code)， which is synchronized with the short PN code, is modulated directly onto an orthogonal carrier also by using BPSK. Then, the two carriers are added together to form a new one. After being up-converted, amplified, the carrier is transmitted through the antenna to the spacecraft. The transponder of the spacecraft, after receiving the uplink signal, firstly accomplishes the acquisition and tracking of the short PN code, and demodulates the command information, then with the aid the short PN code, accomplishes the acquisition, tracking and regeneration of the long PN code. Telemetry information and downlink ranging code are modulated by using the same method as that in the ground station, and transmitted. The BBE processes the downlink signal and gets telemetry data, range and range rate, which is computed from the Doppler frequency.

2.2.3 Base Band Equipment The concept of software-defined radio was widely used in the design of the unified systems of the new-generation tracking ship and the part below 70MHz was integrated into one multifunctional BBE(Base Band Equipment) and the reliability, maintainability and flexibility were greatly improved. The BBE, by loading different software, can be configured with various operation modes, i.e., standard TT&C, SS TT&C, DDT (Digital Data Transmission), PCM/FM.

(1)Standard TT&C

Standard TT&C is just the operation mode of the conventional USB. In this mode, the BBE integrates the functions of ranging, range rate measurement, command, telemetry etc. into an industrial control computer.

(2)Spread Spectrum TT&C

The BBE supports both coherent SS TT&C and noncoherent SS TT&C and in this mode, multiple-object tracking can be accomplished.

(3)Digital Data Transmission

In this mode, the BBE can receive and demodulate signals with various modulations such as PCM - BPSK / QPSK /DQPSK / SQPSK / UQPSK, etc. and accomplishes voice and video communication.

Journal of Physics: Conference Series **679** (2016) 012041 doi:10.1088/1742-6596/679/1/012041

(4)PCM/FM

Compared with land-based ones, a significant feature of the USB onboard the tracking ship is that the PCM/FM receiver for rocket telemetry was incorporated. To combat the fading of the rocket telemetry signal, polarization diversity was adopted.

(5)Spacecraft Simulator and Rocket Telemetry Simulator based on BBE

Spacecraft simulator consists of demodulation/modulation platform and simulation computer, and by using which tests and trainings can be implemented. Rocket telemetry simulator generates the telemetry data with the same time sequences as that of a real mission to test the data processing system.

To fully exploit the current equipment and save the installation space, which is limited on the tracking ship, a spacecraft simulator and rocket telemetry simulator based on the hardware of BBE were proposed and developed. In some sense, these simulators can be seen as BBE's built-in components, so except for the benefit of low development cost, the interfaces were also simplified.

2.2.4 Dual-Frequency Tracking The Chang'E satellite utilizes two antennas, each with a different frequency, for downlink transmission and covered the ground station alternatively. In order not to lose the linkage, a scheme called Dual-Frequency Tracking was proposed for USB. In this scheme, USB demodulates angle error signals from the two frequencies simultaneously, and according to some rules, selects and sends one to the ACU(Antenna Control Unit) to realize seamless switch between the two frequencies. This scheme is actually a sort of frequency diversity, although not for communication but for tracking.

2.3. Ship Attitude and Position Measuring System(SPAMS)

The Ships Position and Attitude Measuring System (SPAMS) is one of the principal systems of the tracking ship. This system comprises subsystems and equipments-integrated to provide data for stabi1ization of the shipboard antenna systems and for conversion of angle information from deck-referenced coordinates to earth-referenced coordinates. The system also provides for measurement of ship's velocity and for monitoring antenna mount flexure.

The functional diagram of the SAPMS of the tracking ship is shown in Fig. 1. The major subsystems of the SPAMS are as follows: INS(Inertial Navigation Subsystem), GNSS(Global Navigation Satellite Subsystem), FMS(Flexure Monitor Subsystem)， DVL(Doppler Velocity Log) and calibration theodolite, etc.

The key subsystem in the SPAMS is the Integrated Navigation System (INS). The INS supplies extremely precise information concerning the ship's latitude and longitude; position North, East,and vertical components of velocity; roll, pitch, and heading.

Inertial components, gyros and accelerometers, incorporated into this system are not error free; thus their outputs will degrade with time. These errors are unacceptable in Shipboard app1ication and must be reduced by a reset. System reset is possible when accurate position information is available from one of several sources: Navigation Satellite, and the Marine Star Tracker.

Three mutually orthogonal axes are required as a reference for measurement of all shipboard equipment tracking angles and are established by the SINS. For this reason it is very important that the ship's structural foundation meet certain requirements. To meet these requirements a specially designed bedplate, supported by rigid ship's structure, is provided.

Being compared with those onboard other vehicles, the INS of the tracking ship cares more about the accuracy of attitude rather than that of position. Positioning is mainly accomplished by GNSS which is independent with INS and also provides time synchronization.

Flexure monitor subsystem provides simultaneous automatic measurement of rotation about three mutually orthogonal axes: X (axis parallel to the roll axis of the ship), Y (axis parallel to the pitch axis of the ship) and Z (axis parallel to the yaw axis of the ship) for both the Unified S-Band and the C-Band radar antenna foundations. These measurements are referenced to the INS bedplate. The purpose of these measurements is to determine the alignment error between the INS bedplate and the antenna foundations due to flexure. This data is provided to the CDPS in digital form for use in correction of trajectory data.

Calibration theodolite is a kind of celestial navigation system, which tracks stars to calibrate the course error of INS or angle error and range error of radio tracking systems.

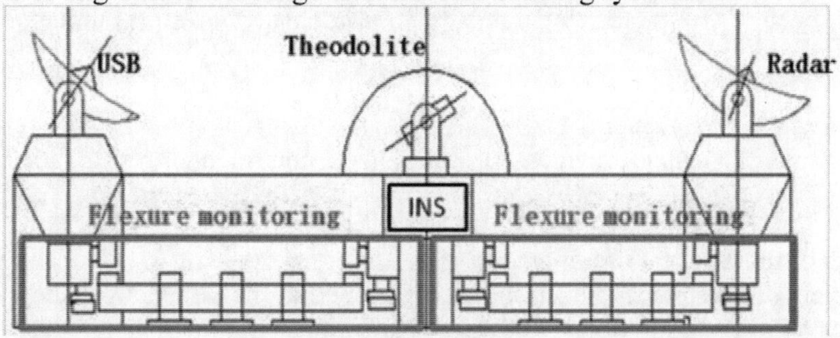

Figure 3.Functional Diagram of SAPMS of the Space Tracking Ship.

2.3.1 GPS Attitude Determination The technique of GPS attitude determination is based on GPS carrier phase measurement. In early 1990s, many institutes and universities began the research on attitude determination and course determination based on GPS carrier phase measurement by theoretical analysis and experiment. Compared with INS, GPS attitude determination has the advantages of low-cost, weather-irrelevant, fast-startup, no-error-accumulation and can work in regions of high latitude.

On the space tracking ship, the accuracy of GPS attitude determination is affected by factors such as complex electromagnetic environment, shadowing of large tracking antennas and ship ' s deformation. After intensive experiments, the feasibility of applying this technique on tracking ship was verified. Design schemes of appropriate GPS antenna array, system calibration and coordinate reference unification was proposed and successfully applied.

2.3.2 Information Fusion of SAPMS INS can provide continuous navigation information of position and attitude, but because of its error accumulation with time, it can't work alone and needs frequent calibration. GNSS can provide high-precision navigation information globally, but is apt to be interfered. Any kind of navigation system alone has its pros and cons and is impossible to meet all the tracking ship's requirements simultaneously. To realize highly precise, highly reliable, global, all-weather, continuous navigation, the new-generation tracking ship adopted an information fusion scheme based on federated Kalman filter and improved chi-square fault detection method, which fully exploits the inherent redundancy in the data and achieves high reliability and fault tolerance.

2.4. Central Data Processing System

The Central Data Processing System (CDPS) accepts digital data from the various systems and presents a processed data to various end equipments for control, display, recording or sends to the operation center through communication system.

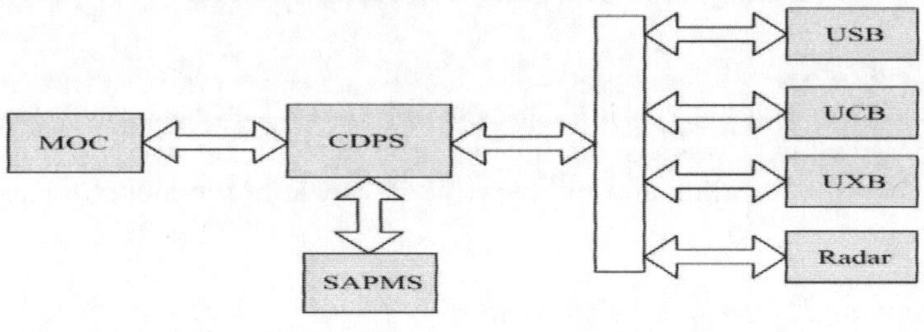

Figure 4. The working principle of Central Data Processing System on tracking ship.

To assist in acquiring the target, CDPS provides acquisition programs which direct the tracking system to the target.

Journal of Physics: Conference Series **679** (2016) 012041 doi:10.1088/1742-6596/679/1/012041

2.5. Communication System
The purpose of the communication system is to provide linkage between the tracking ship and the mission control center, in Beijing or Xi'an for the ship's daily operation and for support of space missions.

2.6. The Operations Control Center
The Operations Control Center provides for centralized control and coordination of all an-board systems.

3. Special Issues of the Space Tracking Ship
Besides the problem of antenna stabilization just mentioned in the last section, there are many special issues and challenges must be addressed in the space tracking ship.

3.1. Electromagnetic Compatibility Design
Located on the top deck of the ship are dozens of antennas with various sizes. In order to eliminate electromagnetic interference, during design and layout, a rational selection is made of the operating frequencies and frequency spectra of the electronic equipment; there is a rational layout of the equipment and antennas, and strict antenna selection and cable-laying. Furthermore, shielding, restricting and filtering technology must be used.

3.2. Measures to Overcome Adverse Environments
The tracking ship sails the oceans for long periods of time and will frequently encounter high temperatures, high humidity, waves, and salt fog. To ensure normal and reliable operation and long life of the electronic equipment on board the ship, it is equipped with waterproof portholes and doors. All the cabins are air-conditioned. Electrical devices use large-scale integrated circuits and are sealed and heated. Exposed portions are treated against moisture, salt fog, and mildew.

3.3. System Calibration
Parameter drift, especially under adversary conditions, will increase the system error and sometimes will even cause failure of a mission, so regular and frequent maintenance, test and calibration is necessary to guarantee the required performance. For example, the zero value of the radar needs regularly check, the amplitude-phase inconsistency of the momopulse tracking channel must be carefully measured before each mission.

4. Future Missions
The development and deployment of China's tracking and data relay satellites brings unprecedented challenge to the space tracking ship, but because of many factors, especially the tracking ship's unique advantage, the space tracking ship is still going to play a critical role in the future.

In the future, the space tracking ship will continue to support various space programs including:

4.1. Space Station Program
As the most complicated space program of China, the objective of the space station program is to build a manned space laboratory to commit scientific research, and it also needs a much stronger ground support. In this mission, the tracking ship might will participate in the launch phase, orbit tracking or as voice/video relay.

4.2. Lunar Exploration and Deep Space Exploration
The ambitious goals of China's lunar exploration probably include astronaut's landing and surveying on the moon. In this and subsequent deep space explorations, the tracking ship might will participate in the launch phase and the return phase.

4.3. BeiDou Navigation Satellite System

The BeiDou Navigation Satellite System (BDS) is planned to be completed before 2020 and it needs the tracking ship to provide service of vehicle telemetry and satellite tracking in launch phase. The tracking ship will continue to support this grand plan in several years.

5. Future Challenges

These and other possible missions propose more rigorous requirements. The challenges that the tracking ship confronts, which are also the development trends of it, include:

5.1. Larger Operation Range

The diameter of the tracking antenna, limited by the size of the tracking ship, can't be very large and will prevent the tracking ship to take part in the deep space exploration of which the downlink signal is very weak due to long range.

Possible approaches to increase the operation range include using high frequency, array of all antennas onboard a single ship, coding and using very-low-temperature amplifier.

The beam of the reflector operating in high frequency is very narrow, and because of the ship motion, the stabilization of the antenna must be paid much effort.

5.2. Higher Tracking Accuracy

Because of operating in a dynamic environment, the tracking ship has much more error sources than land-based station. To further improve the measurement accuracy is a very difficult task and also an important research direction. For this aim, some promising data processing methods have been studied and tested, such as data fusion algorithm and adaptive method.

The accuracy of ship position and attitude measurement affects greatly the final result, and some new advanced techniques such as strap-down inertial navigation system and star-matching-based star sensor are now under consideration.

5.3. Higher data rate

In future missions, video from the space station or observation data of deep space explorer will have much higher data rate, so the tracking ship needs to expand its bandwidth to receive and relay those data.

References

[1] Cong Bo, XI Zhendong, WANG Jue and WU Youxing 2013 New advances of Chinese space tracking ship *International Astronautical Congress* **28** 1219

[2] Zhao Wenhua *et al* 2004 Application of GPS attitude determination on survey ship *Journal of Chinese inertial technology* **12** 394

[3] Feng Hongkui *et al* 2008 An information processing method for the navigation system of space instrumentation ship based on federated Kalman filtering *Journal of spacecraft TT&C technology* **27** 174

[4] Zhang Tongshuang *et al* 2007 An improved Chi-square fault detecting method for space TT&C ship's integrated navigation system *Journal of telemetry ,tracking and command* **28** 361

[5] Xue Guohu, Tao Xiaohong *et al* 2010 Research on key techniques of shipborne X band microwave universal system *Journal of telemetry ,tracking and command* **31** 263

[6] Wang Xiaoming *et al* 2010 Functional expansion of pulse radar based on compounded radar station *Modern Radar* **32** 277

Journal of Physics: Conference Series **679** (2016) 012042 doi:10.1088/1742-6596/679/1/012042

The Fiber Grating Sensors Applied in the Deformation Measurement of Shipborne Antenna Basement

Yong Liu, Jiahong Chen, Wenhua Zhao

Joint Laboratory of Ocean-based Flight Vehicle Measurement and Control, China Satellite Maritime Tracking and Control Department, Jiangyin, Jiangsu Province

E-mail: gfkdly@aliyun.com

Abstract. The optical fiber grating sensor is a novel fibre-optical passive device, its reflecting optical spectrum is linearly related with strain. It is broadly applied in the structural monitoring industry. Shipborne antenna basement is the basic supporting structure for the radar tracking movement. The bending deformation of the basement caused by ship attitude changing influences the antenna tracking precision, According to the structure of shipborne antenna basement, a distributed strain testing method based on the fibre grating sensor is approved to measure the bending deformation under the bending force. The strain-angle model is built. The regularity of the strain distribution is obtained. The finite element method is used to analyze the deformation of the antenna basement. The measuring experiment on the contractible basement mould is carried out to verify the availability of the method. The result of the experiment proves that the model is effective to apply in the deformation measurement. It provides an optimized method for the distribution of the fiber grating sensor in the actual measuring process.

1. Introduction

The optical fiber grating sensor is a novel fibre-optical passive device, its reflecting optical spectrum is linearly related with the strain of the fiber. It is broadly applied in the structural monitoring industry. The working principle of the sensor is shown in figure 1.

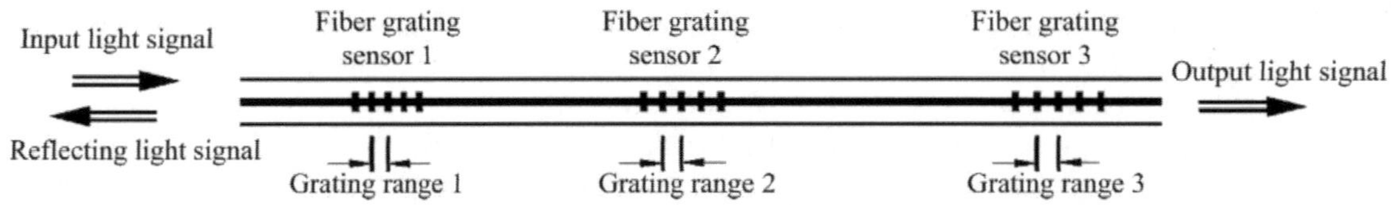

Figure 1. Working principle of optical fiber grating sensor.

Shipborne antenna basement is the basic supporting structure for the radar tracking movement. The bending deformation of the basement caused by ship attitude changing influences the antenna tracking precision. The Shipborne antenna basement of tracking radar is shown in figure 2.

Figure 2. Shipborne antenna basement of tracking radar.

To calculate the value of the bending angle, a distributed strain measuring method based on the fiber grating sensor is approved to measure the bending deformation of the antenna basement.

2. Measuring principle

2.1. Strain model

The bending angel of the antenna is related with the strain and the structural dimensions. θ is the bending angle. ε is the strain on the wall of basement. H is the height, R is the radius of the basement. The schematic of strucureal of bending is shown in figure 3.

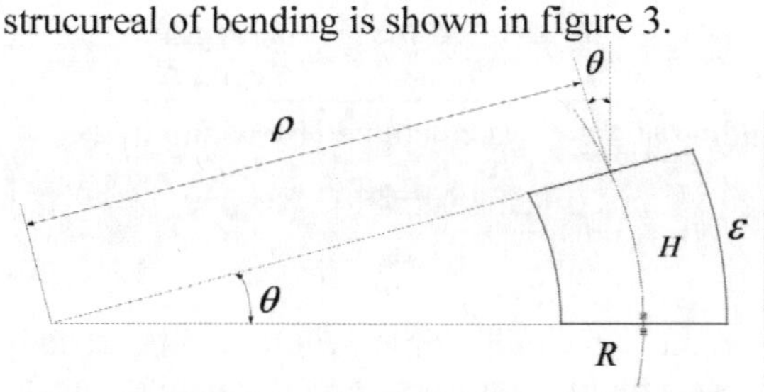

Figure 3. Schematic of the structural bending.

The value of the bending angle can be formatted as:

$$\theta = \frac{H}{\rho} = \frac{H}{R/\varepsilon} = \frac{\varepsilon H}{R} \tag{1}$$

2.2. Calculating method

The bending angle of the entire basement is composed by all the separating parts. So the general value should be accumulated as formula (2):

$$\theta_i = \frac{\Delta H_i}{R_i}\varepsilon_i + \theta_{i-1}$$
$$(i = 1, 2, 3, \cdots n), \Delta H_i = H_i - H_{i-1} \tag{2}$$

3. Simulation analysis

A finite elements model is built to analyze the deformation of the antenna basement. The simulation result of antenna basement is shown in figure 4.

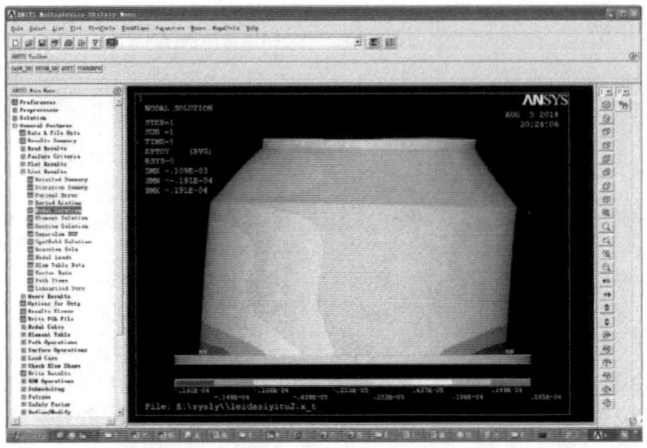

Figure 4. Simulation result of antenna basement.

The result of the simulation is shown in Table 1.

Table 1. Simulation results of antenna basement.

Num.	Location	Train
1	0.006	1.905e-4
2	0.575	1.676e-5
3	1.057	1.234e-5
4	1.884	0.732e-5
5	1.981	0.671e-5
6	2.286	0.537e-5
7	2.443	0.410e-5

The general value of the bending angle θ_s is calculated by formula (2):

$$\theta_s = 20.74''$$

4. Measuring experiment

The measuring experiment on the contractible basement mould is carried out to verify the availability of the measuring method. The volume ratio between the mould and the actual shipborne antenna basement is 1:10.The Measuring experiment on the contractible mould of antenna basement is shown in Figure 5.

Figure 5. Measuring experiment on the contractible mould of antenna basement.

The result of the experiment is shown in Table 2.

The general value of the bending angle θ_e is calculated by formula (2):

$$\theta_e = 23.51''$$

5. Conclusions

Journal of Physics: Conference Series **679** (2016) 012042 doi:10.1088/1742-6596/679/1/012042

The result of the experiment proves that the measuring method is effective to apply in the deformation measurement. It provides an optimized method for the distribution of the fiber grating sensor in the actual measuring process.

Table 2. Result of the measuring experiment.

Num.	Location	Train
1	0.01	2.231e-4
2	0.63	1.852e-5
3	1.22	1.457e-5
4	1.81	1.083e-5
5	1.92	0.962e-5
6	2.21	0.739e-5
7	2.40	0.562e-5

References

[1] Shahriar Negahdaripour, Pezhman Firoozfam 2006 An ROV stereovision system for ship-hull inspection *IEEE Journal of oceanic engineering* **31** 551

[2] Wang Yin, ZHU Zhenyu et al 2012 A method of wing shape predictions for highly flexible UAVs *Computer measurement and control* **20** 2894

[3] Fan Hongchao et al 2008 Vibration intelligent test of large-scale flexible surface *Journal of mechanical engineering* **44** 202

[4] GUO Meng et al 2012 Deformation detection of satellite antenna based on strain testing method *Spacecraft environment engineering* **9** 663

[5] WANG Chang-wu et al 2006 A real-time measuring system of operational deformation of radar antenna *modern radar* **28** 90

[6] SHEN Chun-yuan et al 2011 The force calculation and finite element analysis of stabilizing rotary platform systems of shipborne radars *radar &ECM* **31** 67

Journal of Physics: Conference Series **679** (2016) 012043 doi:10.1088/1742-6596/679/1/012043

Satellite-borne QPSK Direct Modulator for Ka Band

Min Zhang, Changsheng Li

Shanghai Aerospace Electronic Co. Ltd, Shanghai, China

E-mail: zhangminnp@126.com

Abstract. Ka band is referred to a microwave band whose frequency range is from 24.6 GHz to 40 GHz, it shares a wide available bandwidth, high frequency reuse rate and strong ability of anti-jamming. This paper presents a novel method to design a modulator for Ka-band satellite communication. Using QPSK to improve the ability of anti-jamming, using direct modulation to reduce the weight, volume and cost of electronic equipment, using sub-harmonic mixer to cut the LO power leakage, excellent modulation results are obtained.

1. Introduction

Nowadays, in the satellite communication domain requirements of data transfer rate have reached 300Mbit/s in the Tracking and Data Relay Satellite System (TDRSS), while that in the space station is even more than 600 Mbit/s. With constant development of current satellite communication system, the amount of information it carries will be larger and larger, and the request of bandwidth will be higher and higher. Whereas, generic traditional secondary planets are using C band and Ku band, whose maximum available bandwidth is 800 MHz and 1000 MHz respectively, which cannot keep up with the ever-increasing demand of satellite communication date transfer rate. Hence, the focus of world attention has gradually shifted towards the Ka band, which is relatively unoccupied.

The secondary planets for Ka band utilize amounts of narrow spot-beam mostly, which not only improves the value of G/T and EIRP of radiating antenna, reduces the size of the antenna in ground terminal, but also has high frequency reuse rate and strong ability of anti-jamming. In order to enhance the ability to support the measurement control and communication for the MEO/LEO spacecraft, NASA and ESA have made plans to divert into Ka band. Besides， the improvement of components and craft advances the development of satellite communication for Ka band.

2. Methods &Results

In this paper, a new method to design a satellite-borne QPSK direct modulator for Ka band was presented. The schematic is shown in figure 1, which consists of three main parts: coupler, BPSK modulator, power divider.

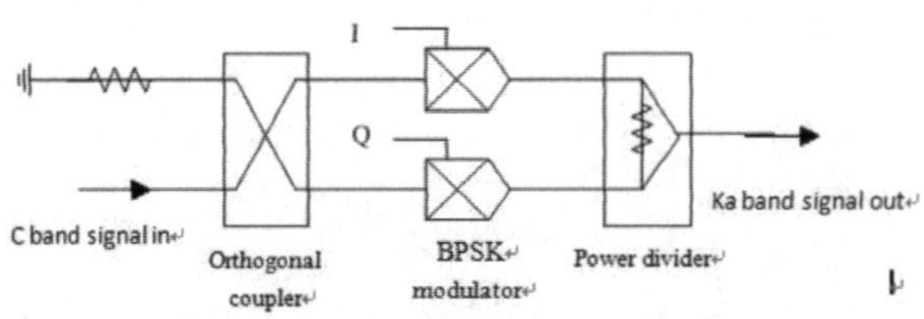

Figure 1. QPSK schematic.

Journal of Physics: Conference Series **679** (2016) 012043 doi:10.1088/1742-6596/679/1/012043

2.1. Orthogonal coupler for C band

Branch line coupler, directional coupler, and Lange coupler are most extensive using coupler in project. Considering the strict demand of processing craft, we choose the branch line coupler to provide 90° phase shift.

The simulation result is shown in figure 2.

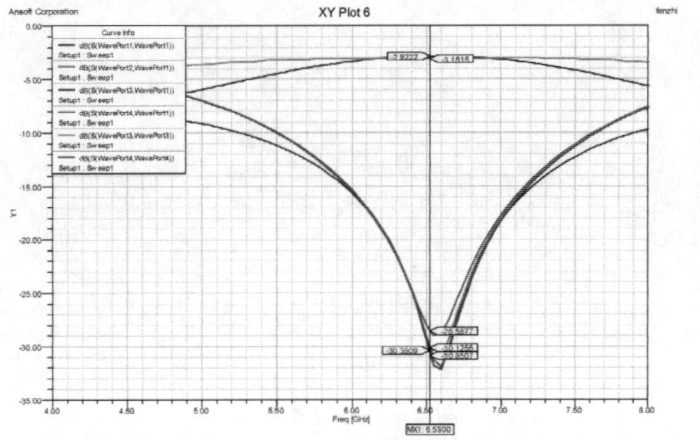

Figure 2. Simulation result of Orthogonal coupler for C band.

2.2. Power divider for Ka band

Power divider is used in QPSK modulator to combine two BPSK signal. In a conventional Wilkinson power divider, all branches are of length $\lambda_g/4$ and chip of 100Ω ensures isolation between output ports. For easy integration of chip resistor, branch length of each power divider is increased to $3\lambda_g/4$, as shown in figure 3.

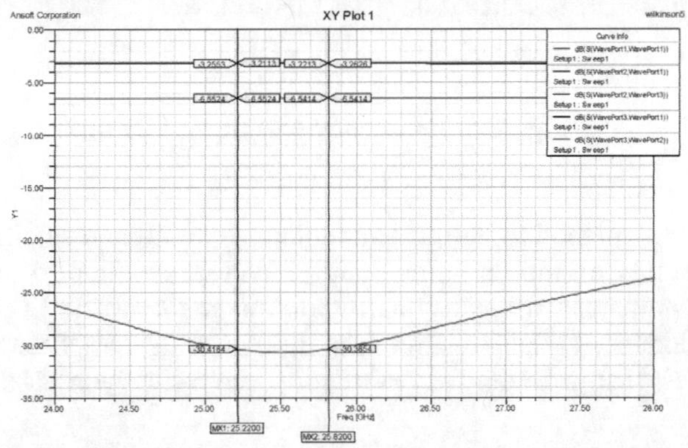

Figure 3. Simulation result of Power divider for Ka band.

2.3. BPSK modulator for Ka band

BPSK modulator is the core of this design, two orthogonal BPSK modulators make up a QPSK modulator. To reduce the LO power leakage and decrease the source's developing difficulty, our design takes sub-harmonic mixer as modulate device.

According to the schematic above, we design a sub-harmonic mixer, its structure and frequency spectrum is shown in figure 4 and figure 5.

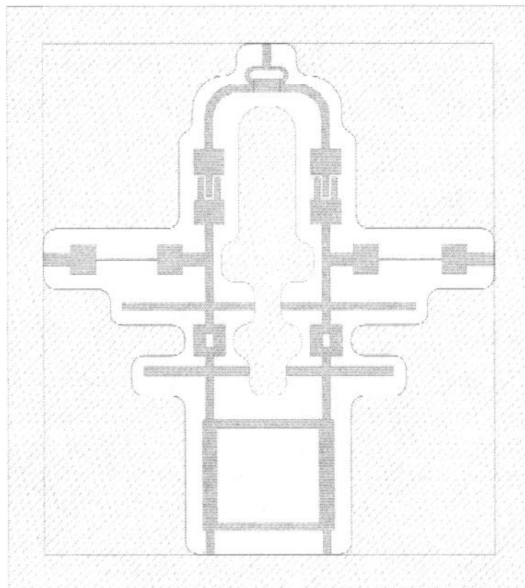

Figure 4. Structure of the modulator.

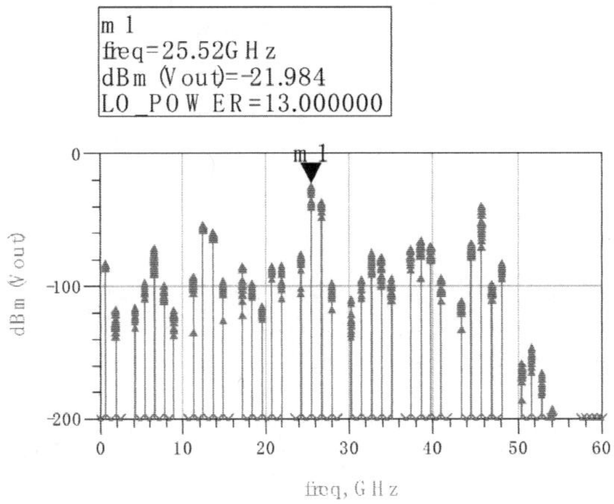

Figure 5. Frequency spectrum after mix.

3. Conclusion

This paper presents a novel method to design a modulator for Ka-band satellite communication. Using QPSK to improve the ability of anti-jamming, using direct modulation to reduce the weight, volume and cost of electronic equipment, using sub-harmonic mixer to cut the LO power leakage, excellent modulation results are obtained at data transfer rate of 600 Mbit/s, the phase error is less than 2.3°, the amplitude deviation is less than 1 dB ,and the minimum conversion gain is less than 12 dB.

References

[1] Chen Qiang *et al* 2014 Global Ka Band Broadband Satellite Services Overview(1) *Digital Communication World* **3** 198

[2] Michael Ludwing, Elena Daganzo and Malcolm Davidson 2013 Ka-Band radar missions for earth observation *Proc. Int. Conf. On Geoscience and Remote Sensing* **5** 316

[3] Tsenes P S, Stratakos G E *et al* 2002 A MMIC QPSK Modulator *Active & Passive Electronic Components* **1** 47

Journal of Physics: Conference Series **679** (2016) 012044 doi:10.1088/1742-6596/679/1/012044

An optical multiplier setup with dual digital micro-mirror array devices

Hui-feng Liu*, Yun-long Ma, San-wen Li and Yong-ming Nie

Satellite Maritime Tracking and Controlling Department, Jiangyin, Jiangsu, China, 214431

* Corresponding author: Liuhuif@mail.ustc.edu.cn

Abstract. An optical multiplier setup is demonstrated. The key of the setup is the dual digital micro-mirror array devices which can afford optical vectors by properly designing. The output light passing through the mirror arrays and lens is collected by a CCD camera which outputs the calculating result.

1. Introduction

Optical computing has very good properties in parallel data process, which can be 103 faster than the electric calculating method and has been widely used in areas such as numerical calculation, information management, process control, supplementary system and artificial intelligence. Early in 1978, J. W. Goodman had proposed theory model of the optical vector matrix multiplier and used in pulse beam shaping, spectral analyzing and filting [1]. Baed on that model many noval setups were demenstrated [2-5]. More than ten years later, E. P. Mosca designed an acousto-optic optical multiplier, which can finish caculating vercor with 128 multiplying 128×128 matrix fastly as 8×105 per second and correctly. The setup is too big and optical efficient is low [6]. Using large scale integrated circuit technology Matthias Gruber designed planar-integrated optical multiplier, which had good properties with low crosstalk [7-9]. In 2004 Lenslet Company generated Elight256 products which are the first digital vector processing with optical heart. The information transmission distortion is small, and the energy consuming is much little. In this manuscript, we designed an optical multiplier setup that can be also used as an optical logic operator is demonstrated. The key of the setup is the dual digital micro-mirror arrays which can afford optical vectors by properly designing. The output light passing through the mirror arrays and lens is collected by a CCD camera which outputs the calculating result. Experimental results indicate that the setup can realize optical logical operator function effectively.

Journal of Physics: Conference Series **679** (2016) 012044 doi:10.1088/1742-6596/679/1/012044

2. Theory analysis

The key theories are the optical vector-matrix multiplier caculating formula and the digital micro-mirror device woring principal, which will be given in the following..

2.1. Theory of the optical vector-matrix multiplier

Based on the model proposed by Goodman as shown in Fig.1 the calculating formula can be written as following.

$$c_i = \sum_{j=1}^{n} a_{ij} b_j, \qquad (i = 1, 2, \cdots m) \tag{1}$$

The setup is composed by the following elements. An optical source array provides the light. A lens collimates the beam. Two cylinder lenses and spatial light modulator and detector are also important devices. Vector B with n-demension is presented as b_j. Matrix A with m×n demensions is presented a_{ij}.

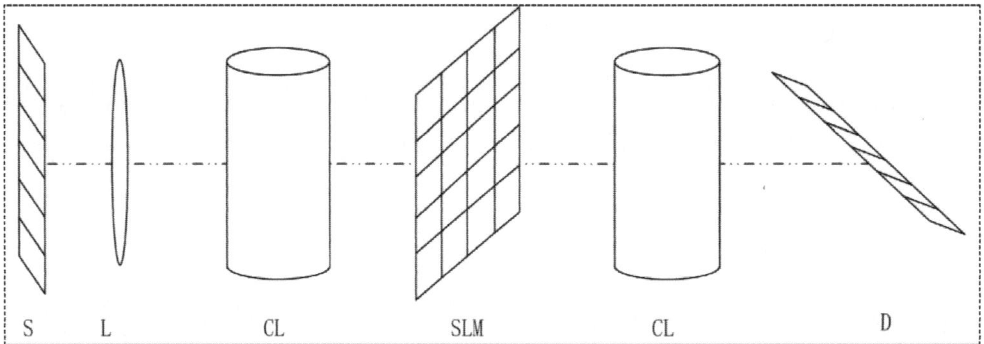

Figure 1. Schematic of optical vector-matrix multiplier, S presents optical source, L present lens, CL presents cylinder lens, SLM presents spatial light modulator and D presents detector.

2.2. Principal of the digital micro-mirror device

The digital micro-mirror device is a novel reflecting spatial modulating with high energy efficiency which may be higher than 25%. The micro structure and diffraction properties are shown in Fig.2.

Journal of Physics: Conference Series **679** (2016) 012044 doi:10.1088/1742-6596/679/1/012044

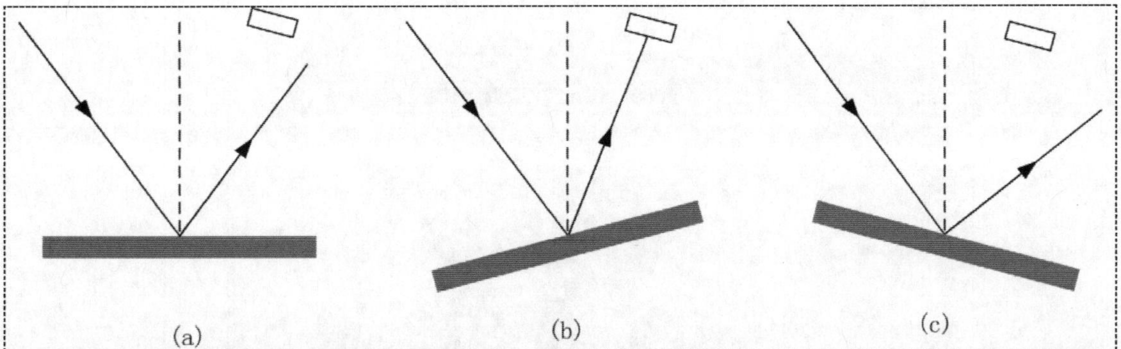

Figure 2. Schematic of single DMD mocro mirror working principal. (a) is normal state, (b) is 'on' state, and (c) is 'off' state.

The light inclined irradiating the mirror, when the DMD is working on 'normal' state, the mirror is horizontal. When the DMD is working on 'on' state, the mirror has angle with horizontal. When the DMD is working on 'off' state, the mirror has angle with horizontal. It is just when the DMD is working on 'on' state the light can reach the object detector. The DMD diffraction optical intensity distribution schematic is shown in Fig.3.

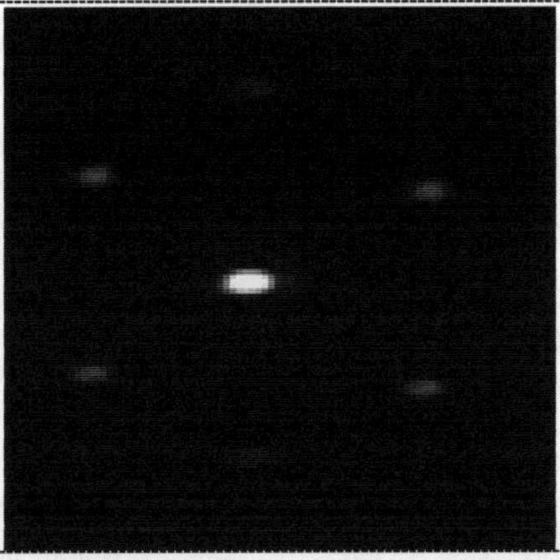

Figure 3. Schematic of the DMD diffraction optical intensity distribution.

3. Experimental results

In order to analyze the feasibility of using DMD as the spatial light modulators in the optical vector matrix multiplier system, on the basis of the primary structure and principle of the digital micro-mirror device and the optical vector matrix multiplier, an optical vector matrix multiplier system which introduces two digital micro-mirror devices is proposed as shown in Fig.4. In the system both the input of the optical vector and the matrix is finished by the digital micro-mirror devices.

Journal of Physics: Conference Series **679** (2016) 012044 doi:10.1088/1742-6596/679/1/012044

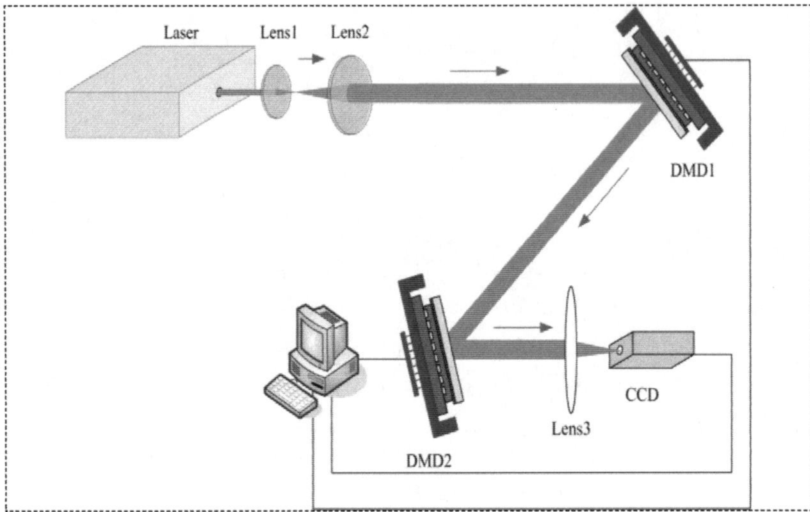

Figure 4. Schematic of the multiplier setup. Lens1 and lens2 compose a beam expander.

The key of the setup is the optical vector designing, which is shown in Fig.5. In this manuscript several kinds of optical vectors and corresponding experiments are given. As shown in Fig.5, the first vector is generating when the light passes through the DMD1 and the first step of the multiplier is accomplished when the light passes through the second DMD. At last, when the output light is collected by the CCD, the result is given.

The key of the setup is the optical vector designing, which is shown in Fig.5. In this manuscript several kinds of optical vectors and corresponding experiments are given. As shown in Fig.5, the first vector is generating when the light passes through the DMD1 and the first step of the multiplier is accomplished when the light passes through the second DMD. At last, when the output light is collected by the CCD, the result is given.

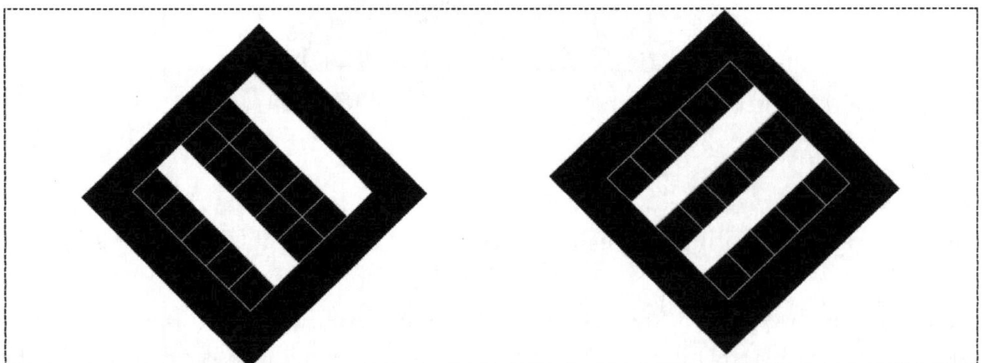

Figure 5. Schematic of the vectors generated with DMD. The left vector presents 01001 and the right vector presents 01010.

The input light is He-Ne laser with 2mm diameter at 632.8 nm central wavelength. After expanded by a pair of lenses the beam diameter is 10mm. Then the light incidents on the DMD1 which is loaded the information as shown in the left part of Fig.5. The reflected light after DMD1 is the incident light to DMD2 which is loaded the information as shown in the right part of Fig.5. At last the output light is collected by a CCD camera.

The theoretical and experimental results are shown in Fig. 6. The experimental results are average values after ten times experimentations.

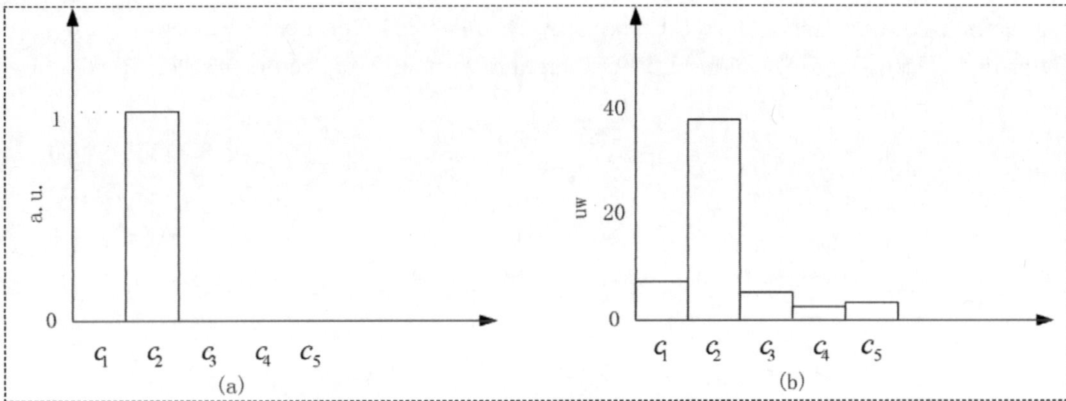

Figure 6. Schematic of the vectors generated with DMD. The left vector presents 01001 and the right vector presents 01010.

Vector 01001 multiplies vector 01010 and the result is 01000 as shown in Fig. 6(a). The experimental results are shown in Fig. 6(b). It can be easily found that though the result is zero, there is still light being detected, which is caused by diffraction by the optical devices. However, when the optical intensity is quantified according to proper scales and properly deal with the dirraction interfering, results with high precision can be obtained.

4. Conclusions

In this manuscript an optical multiplier setup that can be also used as an optical logic operator is demonstrated. The key of the setup is the dual digital micro-mirror arrays which can afford optical vectors by properly designing. The DMD is working in reflecting form, so the energy efficiency is high than normal liquid crystal spatial light modulators. The output light energy efficiency can be about 25%. By theory analyzing and experiments,we gain the optical properties of the DMD and the influence on the system introduced by the DMD.The results we gain are in accordance with the anticipation.It is proved that by matrix-coding and orderselecting the influence of diffraction and interference can be reduced a lot.These conclusions are meaningful for the designation and coding methods of the optical vector matrix multiplier.They are useful for the study of the key applications of the optical multiplier and have some significance for the designation of the optical vector matrix multiplier.

Acknowledgment s

The authors acknowledge Doctor Yun-fei Mao and Jia-li Liao for fruitful discussions.

References

[1] Goodman J W and Diaz A R 1978. Opt Lett **2** 1

[2] Gruber M, Jahns J and Sinzinger S 2000 Appl. Opt. **39** 5367

Journal of Physics: Conference Series **679** (2016) 012044 doi:10.1088/1742-6596/679/1/012044

[3] Atthale R A and Lee J N 1984 Proc. IEEE **72** 931
[4] Caulfield H J, Rhodes W T, Foster M J and Horvitz S 1981 Opt. Commun. **40** 86
[5] Carlotto M and Casasent D 1982 Appl. Opt. **21** 147
[6] Reif J H and Oshida A Y 1993 Appl. Opt. **32** 159
[7] Mosca E P, Griffin R D, Pursel F P and Lee J N 1989 Appl. Opt. **28** 3843
[8] Ibrahim T A, Amarnath K, Kuo L C and GroverVan V 2004 Opt. Lett. **29** 2779
[9] Barros S, Guan S and Alukaidey T 1997 Journal of System Architecture **43** 391

Laser Monitoring System for Crossing Border Based on the Short Message Service

Wen-jin Zhu*, Dao-qing Zhou, and Lu-jun Zhou

China Satellite Maritime Tracking and Controlling Department, Wuxi, Jiangsu, China, 214431

*Corresponding author: z041230@163.com

Abstract. A warning system applied to long-distance crossing-border monitoring by the technology SMS and IR(infrared) laser was introduced. Master station, which was based on PC and module TC35i, would transmit or receive short message by COM communication. Monitoring station watched the forbidden zone produced by an IR warning apparatus. CMOS chip AT89C52 incepted the warning information, and transmitted out a warning message by module TC35i. Circumstance of hardware and software was provided and programming was described also. The system overcame the shortages and restrictions on conventional communication and monitoring system.

1. Preface

Traditionally, radio station was used often to receive and dispatch data or digital command in the system of the wireless monitoring and controlling [1]. Because of its inherent weaknesses such as short transmitting distance (up to 30~50km), complex of debugging and maintenance, costly operating, lack of frequency resources, and the instability of long-term running， the applications of wireless measuring and controlling system in engineering were restricted seriously. Meanwhile, when visible light, camera, and heat-sensor were used generally in the supervision system, the shortcomings of short monitoring distance, small coverage, and low definition owing to low visibility will lead to a high false alarm rate. Therefore, difficulties such as the monitoring range cannot meet the requirements, signal transmission and processing are awkward, and being of high producing cost are generally have to faced with when it comes to cross-border alarm in remote areas, long distance and large range.

Now, we combined the data business of the GSM network with monitoring system properly, and designed a monitoring system for crossing border which based on the technique of message service and IR laser. This system overcame the restrictions on conventional method effectively and had advantages such as long warming distance, large-range signal transmission, accuracy and timeliness, simple operation, and flexibility of assembling and extending.

2. Theory and method

2.1. System theory

The monitoring system introduced in this paper observes the principle and function exhibited as Fig.1.

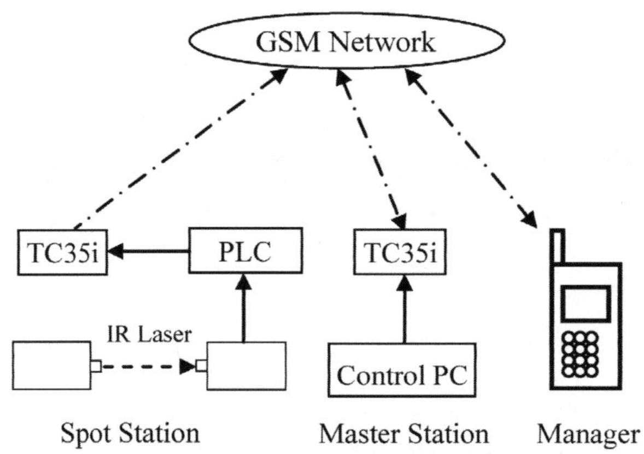

Figure 1. Function of system

Master station system was constituted by the module TC35i and the PC. Spot station, that is IR Laser Monitoring station, was constituted by module TC35i, PLC and laser transmit-receiving equipment. Semi-conductor laser transmitter send out an IR laser light which form an invisible guard line. And laser receiver could receive the IR laser persistently. When someone went through the guard area, laser light was cut off and the signal of receiver changed correspondingly. The change was distinguished by electric circuit, and the information was coded by PLC. Finally, the message was sent to manager or master station by module to display warning consent. So a complete set of monitoring system of supervising and auto warning was realized.

2.2. Extending of monitoring station

2.2.1. Discrimination of Cross-in or Cross-out Border

Cross border detection was realized by a pair of laser transmitter and a pair of laser receiver. The infrared laser transmitter sent out two beams of laser, forming an invisible warning line, and make up a few hundred meters of the warning area. The laser receiver receives the two laser beams. In general situation, the two optical paths is unobstructed. When someone is trying to go through the warning area illegal, laser path would be cut off. Laser receiver could detect the order of block of the two laser beams. Identification circuit would judge the information, and PLC would process and coding the information. Communication module emits the alarm message to master station or manager and display relevant information. Double laser beam also provides a guarantee for reducing the false alarm rate which could cause by leaves and other accidental factors.

2.2.2. Independent monitoring unit

By means of dual laser beams, transmitter No.1 and transmitter No.2 share one receiver which contains a short message sending and receiving device of GSM module controlled by a PLC. Such scheme is conducive to extend system and save costs. One receiver receives four beams simultaneously, analyze the change of lights, and send out warning message based on some judgments. According to the telephone number and message data, manager could acquire the information such as the warning occurring location or person's departing and entering. The sketch of single monitoring unit was displayed as follows:

Figure 2. Monitoring unit

2.2.3. Monitoring system after extending

According to actual situation and demand, monitoring unit can extend freely as the following figure.

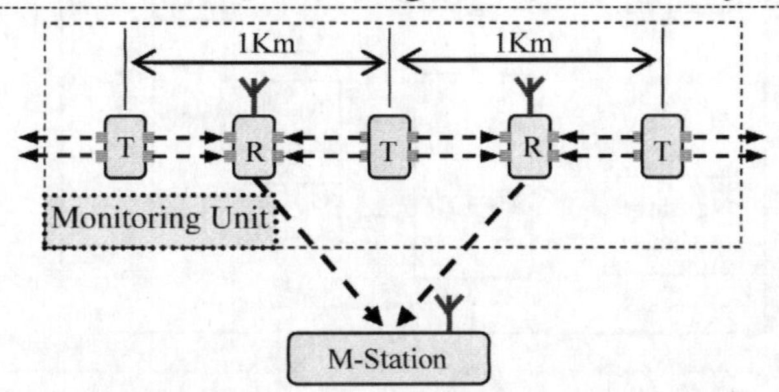

Figure 3. Monitoring system after extending

IR laser signal could be received stably at the scope of about 500 meters with the semiconductor IR laser transmitter of power 15mW. So a single monitoring unit could create a guarding distance of 1 kilometer. If master station runs a system of 40 units, the guarding area will enlarge to 40 kilometers.

3. Electric circuit design

3.1. Circuit realization of master station

Journal of Physics: Conference Series **679** (2016) 012045 doi:10.1088/1742-6596/679/1/012045

Master station adopts the new Siemens wireless module TC35i and its accessories. Around the module, there are some interfaces and conjunctions should be designed respectively such as chip ZIF40, SIM card, power supplying, data communicating connection, and other outer peripheral circuits [2].

The power supply circuit uses the mobile phone charging power, which is input 220V AC and output 5V regulated voltage and maximum 2A current.

In order to make the module normal start to run, a low voltage signal of at least 100ms must be input to "IGT" pin. And the signal's fall time must be less than 1ms. Therefore, the starting circuit should be composed by open drain pole triode and power on reset circuit. After start-up, the pin should remain high voltage.

Emergency shutdown circuit would work only if the module cannot complete soft power-off and other fault conditions. The circuit also uses open drain triode to connect the "PowerDown" pin to the ground and a low voltage signal more than 3.2S is required to put on the pin.

The light-emitting diode circuit is used to indicate the working status of the module such as being shutdown (or sleep) / network logon / standby, etc..

TC35i's serial port uses CMOS level (high level 2.65V), while the PC serial port uses RS232 level. So chip SP3238EEY of Siper could used here to realize the level conversion and serial communication function. SP3238EEY has properties of low power consumption, high data rate, enhanced ESD protection, etc..

3.2. Circuit realization of monitoring station

Monitoring station system is exhibited as follow figure:

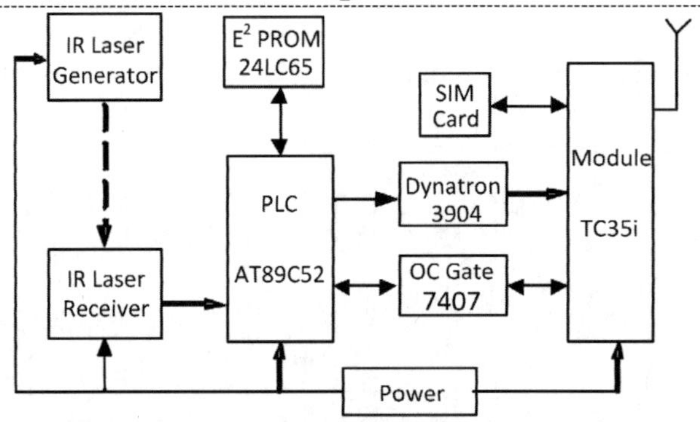

Figure 4. Composing of monitoring station

Monitoring station system is mainly consists of semiconductor IR laser generator, laser receiver, PLC of chip AT89C52, module TC35i, interface circuit, power supply circuit and chips 24LC65, 3904, 7407 etc..

Laser receiver detects the infrared laser from laser generator and changes of optical signal will be input to PLC. PLC will determine whether to send text messages through a simple algorithm.

Chip AT89C52 of PLC connected to module TC35i's data interface through serial port which complete the level conversion by OC gate of chip 7407. Then PLC initializes the module and realizes control of SMS transceiver through the AT command.

Some initial information and the default values of the system are stored in the E2PROM 24LC65 of PLC.

The power supply circuit provides power of requested voltage and current for each component.

4. Software design and programming

4.1. Software and programming of PC

A suit of program which is simple but functional was realized at PC by programming language VC++ 6.0. The program is flexible and easy to extend function when it's necessary. Controller MSCOMM was adopted to realize short messages sending or receiving [3].

Program has following characteristics: supporting hardware of Modem GSM and standard AT instruction [4]; supporting for sending Chinese or English message; supporting for receiving message real-time; extendible function of phone book operation; open AT instruction interface through which users can program custom AT instruction and extend component; attribute, method and event are reasonable and concise and easy to understand and use; event driven mode take up less system resources and running efficiently.

4.2. Software and programming of PLC

Language C51, which was compiled under the Keil μVision2, was used in PLC system. The HEX documents were produced and written into the chip AT89C52 of PLC by compiler SUPERPRO/LX of XELTEK [5].

5. Conclusions

There were successful applications at the border defense in the field of crossing border warning based on radio and visible laser. As monitoring system this paper elaborated is based on the short message service and invisible laser, it has more advantages as follow:

- Long distance and big area of guarding: Tens to hundreds kilometres after system extending.
- Instant and remote supervising: There was no restriction of distance as long as GSM network covers the stations.
- Energy conservation: Module TC35i consumes much less power than radio communication.
- Easy to carry out: This project save work largely of spreading wire out when it's comparing to wire network, and also leave load out of building antenna, obtaining admission of radio frequency when it's comparing to radio station.
- The system is easy to extend or upgrade on the function or scale.

For a monitoring station with single laser generator and receiver, experiment has testified monitoring alarm and short message sending and receiving were all reliable and stable enough to meet the design requirement; If it can be extended and devoted into application, the problem of border guarding to key areas would been solved effectively. All in all, the whole system has large practical value and a favorable applying prospect.

Acknowledgments

This research was guided by Prof. Bin Lin and finished at the State Key Laboratory of Optical Instrument. We acknowledge all staffs of Lab. and thank Dr. Yong-ming Nie for fruitful discussions.

References

Journal of Physics: Conference Series **679** (2016) 012045 doi:10.1088/1742-6596/679/1/012045

[1] Wang Q and Wang J 2002 Design of computer controlled radio laser monitoring system for crossing border Journal of Shenyang University of Technology **24** 152

[2] Nieva T M and Wegmann A A 2002 Conceptual model for remote data acquisition system COMPUTER IN INDUSTRY **47** 215

[3] Liu W 2003 Tutorial of Visual C++ 6.0 Publishing House of Electronic Industry

[4] Rahnema M 1993 Overview of GSM System and Protocol Architecture IEEE Communication **3** 92

[5] Ma Z, Ji S and Zhang K 2004 Applications Design of Single chip by C Language Beijing Aerial Navigation University Publishing House

Ultrafast optical signal pulse descending time controlling with deformable mirror and spatial light modulators

Yong-ming Nie*, Yang Liu and Liu Jun-hui

Satellite Maritime Tracking and Controlling Department, Jiangyin, Jiangsu, China, 214431

*Corresponding author:yimonie@163.com

Abstract. A programmable ultrafast optical signal pulse descending time controlling apparatus is demonstrated based on micro-machined deformable mirrors (MDM) and liquid crystal spatial light modulators (LCSLM), which can smoothly modulate the phase and amplitude of the optical pulse frequency component linearly or nonlinearly.

1. Introduction

Femtosecond pulse shaping, which allows for generating specified temporal optical waveforms by parallel manipulation of the complex optical spectrum, has advanced dramatically since the advent of reliable generation of laser pulses below 100 fs in duration in 1981 with the invention of the colliding pulse mode locked (CPM) ring dye laser [1]. The temporal femtosecond pulse shaping technology has been widely adopted in various applications such as harmonic generation [2], coherent control [3], generation of two-dimension electronic spectra and temporal solutions [4], high-energy isolated attosecond pulse generating [5], compression of optical pulses [6], single cycle pulse generation [7], radio-frequency waveform generation [8], and optical communications [9]. Recently, the application has been further expanded. For example, in 2013, Ferreira P H D and his coworkers used shaped 30 fs pulses to accomplish the enhancement of laser induced Au nano-particle formation [10] and Yi Cai used carrier-envelope phase stabilization techniques to obtain a 1kHz, 90fs, 0.4mJ femtosecond pulse train with a total gain of 2×106 [11]. Bagayev S N obtained super-intense femtosecond multichannel laser system using parametric amplifier to overcome the phase distortions, which may be solved by Femtosecond pulse shaping techniques [12].

In this manuscript, a programmable femtosecond pulse shaping apparatus is designed based on MDM and LCSLM, which changes the phase and the amplitude of the frequency component of the pulse respectively. According to the fact that the time domain signal and its corresponding frequency

Journal of Physics: Conference Series **679** (2016) 012046 doi:10.1088/1742-6596/679/1/012046

domain signal are Fourier transform pair, the pulse can be delayed or advanced by using the phase only MDM that just imposes a linear phase change onto its spectrum.

2. Theory analysis

Ultrafast optical signal pulse descending time controlling with deformable mirror and spatial light modulators refers two key factors which are temporal and spatial controlling.

2.1. Temporal controlling

In the time domain, the incident pulse electric field is denoted by $\mathrm{E}_{in}(t)$ and the output pulse electric field is denoted by $\mathrm{E}_{out}(t)$, so the output of a linear filter can be simply given by the following convolution product:

$$\mathrm{E}_{\mathrm{out}}(t) = \mathrm{R}(t) * \mathrm{E}_{\mathrm{in}}(t) \tag{1}$$

where $\mathrm{R}(t)$ is the impulse response of the pulse shaper. However, for femtosecond pulses, direct shaping in time is quite a hard task and most of the devices operate in the frequency domain and the output pulse can be written as following.

$$\tilde{\mathrm{E}}_{\mathrm{out}}(\omega) = \mathrm{H}(\omega)\,\tilde{\mathrm{E}}_{\mathrm{in}}(\omega) \tag{2}$$

According to the formula (2), the temporal shaping can be accomplished through the spectral phase and amplitude controlling.

2.2. Spatial controlling

Fourier lens and LC-SLM constitute a conventional folded 4f zero-dispersion setup. The frequency components after being phase modulated by the LC-SLM were reflected back through the Fourier lens and recombined into a beam at the diffraction grating, which was anti-parallel by symmetry, and then a beam splitter separated out the retro-reflected beam. Hereto, the temporal shaping is accomplished.

According to the treatment of Kreuzer, the surface distribution of the two aspheric lenses in the Galilean shaping system can be represented as following [13].

$$\begin{aligned} z(r) &= \int_0^r \{(n^2-1)+[\frac{(n-1)d}{h(x)-x}]^2\}^{-1/2}dx \\ Z(R) &= \int_0^R \{(n^2-1)+[\frac{(n-1)d}{h^{-1}(x)-x}]^2\}^{-1/2}dx \end{aligned} \tag{3}$$

The relationship between the surface and phase distribution of the first aspheric lens in Galilean shaping system can be expressed as following [14].

$$f_p(r) = \frac{2\pi(z_e - z(r) + nz(r))}{\lambda} \tag{4}$$

The LC-SLM imprints the phase distribution onto the input beam and the output beam with desired intensity distribution will be realized.

3. Experimental results

We concentrate our work on generating the flat-top profile femtosecond pulse beam in the target plane in the near field. The pulse beam used in all the measurements is generated by the Mai Tai HP Ti: sapphire laser with central wavelength at 707 nm. The beam waist radius after beam expander is 1.2 mm. The phase only LC-SLM with 1920×1080, 9 um×9 um pixels is a product of Holoeye Company. The phase level of each pixel is assigned between 0 and 2π. The CCD camera which has 1392×1040, 6.45 um×6.45 um pixels is a Dolphin F-145B modal 12 bit product of AVT Company. After passing through the shaping setup, the intensity distribution of the output beam is captured by the CCD camera and tested by the pulse checking component by Second Harmonic Generation Frequency Resolved Optical Gating (SHG-FROG) method. FROG is simply a spectrally resolved autocorrelation, which allows the use of a phase-retrieval algorithm to retrieve the precise pulse intensity and phase vs. time. It can measure both very simple and very complex ultrashort laser pulses, and it has measured the most complex pulse ever measured without the use of a reference pulse. The experimental setup is shown in Fig.1.

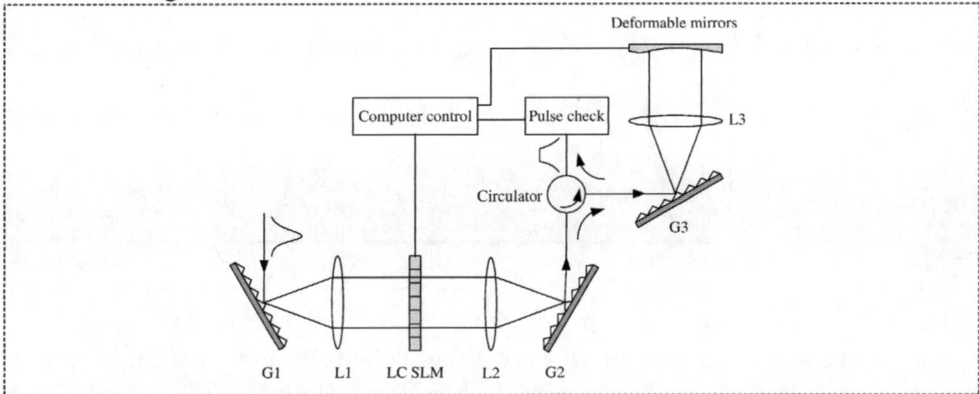

Figure 1. Schematic of the experimental setup. G1, G2 and G3 are three gratings of the same personality. L1, L2 and L3 are three lens of the same personality. LC SLM is liquid crystal spatial light modulations.

In this manuscript, a programmable femtosecond pulse shaping apparatus is designed based on MDM and LCSLM, which changes the phase and the amplitude of the frequency component of the pulse respectively. According to the fact that the time domain signal and its corresponding frequency domain signal are Fourier transform pair, the pulse can be delayed or advanced by using the phase only MDM that just imposes a linear phase change onto its spectrum. The experimental results are shown in Fig. 2 and Fig.3.

Journal of Physics: Conference Series **679** (2016) 012046 doi:10.1088/1742-6596/679/1/012046

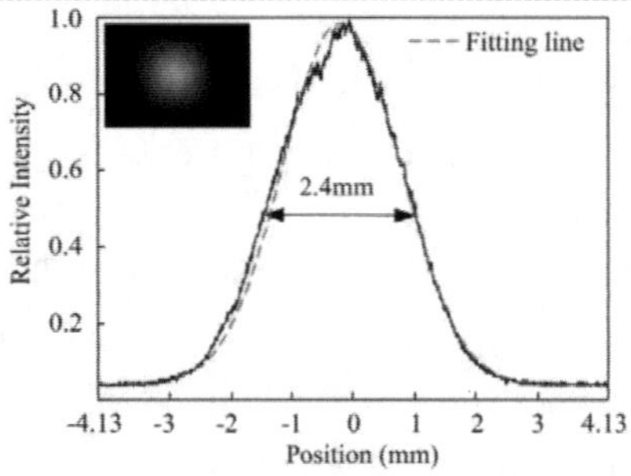

(a)

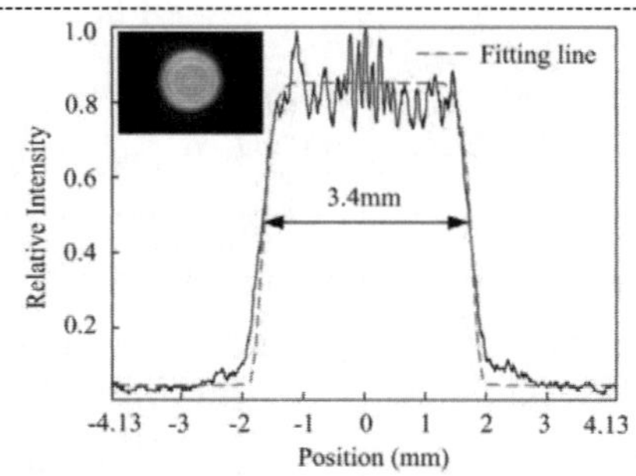

(b)

Figure 2. Spatial profiles of input and output pulse beams, (a) spatial intensity profile of the input Gaussian pulse beam and its gray-scale image (inset), (b) Experimental result of spatial flat-top intensity profile shaping and its gray-scale image (inset)..

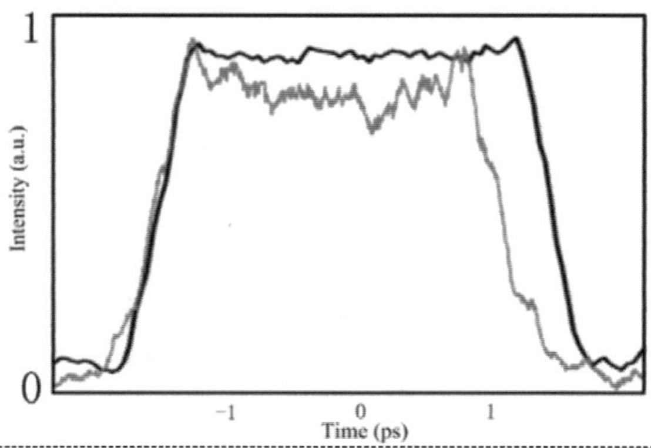

Figure 3. Ultrafast optical signal pulse temporal descending time controlling results

The results above prove that this setup can not only controlling the spatial intensity profile but also the temperoal waveforms.

4. Conclusions

Both theoretical and experimental results indicate that the ultrafast optical signal pulse descending time controlling with deformable mirrors and spatial light modulators demonstrated in this manuscript can controlling the descending time effectively. Moreover, optimization algorithms and deformable mirrors and spatial light modulators with better properties used can give better results.

Acknowledgment s

This research was funded by the National Natural Science Foundation of China (NSFC) under no.61108089, no.61107005. We also acknowledge Doctor Jia-li Liao and Hao-tong Ma for fruitful discussions.

References

[1] Weiner A M 2000 Femtosecond pulse shaping using spatial light modulators Rev. Sci. Instrum. **71** 1929-60.

[2] Bartels R, Backus S and Zeek E 2000 Shaped-pulse optimization of coherent emission of high-harmonic soft X-rays Nature **406** 164-2

[3] Cheng W, Zhang S and Jia T 2014 Quantum coherent control of blue, green and red emissions from codoped lanthanide ions of Er3C/Tm3C/Yb3C by two shaped infrared ultrashort laser beams Laser Phys. **24** 015402-11

[4] Grumstrup E M, Shim S and Montgomery M A 2007 Facile collection of two-dimensional electronic spectra using femtosecond pulse shaping technology Opt. Express **15**16681-9.

[5] Terrari F, Calegari F and Lucchini M 2010 High-energy isolated attosecond pulse generated by above saturation few-cycle fields Nature Photonics **4** 875-9

[6] Karasawa N, Li L M and Suguro A 2001 Optical pulse compression to 5.0fs by use of only a spatial light modulator for phase compensation J. Opt. Soc. Am. B **18** 1742-6

[7] Shverdin M Y, Walker D R and Yavuz D D 2005 Generation of a single-cycle optical pulse Phys. Rev. Lett. **94** 033904-9

[8] McKinney J D, Leaird D E and Weiner A M 2002 Millimeter-wave arbitrary waveform generation with a direct space-to-time pulse shaper Opt. Lett. **27** 1345-7

[9] Saradesai H P, Chang C C and Weiner A M 1998 A Femtosecond Code-Division Multiple-Access Communication System Test Bed L. Lightwave Technol. **16** 1953-64

[10] Ferreira P H D, Silva D L and Siqueira J P 2013 Enhancement of laser induced Au nanoparticle formation by femtosecond pulse shaping Laser Phys. **23** 076004-9

[11] Cai Y, Li J and Lu X al 2013 1 kHz high energy femtosecond pulses with selfstabilized carrier-envelope phases Laser Phys. Lett. **10** 125402-6

[12] Bagayev S N, Trunov V I and Pestryakov 2014 Super-intense femtosecond multichannel laser system with coherent beam combining Laser Phys. **24** 074016-28

[13] Ma H T, Zhou P and Wang X L 2010 Near-diffraction-limited annular flattop beam shaping with dual phase only liquid crystal spatial light modulators Opt. Express **18** 8251-60

[14] Ma H T, Zhang H C and Zhou P 2010 Adaptive conversion of multimode beam to near diffraction-limited flattop beam based on dual phase-only liquid-crystal spatial light modulators Opt. Express **18** 27723-30

Research of Fast 3D Imaging Based on Multiple Mode

Shibing Chen[1,2], Huimin Yan[1,2], Xuxiang Ni[1,2], Xiuda Zhang[1,2],Yu Wang[1,2]

[1] State Key Laboratory of Modern Optical Instrument, Zhejiang University

[2] National Engineering and Technology Research Center for Optical Instrument, Zhejiang University, Hangzhou, 310027, P.R.China

Email:yhm@zju.edu.cn

Abstract. Three-dimensional (3D) imaging has received increasingly extensive attention and has been widely used currently. Lots of efforts have been put on three-dimensional imaging method and system study, in order to meet fast and high accurate requirement. In this article, we realize a fast and high quality stereo matching algorithm on field programmable gate array (FPGA) using the combination of time-of-flight (TOF) camera and binocular camera. Images captured from the two cameras own a same spatial resolution, letting us use the depth maps taken by the TOF camera to figure initial disparity. Under the constraint of the depth map as the stereo pairs when comes to stereo matching, expected disparity of each pixel is limited within a narrow search range. In the meanwhile, using field programmable gate array (FPGA, altera cyclone IV series) concurrent computing we can configure multi core image matching system, thus doing stereo matching on embedded system. The simulation results demonstrate that it can speed up the process of stereo matching and increase matching reliability and stability, realize embedded calculation, expand application range.

1. Introduction

Stereo matching is one of the most active research areas in computer vision. Researchers domestic and overseas have presented a large number of excellent property binocular vision and stereo algorithms over the years[1].Among those ways, using transcendental parameter to constraint the search range when comes to stereo matching, has been proved to be an effective way. As without this limit, not only will it cost tremendous calculations (especially for big size images), but also many algorithms will not perform well[2]. An improved variogram analysis of the maximum expected disparity in stereo images, which is based on statistical correlations[3]. One research estimated the maximum and minimum disparity of stereo image depending on image characteristic[4]. And a hierarchical matching method was applied[5], however, it still need choose proper disparity range factitiously.

The use of a calibrated system, composed of a time-of-flight (TOF) camera and of a stereoscopic camera pair, allows data fusion thus overcoming the weaknesses of both individual sensors[6]. This method could greatly improve the resolution of depth map, of course this is a good try using multiple mode. But it's quite complicated to map the point captured by the TOF camera and stereo image, because he spatial resolution of currently available range sensors is lower than high-definition (HD) color cameras and the two different camera is individual. In order to solve this problem, a dual-mode camera was used to gather binocular vision image and TOF depth maps time-sharing, thus the images own a same spatial resolution[7]. The experimental results demonstrate that it can speed up the process of stereo matching and reduce the error in matching effectively for stereo pairs with large disparity range.

In the meanwhile, a lot hardware system are presented to implement real-time processing. Reference [8] use verilog (a hardware design language) to design sum of absolute differences (SAD) computing units, and dynamic programming algorithm, gaining a fast speed. A preprocessing is applied on each rectified image before looking for the best matching between a left pixel, with a right pixel and different algorithms units are constructed on FPGA with a frequency of 100MHz[9]. Sophisticated local matching algorithms (CA and FLC) that are suitable for FPGA implementation are used to achieve low error rate while maintaining the high processing speed in reference[10] and evaluate the performance of our circuit on Xilinx Vertex-6 FPGAs.

Those method mentioned above have achieved considerably high processing speeds; however, their error rates are not as good as those of software programs. And they also cannot avoid another problems, about the influence of image storage, transport, display on the processing speed, and the restriction of image size. In addition, the configuration of such a hardware system is so complex that it often needs long time to test and the portability is low.

This paper synthesizes those methods advantages, applies the TOF camera and binocular vision with the same spatial resolution as in reference[7]. With the help of TOF depth maps, we could estimate each stereo pixel's disparity range. Simultaneously using field programmable gate array (FPGA, altera cyclone IV series) concurrent computing we can configure multi core image matching system, thus doing stereo matching on embedded system. Our experiment results indicate that it is benefit to use the Nios system to handle stereo imaging.

The remainder of this paper is structured as follows: Section II describes the proposed research algorithm. And the experiment set-up shows in section III. Experimental results on a real dataset and evaluation of the method, are presented in section IV. Finally, section V draws some conclusions.

2. Research method

We use dual mode camera developed in our lab, which capture the grayscale and depth maps with the same spatial resolution in division time. Then we use the data from depth maps to figure out disparity limit factor when comes to stereo matching.

2.1. TOF camera working principle

TOF camera means its working principle is the time of flight method. As Figure.1(a) shows, it is made up of fast gate functional CCD camera and short pulse laser diode. As shows in Figure.1 (b), laser diode emits rectangular wave and the CCD camera use gate technology[10]. A modulated near infrared light from the camera's internal lighting source is reflected by objects in the scene and travels back to the sensor, where its precise time of flight is measured independently at each of the sensor's pixel by calculating the phase delay between the emitted and the detected wave[6]. In the valid ranging scope, the rate of echo pulse falling into the gate time is linear related to the back time. Considering the effect of the background light and echo pulse power, and capturing two frames of background image and when all echo pulse falling into the gate time, normalization intensity of every pixel that have been eliminated background will be in proportion to target point.

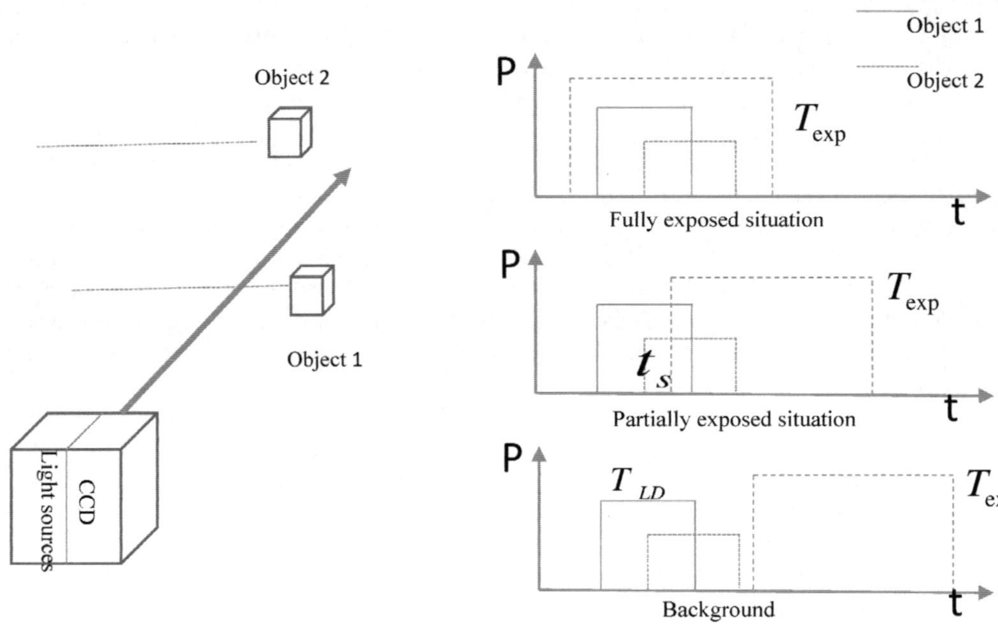

(a) (b)

Figure. 1. The description of TOF camera. The image (a): the structure of the system. The image (b): CCD gate switch form

When all the echo pulse fall into the gate time, the grayscale off each CCD pixel is:
$P_1 = \eta(P_{LD}T_{LD} + P_B T_{exp})$ (1)

(η means the optical-to-electrical efficiency of CCD, P_{LD} and P_B means the echo pulse power and background power, T_{LD} is the pulse width of laser diode, T_{exp} is the gate time of CCD)

When part of the echo pulse fall into the gate time: $P_2 = \eta[P_{LD}(\frac{2z}{c} - t_s) + P_B T_{exp}]$ (2)

(where z means target distance, c means the speed of light, t_s the gate time of CCD)

The background grayscale is: $P_0 = \eta P_B T_{exp}$ (3)

From formula (1)~(3) we could know each pixel corresponding target range is: $z = \alpha r + \beta$ (4)

When we know α and β, we can figure out z according to normalization grayscale. When we calibrate system parameters, we can chose target longitudinal distance to obtain z and r relation matrix and transform the TOF depth maps to initial disparity image, for the sake of reducing hardware image processing procedure.

2.2. Matching Step

The matching process mainly contains three parts, census transformation, cost aggregation and disparity calculation, as we can see in Figure.2. For the purpose of paying key attention to matching algorithm, we ignore the image capture course and camera calibration, and we implement image preprocessing on PC for getting initial disparity maps. On the Nios II embedded system constructed by altera FPGA, we try to run the stereo matching algorithm and finish 3D reconstruction.

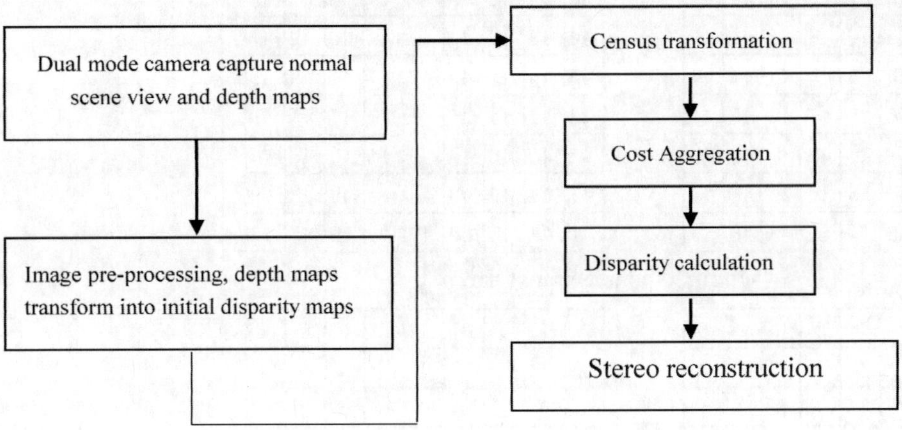

Figure.2 The description of the stereo vision algorithm procedure

2.3. Stereo algorithm

2.3.1. Single point match cost calculation

We chose census algorithm that has been confirmed owning obvious strength in local matching algorithm, as the similarity measurement function with regard to single point match cost calculation. This function make a bit string to express the structure characteristics of the matching points of local area (census transform neighboring area) by using census transform.

And the hamming distance of two match point's bit string represents the match cost. The census method single point match cost C_{Census} calculation procedure is below:

$$C_{Census}(P,d) = d_{Hamming}(Str_R(P), Str_L(P+d)) \quad (5)$$

where $d_{Hamming}$ represents the Hamming distance between two bit string (the number bits of different value at the same bit. P=(x,y) represents the matching pixel coordinate in the reference view, d is the disparity value, Str represents the bit string gained from modified ways based on the foundation of traditional census transform:

$$Str(P) = \underset{[i,j]\in N}{\otimes} \xi(\bar{I}(P), I(P+[i,j])) \quad (6)$$

$\bar{I}$(P) is the average grayscale in the census transform window N(P), and

$$\xi(I,I') = \begin{cases} 1 & I' < I \\ 0 & \text{otherwise} \end{cases}, \otimes \text{ means bit connection}$$

And Figure.3 indicates the single point match cost calculation course based on real experiment data, and the census window is rectangle window with a size of 9×7

The current search point on left view

45	46	45	45	47	46	46	46	46
45	44	45	45	46	48	46	46	48
44	45	44	45	46	45	46	46	45
45	43	44	46	48	47	44	45	44
45	44	43	46	48	46	46	46	46
46	44	47	47	46	46	46	46	46
46	45	45	46	44	46	45	46	49

The point required matching on right view (Census neighboring area)

36	42	42	42	43	46	46	47	45
44	43	45	44	44	49	46	49	47
42	46	41	42	47	47	47	48	49
43	42	44	48	47	49	47	47	49
41	45	46	45	46	44	49	48	46
43	44	44	48	43	44	49	48	49
41	46	48	46	49	48	49	47	48

Modified census transform

1	0	1	0	0	0	0	0	0
1	1	1	1	0	0	0	0	0
1	1	1	1	0	1	0	0	1
1	1	1	0		0	1	1	1
1	1	1	0	0	0	0	0	0
0	1	0	0	0	0	0	0	0
0	1	1	0	1	0	1	0	0

1	1	1	1	1	0	0	0	1
1	1	1	1	1	0	0	0	0
1	0	1	1	0	0	0	0	0
1	1	1	0		0	0	0	0
1	1	0	1	0	1	1	1	1
1	1	1	0	1	1	0	0	0
1	0	0	0	0	0	0	0	0

Str_L=101000000_111100000_111101001_1110X0111_111000000_010000000_011010100

Str_R=111110001_111110000_101100000_1110X0000_110101111_111011000_100000000

$$d_{Hamming}(Str_L, Str_R)=26$$

Figure.3 The bit string of census transform calculation diagram

2.3.2. Cost Aggregation

For accurate local stereo matching, it is important to decide an appropriate local support region for each pixel adaptively. In principle, this local support region should contain only the neighboring pixels from the same depth with the pixel under consideration. A common assumption is that pixels with similar intensity within a constrained area are likely from the same image structure, therefore having similar disparity [11].

In this paper we chose Cross-based aggregation window. The determine ways of the cross-based window shows in Figure.4 First, take the pixel required matching as start point to find the up and down boundry based on disparity similarity code. Then use determined point in the 1D vertical window as start point to search right and left directions in order to find the boundry.

Take the up boundry as an example, search every one pixel every step until the point does not satisfy this three disparity similarity code

$$\begin{cases} 1.D_c(P',P) < \tau_1 \text{ and } D_c(P',P'-(0,1))) < \tau_1 \\ 2.D_s(P',P) < L_1 \\ 3.D_c(P',P) < \tau_2 \text{ if } L_2 < D_s(P',P) < L_1 \end{cases}$$

Journal of Physics: Conference Series **679** (2016) 012047 doi:10.1088/1742-6596/679/1/012047

P′ represent current searching point, Dc(P′,P)=|I(P′)-I(P)| means the grayscale difference value between two point. τ1、τ2、L1 and L2 are constant. We make τ1、τ2、L1 and L2 are 20、6、10 and 5 on the basis of reference[11]

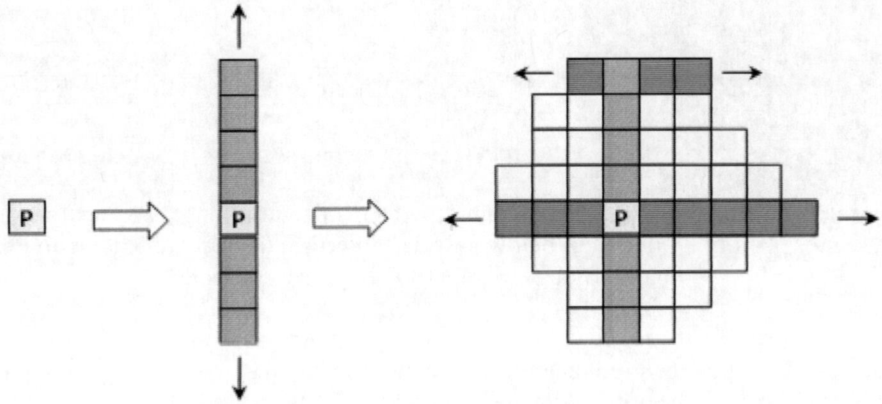

Figure.4 Cross-based aggregation determination method

2.3.3. Disparity Calculation

Now we use WTA (Winner-Takes-All) optimization thoughts for the most optimal operating on every point required matching and chose disparity of single matching point which has the smallest match cost. If we know the TOF depth maps disparity image, we can get the disparity constraint range.

D1= BF * Disp / (BF + DETA * Disp) - Offset + i;

D2 = BF * Disp / (BF - DETA * Disp) - Offset + i;

Where B represents the two camera's distance and F represents the camera focus, Disp represent the TOF disparity value and DETA represents the TOF camera precision. Offset means a rectified constant, and i represents any pixel.

3. Experiment Setup

Our hardware system is based on Altera Cyclone IV E and we can see some key parameters as shows in table.1 The actual working frequency is 100MHz, and the UART is used transport image data to the system. On account of enormous data, SDRAM is taken into consideration. Of course, Matlab is a useful tool to display the processed image.

Table.1 Basic FPGA system parameters used in this paper

Altera	EP4CE15F17C8	Cyclone IV E
CPU	RAM	504kbit
	LE	15408
	Max frequency	200MHz
	Actual frequency	100MHz
	M9K	Six modes
SDRAM	256M (16M*16bit)	Hynix SDRAM Memory
UART	115200bit/s	--
Sub-image Size	32*32 (on FPGA)	640*480 (on PC)

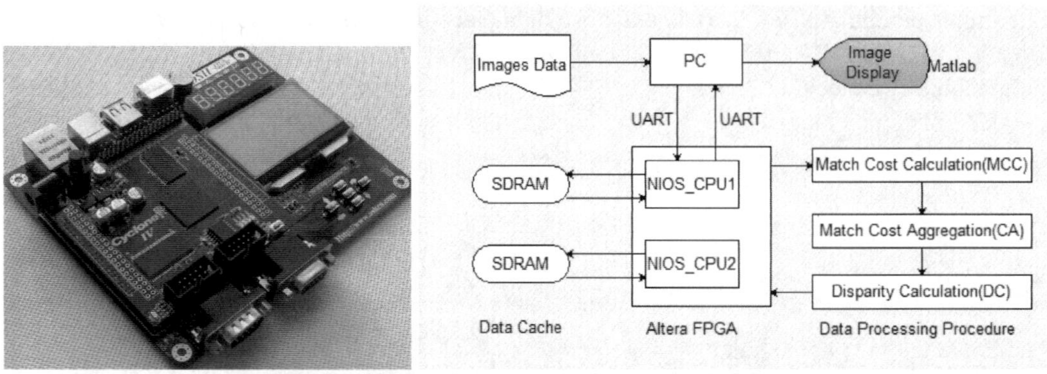

Figure.5 The new-designed FPGA-center platform. The left image: Physical prototype of our FPGA-center platform. On the left below a RS232 interface to communicate with PC. The right image: Block diagram of the system components

4. Results

In the experiment, images required matching were cut into small size sub-image, with a size of 40*30, 96*96, or 160*120, which the original image size is 498*672 as shows in Figure.6. And the stereo match results display in Figure.7. Since the sub-image is so small that we cannot judge the stereo effect macroscopically, we just need to know how long it take to finish the stereo match procedure. And we make a comparison between this system and on PC.

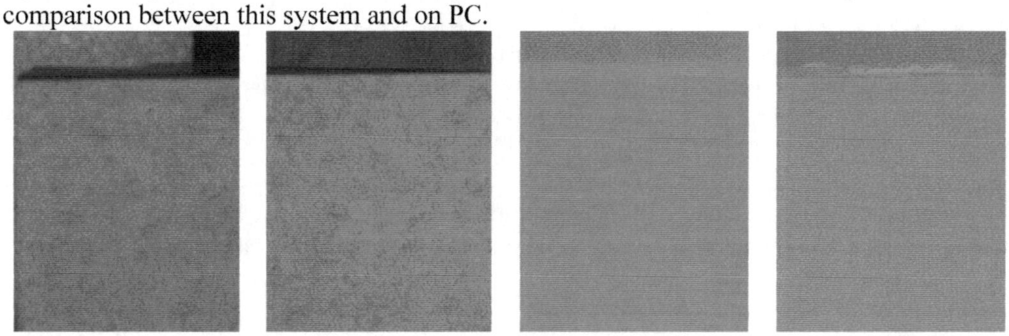

(a) (b) (c) (d)

Figure.6 Results on Sub-image size of 160*120 (a) the left view, (b) the right view, (c) the coarse TOF depth map (d) the disparity image. The (a)···(b) represent the same meaning in Figure.7 and Figure.8

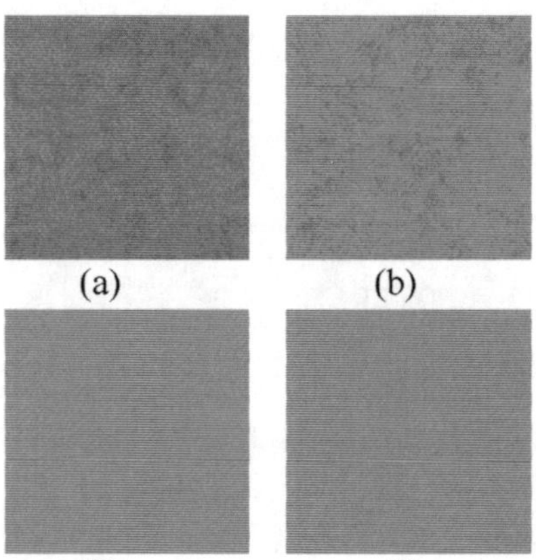

(a) (b)
(b) (d)

Figure.7 Results on Sub-image size of 96*96

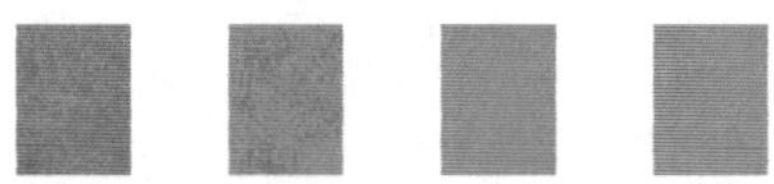

(a) (b) (c) (d)

Figure.8 Results on Sub-image size of 40*30

Match cost time		
Sub_image size	Altera FPGA NiosII	Personal Computer
	100Mhz	2.0GHz
40*30	2.1s	1.1s
96*96	5.3s	2.6s
160*120	9.6s	5.2s

Table.2 Comparisons of the performances on PC and FPGA system

From table.2 we can see, it took a little more time to deal a sub-image on our hardware system. This is of course right and reasonable because of the shortage parallel Nios II cpu. Only one cpu has been tested to calculate the sub-image data. However, take the sub-image 40*30 for an example, it spent about 2.1s to calculate, we could judge that it will spend 35.28s at least to deal with big image owning the size of 672*498, in comparison to at least 84 second on PC, which indicates that using the FPGA and Nios II system can do speed up the matching time.

5. Conclusions

Applying the combination of binocular vision and TOF camera, using FPGA to construct embedded match method, this paper accomplish a effective transplant which bring the same algorithm on PC to Nios II. Although, the simulation results demonstrate that it cannot speed up the process of stereo matching only using one nios cpu to undertake the hug image with the enormous calculation. Under the single Nios CPU, fast importing the Sub-image data divided from original image required matching to this system, we gain a evident speed up for matching. From the comparison between altera FPGA and PC, the speed can be fasten evidently and the effect will be fine. And this paper develop a method to quickly realize stereo match on mobile device, which also avoid complicated hardware programmer to construct algorithm unit.

Of course what's most efficient method is to construct the calculation module using verilog language and implement real parallel computing.

Acknowledgment
The authors would like to thank State Key Laboratory of Modern Optical Instrument and National Engineering and Technology Research Center for Optical Instrument (NERC), for funding this research project.

References

[1] Scharstein, Daniel, and Richard Szeliski. "A taxonomy and evaluation of dense two-frame stereo correspondence algorithms." International journal of computer vision 47.1-3 (2002): 7-42.
[2] Kostkova, J., and R. Sara. "Automatic disparity search range estimation for stereo pairs of unknown scenes." Proc. CVWW (2004): 1-10.
[3] Cyganek, Bogusław, and Jan Borgosz. "An improved variogram analysis of the maximum expected disparity in stereo images." Image Analysis. Springer Berlin Heidelberg, 2003. 640-645.
[4] Min, Dongbo, et al. "Disparity search range estimation: enforcing temporal consistency." Acoustics Speech and Signal Processing (ICASSP), 2010 IEEE International Conference on. IEEE, 2010.
[5] Li, Rueihung, et al. "Disparity search range estimation based on dense stereo matching." Industrial Electronics and Applications (ICIEA), 2013 8th IEEE Conference on. IEEE, 2013.
[6] Gandhi, Vineet, Jan Čech, and Radu Horaud. "High-resolution depth maps based on TOF-stereo fusion." Robotics and Automation (ICRA), 2012 IEEE International Conference on. IEEE, 2012.
[7]Wei Shaopeng, Yan Huiming, Zhang Xiuda. "A Stereo Matching Method Based on the Combination of Depth Camera and Stereo Vision" [J].Opto-Electronic Engineering, 2015.
[8]John Kalomiros and John Lygouras. "Comparative study of local SAD and dynamic programming for stereo processing using dedicated hardware." (2009).
[9]Michel Devy and Jean-louis Boizard. "Stereovision algorithm to be executed at 100hz on a FPGA-based architecture. In Advances in Theory and Applications of Stereo Vision". Asim Bhatti Ed., Chapter 17,2011
[10] Jin, Minxi, and Tsutomu Maruyama. "Fast and accurate stereo vision system on FPGA." ACM Transactions on Reconfigurable Technology and Systems (TRETS) 7.1 (2014): 3.

[11] Xiuda, Zhang, Yan Huimin, and Jiang Yanbing. "Pulse-shape-free method for long-range three-dimensional active imaging with high linear accuracy." Optics letters 33.11 (2008): 1219-1221.
[12] Zhang, Ke, Jiangbo Lu, and Gauthier Lafruit. "Cross-based local stereo matching using orthogonal integral images." Circuits and Systems for Video Technology, IEEE Transactions on 19.7 (2009): 1073-1079.

Journal of Physics: Conference Series **679** (2016) 012048 doi:10.1088/1742-6596/679/1/012048

Application of the Strong Tracking UKF in the Maneuvering Target Tracking

GUO Caifa, DAI Zhengxu, YANG Lei, WANG Xuliang

China Satellite Maritime Tracking and Controlling Department, Jiangyin, China

guo_caifa@sina.com

Abstract. Maneuvering target tracking is one of the key technologies in the space tracking system. The strong tracking unscented Kalman filter is presented for the data processing which uses a fading factor in the state estimation. Using the current statistical model, the strong tracking UKF is tested with the real data obtained in a mission measured by the space tracking ship. Simulation results show the good performance of the proposed method.

1. Introduction

The maneuvering target tracking has been an issue for a long time as the trajectory target has a high maneuverability and wide range of dynamics as well as a complex measurement environment [1,2]. For the space tracking ship, the tracking process is under dynamic conditions including swing, rotation, vibration etc, which the measurement environment is even worse than the other platform. Hence it's necessary to find an effective method to meet the requirements of the maneuvering target tracking.

With the development of the instrument and the data processing, more and more new methods are developed in the maneuvering target tracking [3]. The extended Kalman filter (EKF) is used widely in the state determination of the target. The problem of the EKF is the linearization of the nonlinear system, which may cause large error when the system has high nonlinearity, like the tracking model of the ship [4]. Unscented Kalman filter (UKF) uses the sigma points to approximate the target state instead of the linearization process. This character makes it more adaptive than EKF in dealing with the nonlinear estimates [5,6].

Based on the principle of the strong tracking, this paper presents an improved UKF to actualize the maneuvering target tracking, which is tested with the real tracking data in a mission measured by the space tracking ship.

2. The current statistical model

The current statistical (CS) model is a common tool for the tracking target modeling. In this paper, an improved CS model is used to build the tracking model, which is

$$\begin{cases} \ddot{x} = a_1(t) \\ \dot{a}(t) = -\alpha a_1(t) + \alpha \bar{a} + \upsilon(t) \end{cases} \tag{1}$$

$$X(K+1) = F(K)X(K) + G(K)\bar{a} + V(K) \tag{2}$$

Journal of Physics: Conference Series **679** (2016) 012048 doi:10.1088/1742-6596/679/1/012048

where $\bar{a}$ is the mean value of the acceleration $a_1(t)$, α is the reciprocal of the time constant for the acceleration, $\upsilon(t)$ is the Gaussian white noise, X_K is n dimensional state vector at time K, $G(K)$ is the input control matrix, $F(K)$ is given by

$$F = \begin{bmatrix} 1 & T & (\alpha T - 1 + e^{-\alpha T})/\alpha^2 \\ 0 & 1 & (1 - e^{-\alpha T})/\alpha \\ 0 & 0 & e^{-\alpha T} \end{bmatrix} \tag{3}$$

$V(K)$ is the dynamic noise matrix which the covariance is

$$Q(k) = E[V(k)V'(k)] = 2\alpha\sigma_\alpha^2 \begin{bmatrix} q_{11} & q_{12} & q_{13} \\ q_{12} & q_{22} & q_{23} \\ q_{13} & q_{23} & q_{33} \end{bmatrix} \tag{4}$$

where

$$\begin{cases} q_{11} = (1 - e^{-2\alpha T} + 2\alpha T + \dfrac{2\alpha^3 T^3}{3} - 2\alpha^2 T^2 - 4\alpha T e^{-\alpha T})/2\alpha^5 \\ q_{12} = (e^{-2\alpha T} + 1 - 2e^{-\alpha T} + 2\alpha T e^{-\alpha T} - 2\alpha T + \alpha^2 T^2)/2\alpha^4 \\ q_{13} = (1 - e^{-2\alpha T} - 2\alpha T e^{-\alpha T})/2\alpha^3 \\ q_{22} = (4e^{-\alpha T} - 3 - e^{-2\alpha T} + 2\alpha T)/2\alpha^3 \\ q_{23} = (e^{-2\alpha T} + 1 - 2e^{-\alpha T})/2\alpha^2 \\ q_{33} = (1 - e^{-2\alpha T})/2\alpha \end{cases} \tag{5}$$

$$\sigma_\alpha^2 = \begin{cases} \dfrac{4-\pi}{\pi}[a_{\max} - \bar{a}]^2, \bar{a} \ge 0 \\ \dfrac{4-\pi}{\pi}[\bar{a} - a_{\min}]^2, \bar{a} < 0 \end{cases} \tag{6}$$

As a parameter with no definite physical meaning, the main function of the parameter $\bar{a}$ is to determine the variance of the acceleration. The introduction of $\bar{a}$ into the state prediction described in Eq. 2 makes the model lose the character of exponential autocorrelation. Therefore, $X_{K+1} = F(K)X(K)$ is used in the state prediction of the filtering process instead of $X_{K+1} = F(K)X(K) + G(K)\bar{a}$, which can actualize the adaptive filtering of the mean value and variance [7].

3. The strong tracking UKF method

Using the CS model as the state function in the UKF, the measurement function can be written as

$$Z_K = H(X_K) + \upsilon_K \tag{7}$$

where Z_K is m dimensional measurement vector at time K, H is measurement function, υ_K is the measurement noise.

In order to improve the performance of the nonlinear filtering, Julier proposed the UKF based on the unscented transformation. In this paper, the strong tracking filter is integrated with the UKF to decrease the impact brought by the high maneuverability and the complex environment. The strong filtering has three advantages, the first is its robustness for the model's uncertainty, the second is the strong tracking ability while the jumping phenomenon happened, the last is the moderate calculation complexity. The major difference of the UKF and the strong tracking UKF is the introduction of the fading actor. The filtering process includes 3 steps.

1) The sigma points calculation

$$w_0^{(m)} = \lambda/(n+\lambda) \tag{8}$$

$$w_0^{(c)} = \lambda/(n+\lambda) + (1 - \alpha^2 + \beta) \tag{9}$$

$$w_i^{(m)} = w_i^{(c)} = 0.5/(n+\lambda),\ i = 1,\cdots,2n \tag{10}$$

$$X_K^{(0)} = \hat{X}_K \tag{11}$$

$$X_K^{(i)} = \hat{X}_K + \left(\sqrt{(n+\lambda)P_K}\right)_i, \ i = 1,\cdots,n \tag{12}$$

$$X_K^{(i)} = \hat{X}_K - \left(\sqrt{(n+\lambda)P_K}\right)_i, \ i = n+1,\cdots,2n \tag{13}$$

2) The state prediction and update

$$X_{K+1}^{(i)} = F(K)X_K^{(i)} \tag{14}$$

$$r_{K+1}^{(i)} = H(X_K^{(i)}) \tag{15}$$

$$\hat{X}_{K+1|K} = \sum_{i=0}^{2n} w_i^{(m)} X_{K+1}^{(i)} \tag{16}$$

$$\hat{Z}_{K+1|K} = \sum_{i=0}^{2n} w_i^{(m)} r_{K+1}^{(i)} \tag{17}$$

$$P_{X,K+1} = \mu_{K+1} \sum_{i=0}^{2n} w_i^{(c)} (X_{K+1}^{(i)} - \hat{X}_{K+1|K})(X_{K+1}^{(i)} - \hat{X}_{K+1|K})^T + Q_{K+1} \tag{18}$$

3) The measurement update

$$P_{Z,K+1} = \mu_{K+1} \sum_{i=0}^{2n} w_i^{(c)} (r_{K+1}^{(i)} - \hat{Z}_{K+1|K})(r_{K+1}^{(i)} - \hat{Z}_{K+1|K})^T + R_{K+1} \tag{19}$$

$$P_{XZ,K+1} = \mu_{K+1} \sum_{i=0}^{2n} w_i^{(c)} (X_{K+1}^{(i)} - \hat{X}_{K+1|K})(r_{K+1}^{(i)} - \hat{Z}_{K+1|K})^T \tag{20}$$

$$\hat{X}_{K+1} = \hat{X}_{K+1|K} + K_{K+1}(Z_{K+1} - \hat{Z}_{K+1|K}) \tag{21}$$

$$K_{K+1} = P_{XZ,K+1} P_{Z,K+1}^{-1} \tag{22}$$

$$P_{K+1} = P_{X,K+1} - K_{K+1} P_{Z,K+1} K_{K+1}^T \tag{23}$$

where μ is the fading factor, which is given by

$$\begin{cases} \mu_{K+1} = \begin{cases} \mu_0, \mu_0 \geq 1 \\ 1, \mu_0 \leq 1 \end{cases} \\ \mu_0 = \dfrac{\mathrm{tr}[N_{K+1}]}{\mathrm{tr}[M_{K+1}]} \end{cases} \tag{24}$$

where

$$N_{K+1} = V_{K+1} - (P_{XZ,K+1}^a)^{\mathrm{T}} (P_{K+1/K}^a)^{-1} Q_{K+1} (P_{K+1/K}^a)^{-1} P_{XZ,K+1}^a - R_{K+1} \tag{25}$$

$$M_{K+1} = H_K (P_{K+1/K}^a - Q_{K+1}) H_K^{\mathrm{T}} = H_K P_{K+1/K}^a H_K^{\mathrm{T}} - H_K Q_{K+1} H_K^{\mathrm{T}} = P_{Z,K+1}^a - V_{K+1} + R_{K+1} + N_{K+1} \tag{26}$$

In Eq. 25 and 26, the matrix with superscript a means the one in the typical UKF without the introduction of μ, which is given by

$$P_{K+1/K}^a = E[(X_{K+1} - \hat{X}_{K+1/K})(X_{K+1} - \hat{X}_{K+1/K})^{\mathrm{T}}] \tag{27}$$

$$P_{Z,K+1}^a = H_K P_{K+1,K}^a H_K^{\mathrm{T}} \tag{28}$$

$$P_{XZ,K+1}^a = P_{K+1,K}^a H_K^{\mathrm{T}} \tag{29}$$

4. Simulation test

With the measurement information, simulation test is implemented with the strong tracking UKF and the typical UKF separately. As it's hard to estimate the acceleration with the space tracking ship, the filtering results of the position and velocity are selected to evalute the performance of the proposed method in this paper.

4.1. State equation

The proposed method is tested with the real data from the space tracking ship. In the filtering, the state vector is a 9 dimensional vector including the position, velocity and the acceleration component in the 3 axes of the geocentric coordinate system. From the Eq. 3 and 14, we have

$$F(k+1)=\begin{bmatrix} F_1 & 0 & 0 \\ 0 & F_2 & 0 \\ 0 & 0 & F_3 \end{bmatrix} \tag{30}$$

where

$$F_i=\begin{bmatrix} 1 & T & (\alpha T-1+e^{-\alpha T})/\alpha^2 \\ 0 & 1 & (1-e^{-\alpha T})/\alpha \\ 0 & 0 & e^{-\alpha T} \end{bmatrix}, \alpha=1 \tag{31}$$

In the Eq. 6, the parameters are set as $[a_{x\min,}a_{x\max,}a_{y\min,}a_{y\max,}a_{z\min,}a_{z\max}]=[-10,10,-10,10,-10,10]$.

4.2. Measurement equation

The measurement information is the range R, azimuth A and elevation E which can be obtained directly by the ship-borne devices. The measurement equations are

$$\begin{aligned} R &= \sqrt{x^2+y^2+z^2} \\ E &= \arcsin(y/R) \\ A &= \arctan(z/x)+\begin{cases} 0, if\ x>0, z\ge 0; \\ \pi, if\ x<0; \\ 2\pi, if\ x>0, z<0; \end{cases} \end{aligned} \tag{32}$$

The initial state X_0 is set as the 20th point of the reference orbit from a ground tracking station which can be considered as the true value of the target state. The initial covariance is given by $P_0=diag(1.0\times10^9,1.0\times10^7,1.0\times10^6,1.0\times10^9,\ 1.0\times10^7,1.0\times10^6,1.0\times10^9,1.0\times10^7,1.0\times10^6)$. The unscented transformation parameters are assumed as $\alpha=0.5, \beta=2,\ \kappa=-6$. Using the strong tracking UKF and the typical UKF separately, the test results are shown in Figure 1 and 2.

It's easily to see that the strong tracking UKF method has better performance than the typical UKF. For the position estimation, both of the two methods have good results. In detail, the strong tracking UKF method has better quality. For the velocity estimation, the advantage of the strong tracking UKF is significant. With the introduction of the fading factor μ, the improved UKF method has stronger tracking ability while the jumping phenomenon happened.

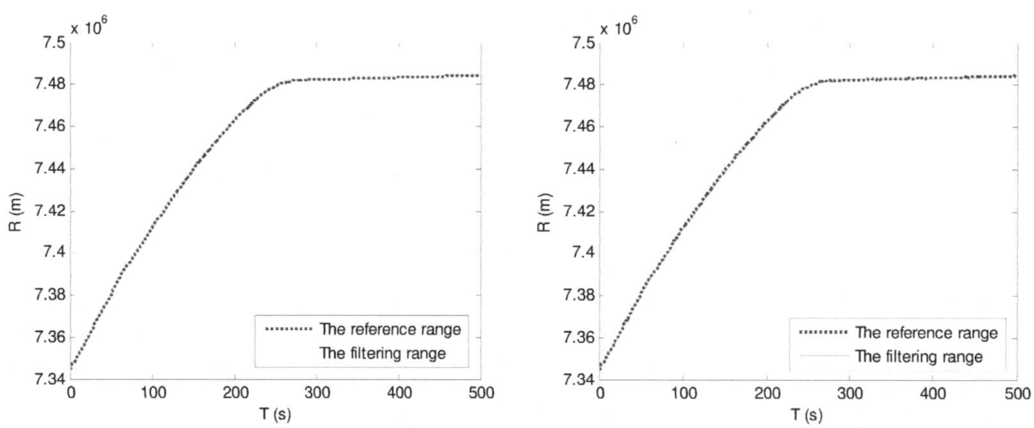

Figure 1. The comparison analysis between the reference range and the filtering result (a) the strong tracking UKF method (b) the typical UKF

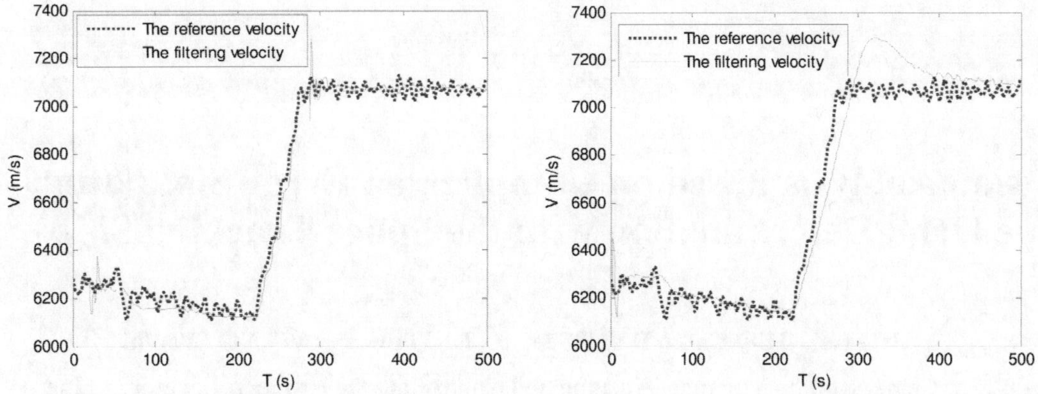

Figure 2. The comparison analysis between the reference velocity and the filtering result (a) the strong tracking UKF method (b) the typical UKF

5. Conclusion

The strong tracking UKF method is tested with the real data in this paper. Comparing with the typical UKF, the strong tracking UKF has better performance especially in the estimation of the velocity. This work can be applied in the maneuvering target tracking with the space tracking ship.

References

[1] Jin C B, Du J J and Bao J 2013 Maneuvering target tracking by adaptive statistics model *J. China Universities of Posts and Telecommunications.* **20**, 108-114

[2] Liu Y 2011 *The robust and high accurate fusion filter for trajectory target real time tracking* (The graduate school of National University of Defense Technology) pp 1-16

[3] Vasuhi S, Vaidehi V and Rincy T 2011 *IMM estimator for maneuvering target tracking with improved current statistical model* Pro. the International Conference on Recent Trends in Information Technology (ICRTIT'11). Chennai, India.

[4] Choi E J, Yoon J C, Lee B S, Park S Y and Choi K H 2010 Onboard orbit determination using GPS observations based on the unscented Kalman filter *Advances in Space Research.* **46** 1440-1450

[5] Cheng H B, Ni S H, Huang G R, Liu H W and Jiang Z Y 2011 Application of UKF strong Tracking filter in fault detection of integrated navigation system *Piezoelectrics & Acoustooptics.* **33** 901-905 (in Chinese)

[6] Julier J S and Uhlmann K J 2004 *Unscented filtering and nonlinear estimation* Pro. the IEEE. Jefferson City, USA

[7] Liu B G, Tao Q C and Pan M H 2012 An improved tracking algorithm based on current statistical model for highly maneuvering targets *Radar & ECM.* **32** 28-30 (in Chinese)

Journal of Physics: Conference Series **679** (2016) 012049 doi:10.1088/1742-6596/679/1/012049

Precision Analysis Based on Complicated Error Simulation for the Orbit Determination with the Space Tracking Ship

YANG Lei, GUO Caifa, DAI Zhengxu, LI Xiaoyong, WANG Shaolin

China Satellite Maritime Tracking and Controlling Department, Jiangyin, China

yanglei831110@163.com

Abstract. The space tracking ship is a moving platform in the TT&C network. The orbit determination precision of the ship plays a key role in the TT&C mission. Based on the measuring data obtained by the ship-borne equipments, the paper presents the mathematic models of the complicated error from the space tracking ship, which can separate the random error and the correction residual error with secondary low frequency from the complicated error. An error simulation algorithm is proposed to analyze the orbit determination precision based on the two set of the different equipments. With this algorithm, a group of complicated error can be simulated from a measured sample. The simulated error groups can meet the requirements of sufficient complicated error for the equipment tests before the mission execution, which is helpful to the practical application.

1. Introduction

The space tracking ship is a very important part of the TT&C system in China, especially when the spacecraft is far away from our airspace. With the continuous improvement of the precision requirement of the orbit determination, the development of the measurement technology of the space tracking ship has also been greatly improved.

In the TT&C mission, both of the two set of the ship-borne equipments can track the target and accomplish the orbit determination. In order to choose the better result, it's necessary to do the precision analysis before the mission execution. The key point is the simulation for the complicated error of the equipments.

The space tracking ship is a moving platform, which the tracking process is under dynamic conditions including swing, rotation, vibration etc. Therefore, the target tracking data needs to be corrected with the corresponding ship's position and attitude. Except the system error and the random error, the tracking data includes the residual error generated by the correction process of ship-pose, hull's deformation and wave refraction, which makes the main difference between the results measured by the ship and the ground stations.

Based on the analysis of the error character, the simulation of the complicated error is presented in this paper with spline function, wavelet technology and time series modeling technique. Furthermore, the comparison analysis between the two set of equipments is studied for the orbit determination precision, which can offer the reasonable reference for the result selection in the mission.

The remainder of this paper consists of three sections plus conclusions. Section 2 introduces the orbit determination method of the space tracking ship. The simulation method of the complicated error is presented in section 3. Section 4 discusses the precision comparison analysis of the two different equipments. Conclusions are resumed in section 5.

2. The Orbit Determination Method

Using the ship's location (λ, B, H) and the orbit elements $(a_0, e_0, i_0, \Omega_0, \omega_0, M_0)$ at one time t_f, the orbit determination can be accomplished based on the following dynamic models.

The orbit elements described in 2000.0 geocentric celestial coordinate can be obtained by orbit propagation when the satellite state parameters are given. In order to get the initial simulated measured data of (R_0, E_0, A_0), the results need to be transferred into the inertial system of the space tracking ship firstly. Then the measured data of the two sets of the equipment (R_a, E_a, A_a) and (R_b, E_b, A_b) can be obtained by adding the corresponding complicated error to the initial (R_0, E_0, A_0). Using the measured data, the orbit elements $(a_a, e_a, i_a, \Omega_a, \omega_a, M_a)$ and $(a_b, e_b, i_b, \Omega_b, \omega_b, M_b)$ can be calculated with the Unit Vector Method (UVM). Hence it's easily to do the comparison analysis of the orbit determination. The flow chart is shown in Figure 1.

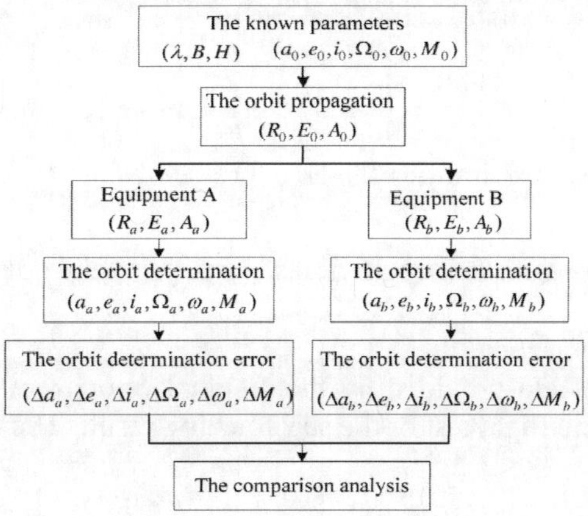

Figure 1. The flow chart of the orbit determination analysis

3. The Simulation Method for Complicated Error

3.1. The modeling for the complicated error

The complicated errors of the tracking data obtained by the ship include the system error and random error. The random error is one kind of the high frequency error which can be described by the auto-regressive model. The system error can be divided into two categories, one can be modeled in detail, such as the constant error, time-delay error and dynamic lag error. The other is with some uncertainty which complicates the tracking error, like the residual error generated by the correction of the hull's deformation and the ship-swaying.

The uncertain system error is one kind of low frequency errors which is hard to describe with Monte-Carlo method. In this paper, spline function is used to separate the uncertain system error from the tracking data. The separate result always comes with the random error as it has some relativity with the uncertain system error. The two kinds of errors can be identified through the Wavelet frequency-division process. Wavelet frequency -division means to decompose the separate results into several different frequency sections by using wavelet transform (WT). The window function of the WT cuts the data into different sections and then studies the specific frequency component of interest. The data processing includes 3 steps which are WT, nonlinear processing of the different components and wavelet reconstruction. The principle of the process is to keep the low frequency component and make wavelet reconstruction.

The complicated error $z(t)$ is given by

$$z(t) = s(t) + c(t) + e(t) \tag{1}$$

where $s(t)+c(t)$ is the system error which needs processing with the spline function and wavelet frequency-division. $s(t)$ can be approximated by spline function with sparse equidistant nodes where $c(t)$ is approximated by spline function with dense equidistant nodes instead. $e(t)$ is the random error which can be approximated by the auto-regression model.

In this paper, the cubic B-spline is selected, which is given by

$$s(t) = \sum_{j=-1}^{m+1} a_j M_4\left(\frac{t-T_j}{h}\right) \tag{2}$$

where a_j are the model coefficients which can be estimated by the least square method, M_4 is the basis function, which has the form of

$$M_4(t) = \begin{cases} 0, |t| \geq 2 \\ \dfrac{|t|^3}{2} - t^2 + \dfrac{2}{3}, |t| < 1 \\ -\dfrac{|t|^3}{6} + t^2 - 2|t| + \dfrac{4}{3}, 1 \leq |t| < 2 \end{cases} \tag{3}$$

The auto-regression model of p th-order can be expressed by

$$e_t = \varphi_1 e_{t-1} + \ldots + \varphi_p e_{t-p} + \varepsilon_t \tag{4}$$

where $\varphi_1, \varphi_2, \ldots, \varphi_p$ are the auto-regression coefficients, ε_t is the white noise.

3.2. The complicated error simulation

The complicated error samples are provided by the residual error from the intercomparison analysis between the precise orbit and the flight test or the star tracking result. The error fitting model is

$$\tilde{z}(t) = \tilde{s}(t) + \tilde{c}(t) + \tilde{e}(t) = \sum_{j=-1}^{m+1} \hat{a}_j M_4\left(\frac{t-T_j^s}{h^s}\right) + \sum_{j=-1}^{n+1} \hat{b}_j M_4\left(\frac{t-T_j^c}{h^c}\right) + \hat{\varphi}_1 e_{t-1} + \ldots + \hat{\varphi}_p e_{t-p} + \varepsilon_t \tag{5}$$

where $\hat{a}_{-1}, \hat{a}_0, \cdots, \hat{a}_{m+1}$ can be considered as the sample data of the random variable ω. Assuming $\omega \sim U(t_1, t_2)$ where U means uniform distribution, t_1 and t_2 can be estimated with the maximum likelihood function, which are $\hat{t}_1 = \min_j a_j$ and $\hat{t}_2 = \max_j a_j$. In the simulation process of $s(t)$, $\hat{a}_{-1}, \hat{a}_0, \cdots, \hat{a}_{m+1}$ are given by the sampling from $\omega \sim U(t_1, t_2)$. Consequently, the complicated error simulation can be completed by the following steps:

Step 1: establish the fitting model $\tilde{z}(t)$ after the spline fitting and wavelet frequency-division of the sampled error data.

Step 2: calculate the parameters $\hat{a}_{-1}, \hat{a}_0, \cdots, \hat{a}_{m+1}$, $\hat{b}_{-1}, \hat{b}_0, \cdots, \hat{b}_{n+1}$ and $\hat{\varphi}_1, \ldots, \hat{\varphi}_p$.

Step 3: calculate $\hat{t}_1$ and $\hat{t}_2$, and get $a_{-1}, a_0, \cdots, a_{m+1}$ by sampling from $\omega \sim U(t_1, t_2)$.

Step 4: simulate $s(t)$ by using the equation $s(t) = \sum_{j=-1}^{m+1} a_j M_4\left(\frac{t-T_j}{h}\right)$.

Step 5: simulate $c(t)$ which the principle is the same as the simulation of $s(t)$.

Step 6: simulate $e(t)$ with $\hat{\varphi}_1, \ldots, \hat{\varphi}_p$ based on the auto-regression model of p th-order.

Step 7: Finish the complicated error simulation with the equation (1).

4. The Simulation Method for Complicated Error

Journal of Physics: Conference Series **679** (2016) 012049 doi:10.1088/1742-6596/679/1/012049

Taking the measured data from the equipment A in a mission as the test data, the real error and the simulated error is shown in figure 1.

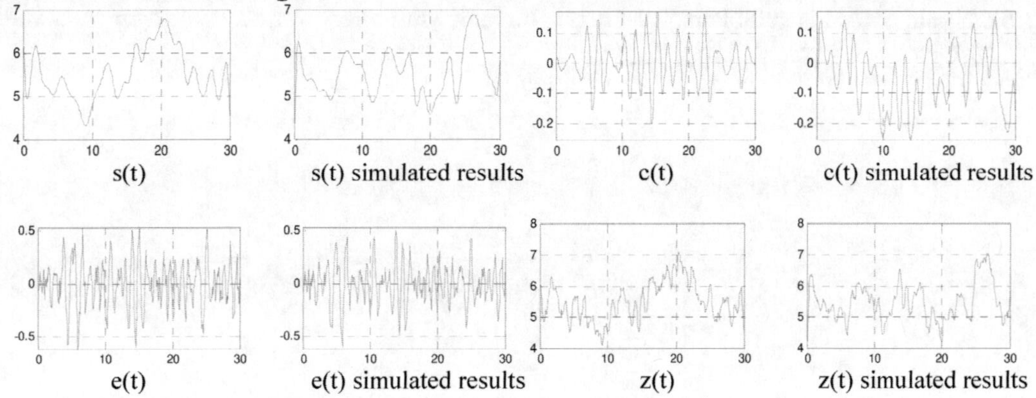

Figure 2. The real error and the simulated results

According the calculating process showed in Figure 1, the precision analysis is carried out with the residual error sample from the equipment A and B in a ship's mission. The results are presented in Table 1.

Table 1. The Precision Analysis Comparison Between Equipment A and B

	Equipment A		Equipment B	
$\triangle a/(m)$	363.33	-330.31	-271.99	-292.56
$\triangle e$	8.17e-5	-8.29e-5	-4.24 e-5	-4.27 e-5
$\triangle i/(°)$	0.0443252	0.045014	0.0463275	0.0466115
$\triangle \Omega_G/(°)$	-0.077497	-0.078611	-0.0703	-0.070932
$\triangle \lambda/(°)$	0.093879	0.098608	0.057	0.056
$\triangle R/(m)$	2833.57	2886.68	2014.44	2036.25
$\triangle V/(m/s)$	7.9	7.8	6.4	6.5

5. Conclusion

The ship-swaying and hull's deformation make the tracking error of the space tracking ship much more complicated than the one of the ground station. In the near future, the space tracking ship still has the irreplaceable advantages in our TT&C system. Hence it's helpful to the engineering application with the complicated error simulation. In this paper, the errors of the different frequency are separated successfully with the spline function, wavelet frequency-division and auto-regression model. Based on this, the complicated error is simulated by modeling of the different error components. The orbit determination simulation shows the method's validity, which could be used for equipments' evaluation in the mission preparation phase.

References

[1] YANG L, WU M D, ZHANG Z H and LI X Y 2008 A Error Separation and Analysis of Instrumentation Ship-Measured Data *J. Spacecraft TT&C Technol.* **27** 51-55 (in Chinese)

[2] HU S L, XU A H and GUO X H 2007 *Data processing technology applied to pulse radar tracking* (Beijing: National Defense Industry Press) (in Chinese)

[3] ZHAO G, ZHOU X H and WU B 2013 Precise orbit determination of Haiyang-2 using satellite laser ranging *Chinese Science Bulletin.* **58** 589-597

[4] ZHANG H Q, LI Y J and ZHANG K 2011 A novel method of periodic orbit computation in circular restricted three-body problem *Science China-Technological Sciences.* **54** 2197-2203

[5] YANG Y, LI Z G, YANG X H, FENG C G and CAO F 2012 Satellite orbit determination by transfer with differenced ranges *Chinese Science Bulletin.* **57** 4701-4706
[6] LI Z Y and SHI X L 2002 Parseval Frame Wavelet Multipliers in L~2(R~d) *Chinese Annals of Mathematics-Series B.* **33B** 949-960
[7] Young M 1989 *The Technical Writer's Handbook.* (CA: University Science)

Measurement of Nonlinear Parameter of an Optical Fiber by Characterizing of Self-phase Modulation based on BOTDA

Hongying Zhang[1], Shaopeng Zhang[1], Zhijun Yuan[1], Ziye Liu[1], Wei Gao[1], and Yongkang Dong [2,*]

1 Institute of Photonics and Optical Fiber Technology, Harbin University of Science and Technology, Harbin, China 150080, zhy_hit@163.com
2 National Key Laboratory of Science and Technology on Tunable Laser, Harbin Institute of Technology, Harbin, China 150080
*aldendong@gmail.com

Abstract. We propose a novel method to measure the nonlinear parameter of an optical fiber by characterizing the evolution of self-phase modulation (SPM) based on Brillouin optical time-domain analysis (BOTDA). The nonlinear parameter of an optical fiber is measured to be 2.03 W-1km-1, which agrees well with previously reported result.

1. Introduction

Due to the long interaction length and the confinement of light in the small core over the entire fiber, the efficiency of the nonlinear processes can be tremendous enhanced in a single-mode fiber, which makes an optical fiber a suitable nonlinear medium to observe a wide variety of nonlinear effects at relative low power levels [1]. The growing applications of the nonlinear effects especially in optical communication drive the interest in the measurement of nonlinear parameter of an optical fiber. Several different nonlinear effects have been proposed to measure the nonlinear parameter, including SPM, XPM, FWM and modulation instability. Among these methods, the SPM was widely used due to its simple structure, where the optical spectrum of a pulse at the fiber output exhibits peaks at multiples because of the SPM-induced phase modulation and the ratio of the peak heights depends only on the nonlinear phase shift and can be used to deduce the nonlinear parameter. However, ps-order or fs-order pulses from a mode-locked Ti:sapphire laser are usually used as the light source and a grating-based optical spectrum analyzer is used to record the SPM-induced broadened spectrum, which makes the measurement system complicated and expensive.

Stimulated Brillouin scattering in optical fibers has been extensively studied and found applications in many fields, such as light storage, slow light, all-optical signal processing, microwave photonics, and optical fiber sensing. The dependence of the Brillouin frequency shift on temperature and strain shows a linear relationship, which has been utilized to develop a distributed optical fiber sensor [2]. BOTDA scheme is widely used for distributed temperature and strain sensing, where the Brillouin spectra over the sensing fiber can be obtained by sweeping the frequency offset between a pump pulse and a CW probe (Stokes wave). The SPM-induced spectral broadening of the pump pulse has a considerably detrimental effect in the measurement for a long-range sensing in a BOTDA system [3], so that the pump pulse power should be carefully controlled to avoid the SPM. In this paper, we propose a method to use an ns-order pulse for measuring the nonlinear parameter of an optical fiber by characterizing the evolution of SPM based on BOTDA, where the Brillouin spectrum is the convolution of the intrinsic Brillouin gain spectrum and the pulse spectrum, subsequently reflecting

Journal of Physics: Conference Series **679** (2016) 012050 doi:10.1088/1742-6596/679/1/012050

the pulse spectrum broadening induced by the SPM. This method features a simple setup by using an ns-order pulse instead of ps-order or fs-order pulses from a complicated mode-locked Ti:sapphire laser.

2. Experimental setup

BOTDA system consist of a pump pulse and a CW probe, where the Brillouin spectra over the fiber can be obtained by the scanning the frequency offset between the pump and probe waves in the vicinity of the Brillouin frequency shift. The experimental setup is shown in Fig. 1. A narrow linewidth (3 kHz) fiber laser operating at 1550 nm is used to provide the light source and its output is divided into two parts to provide the pump and probe waves. A microwave source is applied on a Mach-Zehnder electro-optic modulator performing a carrier-suppressed two sidebands modulation, where the frequency difference between the carrier and sidebands equals to the microwave frequency. A narrow-band FBG with a 3-dB bandwidth of 0.03 nm is used to extract the first-order lower sideband working as the probe wave. The frequency offset between the pump and probe waves can be scanned by changing the microwave frequency. The other part of the laser is launched into a high extinction-ratio (ER) electro-optic modulator (EOM) to create a pump pulse with the ER higher than 45 dB. Before launched into the fiber under test, the pump pulse is amplified by an erbium-doped fiber amplifier (EDFA). A polarization scrambler is used to continuously change the polarization state of the probe wave to reduce the polarization-dependent fluctuation on the signal by averaging a large number of signals, where an averaging number of 2000 is used in our experiment.

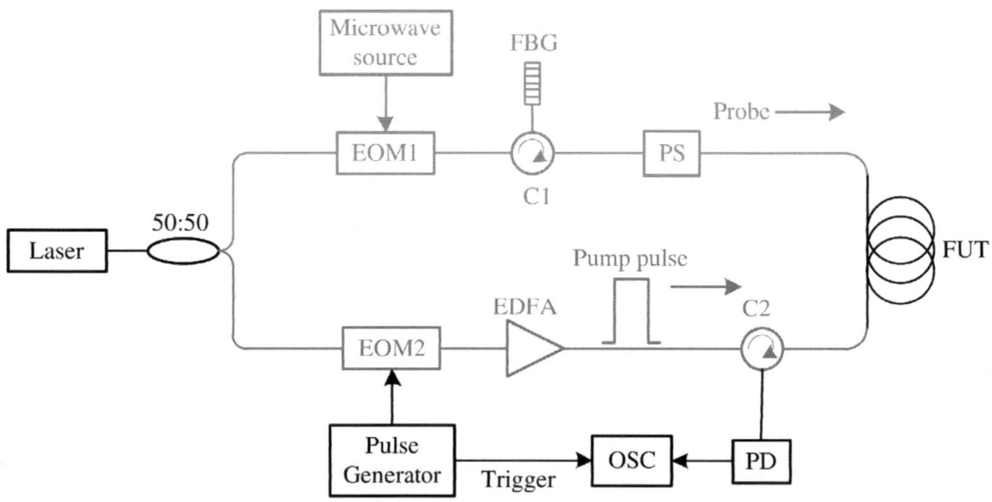

Figure 1. Experimental setup. EOM: electro-optic modulator, FBG: fiber Bragg grating, PS: polarization scrambler, C: circulator, PD: photo-detector, FUT: fiber under test, EDFA: Erbium-doped optical fiber amplifier, OSC: oscilloscope.

3. Results and discussions

A 9.72-ns Gaussian pulse with a peak power of 2.28 W was used as the pump pulse. A 2-km standard single-mode fiber was used as the fiber under test and the frequency offset between the pump and probe was scanned from 10.45 to 11.16 GHz and the measured 3-D Brillouin spectra over the entire fiber is plotted in Fig. 2. The evolution of the Brillouin spectra over the distance clearly shows the SPM-induced pulse spectrum broadening and the generation of multi-peak spectra.

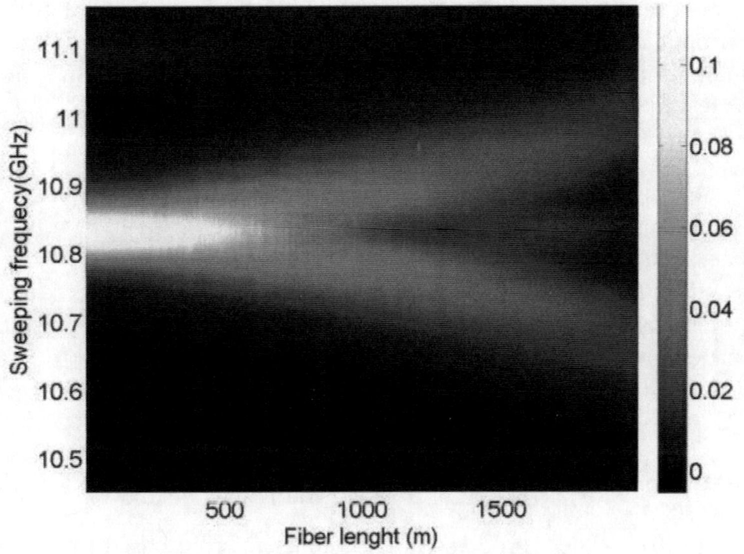

Figure 2. The evolution of Brillouin spectra over 2-km fiber with the SPM.

The simulated Brillouin spectrum under SPM with the nonlinear phase shift of 1.5 π is plotted in Fig. 3, where the measured 9.72-ns pulse shape is chosen as the pump pulse in the simulation. The simulated Brillouin spectra exhibit asymmetric, which is caused by the slightly rising edges deviation from the ideal Gauss curves. Through comparison, it is found that the measured Brillouin spectrum at the position of 1056.8 m in Fig. 2 matched very well with the simulations for the case of 1.5 π . Using a fiber loss of α=0.2dB/km and ignoring the Brillouin loss, the effective length is calculated to be 1031.5 m. Considering the peak power of 2.28 W and according to $\phi_{\max} = \gamma P_0 L_{eff}$, the nonlinear parameter is calculated to be 2.03 $W^{-1}km^{-1}$, which agrees very well with previous reported results[4].

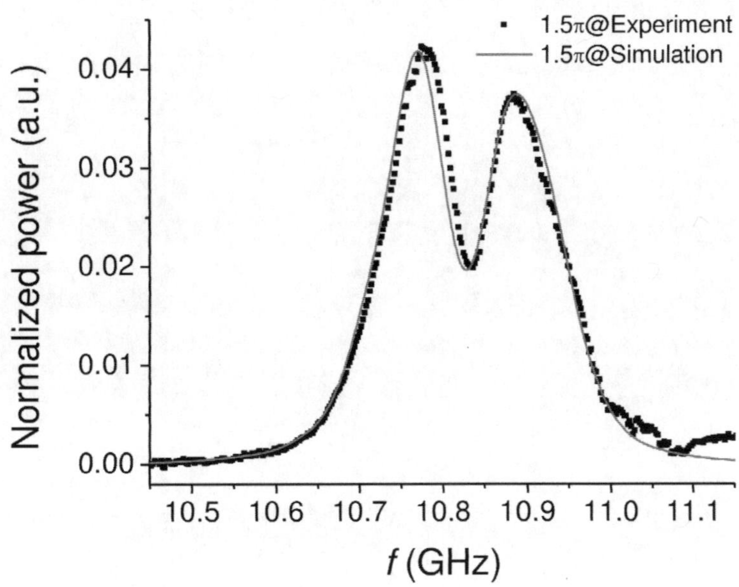

Figure 3. Comparison of measured and simulated Brillouin spectra the nonlinear phase shift of 1.5 π .

4. Conclusions

To summarize, we propose to use BOTDA to characterize the evolution of SPM of an ns-order Gaussian pulse in an optical fiber, where the nonlinear parameter of the fiber can be measured by comparing the Brillouin spectra at a specific phase shift. In experiment, a 2-km standard single-mode fiber was used as the test fiber, and a 9.72-ns Gaussian pump pulse with a peak power of 2.28 W was used as the pump pulse; the nonlinear parameter is measured to be 2.03 W-1km-1 through comparing the Brillouin spectra.

Acknowledgement

National Natural Science Foundation of China (NSFC) (61308004, 61205073, 61378003); Natural Science Foundation of Heilongjiang Province (QC2015087); Scientific Research Fund of Heilongjiang Provincial Education Department (12531093).

References

[1] Agrawal G P 2011 Nonlinear fiber optics: its history and recent progress J. Opt. Soc. Am. B 28(12), p A1–A10
[2] Bao X and Chen L 2011 Recent progress in Brillouin scattering based fiber sensors Sensors 11, p4152-4187
[3] Foaleng S M Barrios F R Lopez S M Herraez M G and Thevenza L 2011 Detrimental effect of self-phase modulation on the performance of Brillouin distributed fiber sensors Opt. Lett. 36, p97-99
[4] Agrawal G P 2006 *Nonlinear Fiber Optics* 4th ed. (New York: Academic Press)

Journal of Physics: Conference Series **679** (2016) 012051 doi:10.1088/1742-6596/679/1/012051

Daily Variation Analysis of Atmospheric Turbulence from Inland to Open Sea

S Y Shao[1], X B Li[1],Y J Li[2],W Y Zhu[1],D Y Kang[2], C Y Fan[1] and N Q Weng[1]

[1]Key Laboratory of Atmospheric Composition and Optical Radiation, Anhui Institute of Optics and Fine Mechanics, Chinese Academy of Sciences, Hefei , Anhui 230031, China
[2]China Satellite Maritime Tracking and Control Department, Jiangyin, Jiangsu 214400,China
Email: shaoshiyong@aiofm.ac.cn

Abstract. Random fluctuation of turbulence brings random fluctuation of refractive index, which makes atmosphere become a random fluctuation medium and destroys the coherence of light wave especially laser transferring in it. Exploration of atmospheric turbulence is essentially investigation of atmospheric refractive index. The atmospheric structure constant of refractive index is a basic parameter of expressing atmospheric turbulence, and was measured using HTP-2 micro-thermal meter at different areas from inland to open sea. It is analysed that the relation of atmospheric structure constant of refractive index with corresponding temperature and wind speed. The conclusion of turbulence and main influencing factors is to deepen the research in atmospheric optical transmission, and to provide data support for the siting of ship board photoelectric systems.

1. Introduction

With the rapid development of modern advanced photoelectric technology, the capability of their various components, which restricts the overall performance of photoelectric system, has been close to the ideal limit level. The overall performance of different photoelectric systems worked in atmospheric envirionment lies on atmospheric medium to a considerable degree, and atmospheric turbulence is one of the important factors.

Atmosphere is always in turbulence motion, the reason is that wind shear caused by gas flow drag due to earth surface, heating difference at different areas irraidated by sun, or heat convection induced by radiation from earth surface. Turbulence structure and corresponding physical parameters random change at various spatial-temporal dimensions. Enegy, momentum and meterials in atmosphere are transported by turbulence, whose transmission speed is several orders than molecular thermal motion. The transform of optical wave in atmosphere is seriously affected by turbulence[1,2]. The understanding of turbulence developing law and its main influencing factors is to deepen the research of atmospheric optical transmission.

2.Measureing principle and technique

Small size turbulence motion is emphases while its effect is assessed for transform of light wave

Journal of Physics: Conference Series **679** (2016) 012051 doi:10.1088/1742-6596/679/1/012051

and electromagnetic wave in atmosphere. Kolmogorov introduced structure function to explore statistic structure of turbulence which accorded with local uniform and isotropy assumption[3-5]. Due to dimensional analysis, turbulence spectra in the inerial subrange conforms to "2/3" power law[6,7]. Fluctuation of atmospheric temperature and refractive index can be expressed as

$$D_n(r)=\left\langle\left[n(r_1+r)-n(r_1)\right]^2\right\rangle=C_n^2 r^{2/3} \tag{1}$$

$$D_T(r)=\left\langle\left[T(r_1+r)-T(r_1)\right]^2\right\rangle=C_T^2 r^{2/3} \tag{2}$$

$D_n(r)$ and $D_T(r)$ are respectively structure function of refractive index and temperature. C_n^2 and C_T^2 are structure constant of refractive index and temperature. To local uniform and isotropy turbulence, structure constant can be indicated intensity of turbulence, which usually be represented by atmospheric structure constant of refractive index C_n^2.

To light wave and electromagnetic wave, refractive index n can be written as function of atmospheric pressure P and temperature T.

$$n-1=79\frac{P}{T}\times10^{-6} \tag{3}$$

refractive index n, atmospheric pressure P and temperature T can be decomposed average term and fluctuation term, respectively written as $n=\overline{n}+n^{'}, T=\overline{T}+T^{'}, P=\overline{P}+P^{'}$, so the fluctuation term of refractive index is[8].

$$n^{'}=79\left(\frac{P^{'}}{\overline{P}}-\frac{T^{'}}{\overline{T}}\right)\frac{\overline{P}}{\overline{T}}\times10^{-6} \tag{4}$$

$\frac{P^{'}}{\overline{P}}\ll\frac{T^{'}}{\overline{T}}$ in true atmosphere, that is to say, the fluctuation term of refractive index is main affected by temperature without considering the fluctuation of pressure etc under general circumstance[9].

$$n^{'}=-79\frac{\overline{P}}{\overline{T}^2}T^{'}\times10^{-6} \tag{5}$$

$$C_n^2=\left(79\frac{P}{T^2}\right)^2 C_T^2\times10^{-12} \tag{6}$$

Formula (6) reveals the method measuring atmospheric structure constant of refractive index C_n^2.

It is realized that the measurement of atmospheric structure constant of temperature C_T^2 using HTP-2 micro-thermal meter, whose detector consists of two spacing 1m platinum wires, each platinum wire 10 μm diameter and 2.4 cm length. The temperature difference $(T_1 - T_2)$ of above mentioned two platinum wires is measured under frequency 0.05~30 Hz and system noise 0.002 K [10,11].

$$C_T^2 = (T_1 - T_2)^2 / r^{2/3} \tag{7}$$

Over the past ten years, many times field experiments, pointed at the atmospheric structure constant of refractive index C_n^2, have been implementing at multiple locations from inland to open sea. A part of measured results in 2013 and 2014 are selected for analysis on characteristics and influencing factors. The evolution of C_n^2 with atmosphere temperature and wind speed will be discussed below.

3. Results and discussion

Firstly, experiment data of four inland fields at sunny day are displayed, and the relation among C_n^2, temperature fluctuation and wind speed shear is explored here. The measurement equipments are all fixed at top of a pole from the ground 4 to 6 meters high.

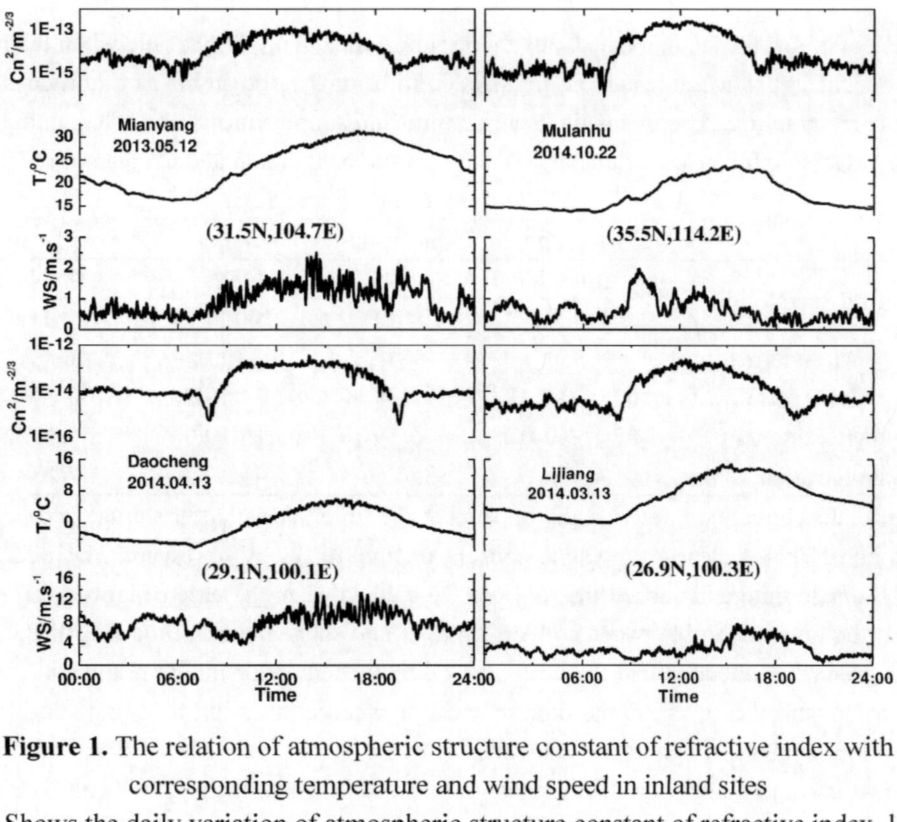

Figure 1. The relation of atmospheric structure constant of refractive index with corresponding temperature and wind speed in inland sites

Figure 1 Shows the daily variation of atmospheric structure constant of refractive index, local temperature and wind speed, which includes four inland measurement sites, respectively Mianyang

Journal of Physics: Conference Series **679** (2016) 012051 doi:10.1088/1742-6596/679/1/012051

and Daocheng in Sichuan province, Mulanhu in Hubei province, and Lijiang in Yunnan province. Temperature fluctuation value of inland areas is mainly caused by sun irradiation, which is regular according to sun movement. Usually, it is weak that the temperature fluctuation at morning, dusk and night in a day, beyond question, the most intense time interval appear at front and back of midday. On the basis of formula (6), the change of daily variation of atmospheric structure constant of refractive index is also like this. In other words, daily variation of atmospheric structure constant of refractive index C_n^2 directly relates to temperature fluctuation from sun irradiation. It also can be seen from Figure 1 that wind speed is impacted by temperature fluctuation, since its maximal and most dramatic change time interval often overlap with corresponding temperature fluctuation. Certainly, the degree of consistency of both is also affected by the underlying surface. The analysis shows that cement and soil underlying surfaces are more susceptible to atmospheric temperature fluctuations, such as Mianyang and Daocheng, and grass or grass around the cement ground underlying surfaces are not easy related to temperature fluctuations, for example Mulanhu and Lijiang. The reason is that the thermal capacity of cement or soil floor is relatively small, air above them easy to form convection current by temperature fluctuation. Comparatively, the thermal capacity of grass floor is so huge like a energy absorption black body that the corresponding affection can be ignored.

The other feature of above inland sites C_n^2 is distinct conversion time, which represents the weakest moment of turbulence in a day. Conversion time appears at twilight time when the heat from sun irradiation and earth surface heat radiation is in equilibrium. In the premise of latitude and longitude of corresponding experiment site, sunrise time and sunset time can be calculated.

Table 1. Geographic information, sunrise/sunset time calculation and Conversion Time of Turbulence to Different Experiment Sites

	Mulanhu	Mianyang	Lijiang	Daocheng
Latitude and longitude	30.5N 114.2E	31.5N 104.7E	26.9N 100.3E	29.1N 100.1E
Altitude(m)	55	531	3200	4692
Calculation	07:05/17:41	06:11/19:43	07:34/19:23	07:00/19:41
Measurement	07:15/17:07	06:46/19:17	08:15/19:06	07:57/19:28
Time Difference(min)	10/34	35/26	41/17	57/13

By literature, the conversion time usually appears at the time interval after sunrise and before sunset for about an hour. This is clearly too vague concept of time to scientific research. Using astronomy Julian angel, sunrise time and sunset time are precise calculated in the case of latitude and longitude. From Table1, the time interval between conversion time and sunrise/sunset time is influenced by local altitude. It is gradually extended that the time interval between conversion time and sunrise time with altitude elevation, and vice versa to the time interval between conversion time and sunrise time with altitude elevation. Such as Mulanhu and Daocheng, altitude respectively 55m and 4692m, the intervals of conversion time and sunrise are 10minutes and 57minutes, lengthening with altitude rising, but corresponding intervals between conversion time and sunset are 34 minutes and 13minutes, shortening with altitude rising. In a word, it is obvious that the precision of time interval between conversion time and sunrise/sunset time can be improved to minutes.

Secondly, measurement data of four offshore experimental sites at sunny day is picked out for analysis. The Qingdao site is port, whose underlying surface is sea water. The distance of Maoming , Zhanjiang and Sanya sites to corresponding seacoast respectively is 6 nautical miles, 15 nautical miles and 20 nautical miles. The measurement equipments at Qingdao were fixed at the top of a pole from the ground 4m meters high. The same equipments at Maoming were fixed on a place 14m high at an iron tower which lied in sea water. The ones at Zhanjiang and Sanya were fixed on each 4m high pole which seated ship deck.

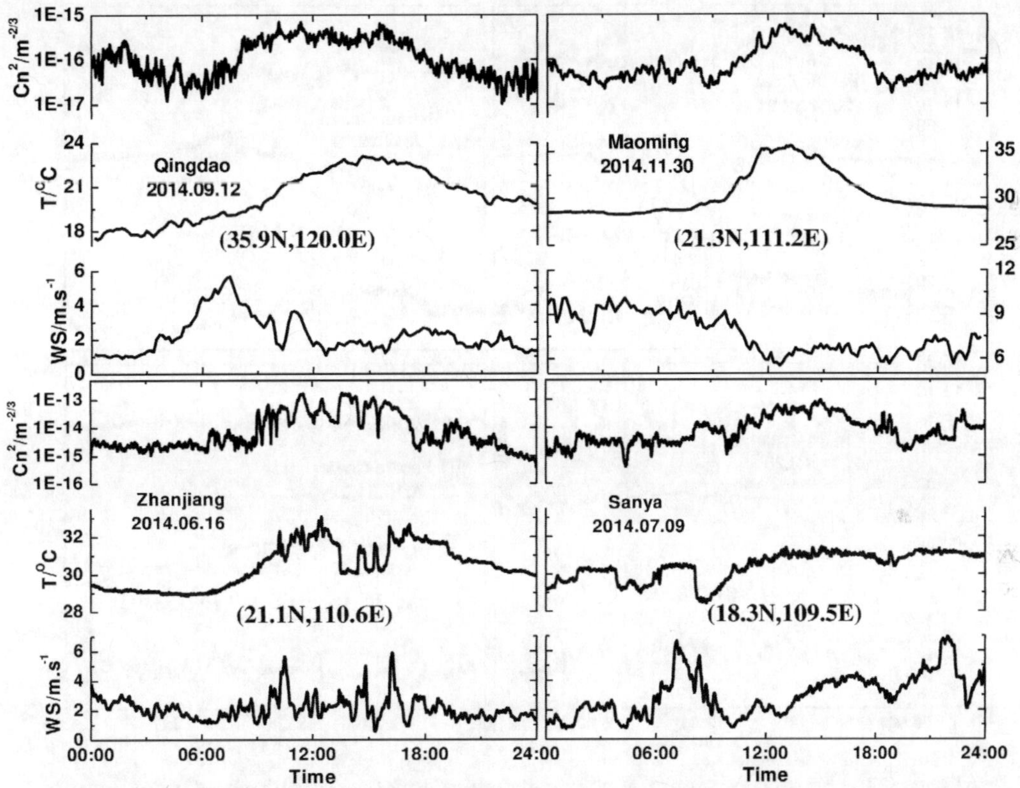

Figure 2. The relation of atmospheric structure constant of refractive index with corresponding temperature and wind speed in offshore areas

Figure 2 displays the development trend of offshore C_n^2, atmospheric temperature and wind speed with time. In general, the value range of C_n^2 in offshore areas is general smaller than in inland. It is obvious that the daily variation characteristics of C_n^2 still retains in Qingdao and Maoming, at the same time, C_n^2 is also mainly affected by temperature fluctuation although the impact of wind speed shear has begun to show, but the relation is weak between temperature and wind speed owing to underlying surface and surrounding environment. In comparison, the daily variation characteristics of

C_n^2 in Zhanjiang and Sanya are fuzzy without distinct conversion time. The C_n^2 of Zhanjiang and Sayan is gradually affected by both temperature fluctuation and wind speed fluctuation, in addition, the influence degree of wind speed fluctuation is seemingly to increase as the offshore distance more and more far.

Finally, it is selected that four sets experiment data of open sea sites in Pacific Ocean and Indian Ocean. In period of experiment, the HTP-2 micro-thermal meter was fixed 2m away on starboard side of the cockpit, where about 20m height above sea water.

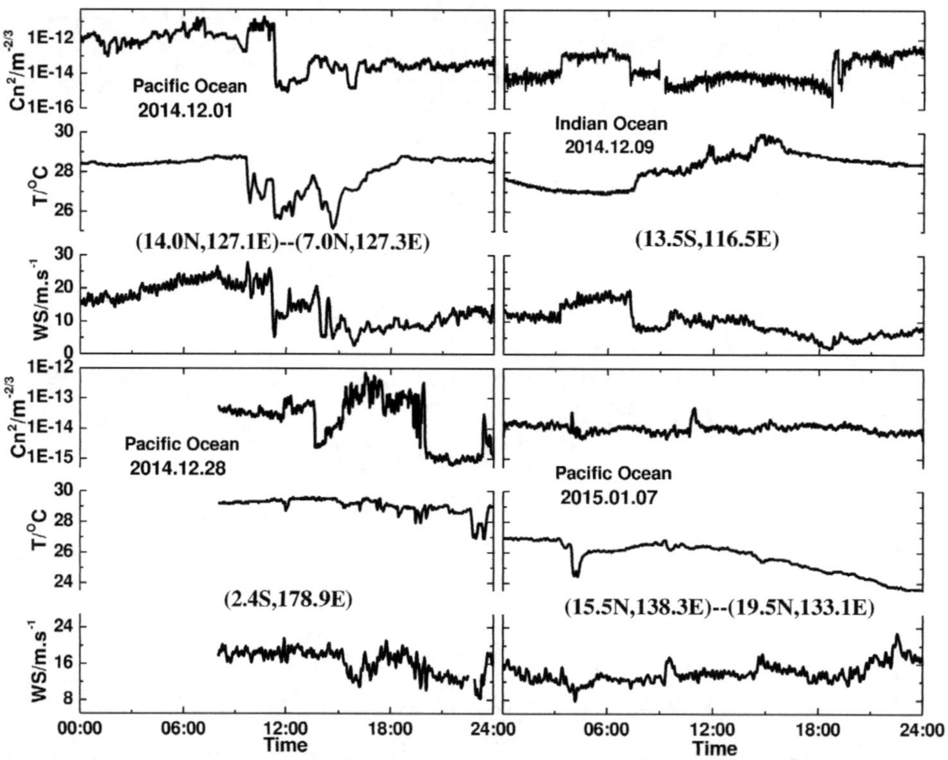

Figure 3. The relation of atmospheric structure constant of refractive index with corresponding temperature and wind speed in open sea

The wind speed used in above chart is absolute wind speed after ship speed subtracted, and the time also Bejing time during the whole period of experiment. The ship routes contained in upper left corner and lower right corner in Figure 3 are all straight line navigation between two sites, which represented by longitude and latitude. There is no way to forecast the weather in open sea, so the sudden change of temperature mostly originates in sudden rain, sudden thick cloud cover or sudden sun irradiation. What need to be explained is the steady decline of temperature after 10:00 in 7 Jan 2015 attributes the ship towards the Chinese East Coast where in cold winter. As Figure 3 shows, whether Pacific Ocean or Indian Ocean, the change of C_n^2 without any regular pattern at all, totally different to inland. The fluctuation of C_n^2 is mainly affected by wind speed fluctuation by each other compared. That is, the temperature fluctuation act on turbulence in open sea maybe caused by wind speed shear not by change of temperature itself. The value of C_n^2 is likely proportional to wind speed shear, in other words, the temperature fluctuation act on turbulence is greater when the wind speed is higher. Figure 3 also reveals that, in open sea, the correlation between C_n^2 and wind speed shear has nothing to do with the ship motion or not. The ship in 7 Jan 2015 is actually in the East China Sea, China's inland sea. The conclusion of offshore sea C_n^2 is also applicative here as mentioned above, that is, the value range of C_n^2 is small.

4. Summary

By comparison of C_n^2, temperature and wind speed at different experiment sites from inland to open sea, the conclusion can be brought out here on the daily variation.

1) The turbulence of inland sites have obvious daily variation characteristic, and conversion time clear. The sunrise time and sunset time can be calculated according to corresponding longitude and latitude. The time interval between conversion time and sunrise/sunset time is effected by local altitude. The time interval between conversion time and sunrise time is gradually extended with elevation, and the time interval between conversion time and sunset time is vice versa. With the distance to the land increase, the conversion time of turbulence of offshore sites is more and more indistinct and finally disappear. In open sea, there is no any regular pattern for turbulence;
2) The turbulence structure of inland is mainly affected by temperature fluctuation, and the wind speed fluctuation tends to agree with temperature fluctuation for small thermal capacity underlying surface, such as soil and cement floor. The consistence declines to high thermal capacity underlying surface, for example grass. The turbulence structure of open sea is mainly decided by wind speed fluctuation, and the turbulence value is like to proportional to wind speed shear. The turbulence structure of offshore areas is influenced by both temperature and wind speed fluctuation, and the influence degree of wind speed shear is more and more high with the distance to the land increase；
3) The numerical range of turbulence first decreases and then increases from inland to open sea, and the weakest turbulence fluctuation appears in inland seas where tens of nautical miles . That is, in the process from land to ocean, there may be a sea area where the effect of wind speed and temperature fluctuation on the turbulence formation is cancel out each other, in other words, small thermal capacity land and large thermal capacity sea water on the atmosphere are equilibrium and atmosphere is the most quiet. This maybe provide data support for the siting of ship board photoelectric systems.

Acknowledgments

This work was supported by the National Natural Science Foundation of China under Grant Nos.

Journal of Physics: Conference Series **679** (2016) 012051 doi:10.1088/1742-6596/679/1/012051

41105020 and 41475024. We thank China Satellite Maritime Tracking and Control Department for financial support under No. FOM2015OF002 and technological support during ocean experiment.

References

[1] Yahel RZ 1990 *Applied Optics* **29** 3088-95
[2] Sasiela RJ 1994 *Electromagnetic wave propagation in turbulence. Evaluation and application of Mellin transforms* (Berlin: Springer-Verlag)
[3] Kolmogorov AN 1962 *Journal of Fluid Mechanics* **13** 82-5
[4] Andrews LC and Phillips RL 1998 *Laser beam propagation through random media* (Bellingham, Washington: SPIE-The international society for optical engineering) 57-80
[5] Kolmogorov AN 1991 *Proc. R. Soc. Lond* **434** 9-13
[6] Song ZF 1990 *Applied Atmosphere Optics* (Beijing: China Meteorological Press) 67-70
[7] Rao RZ 2012 *Modernized Atmospheric Optics* (Beijing: Science Press) 415-75
[8] Zhou XJ, Tao SC and Yao KY 1991 *Senior atmospheric physics* (Beijing： China Meteorological Press) 252-346
[9] Weng NQ, Zeng ZY and Xiao LM 2004 *High Power Laser and Particle Beams* **16** 1101-5
[10] Xiao LM, Weng NQ and Ma CS 2004 *Journal of Hefei University of Technology* **27** 179-82
[11] Sun G, weng NQ and Xiao LM 2008 *High Power Laser and Particle Beams* **20**1 83-8

Distortion measurement for shipboard radar antenna by photogrammetry

Wenhua Zhao[1], Haibo Liu [2], Xinming Liu[1] and Liang Pan[1]

1 China Satellite Maritime Tracking and Control Department, Jiangyin, Jiangsu, China
2 National University of Defense Technology, Changsha, China

E-mail: qianyan99@126.com

Abstract. An approach based on photogrammetry has been developed for the measurement of shipboard radar antenna deformation. In the proposed approach, the images of the radar antenna with diagonal reflecting mark points are captured at a few (at least two) different orientations by the calibrated cameras. Then, the mark points in different images are detected and matched automatically with the information of reference points. At last, the optimum spatial coordinates of diagonal marks and measuring cameras external parameters are achieved via bundle adjustment algorithm. The theoretical analysis indicates that the proposed method precision can reach 1.0mm and has the advantages of high precision, convenient and flexible operation. This method can also be employed in deformation measurement of other large deployable structures.

1. Introduction

The antenna is a key component of the whole radar device, which has important effect on the overall performance of it. The precision of antenna reflector surface affects the quality of the received echo pulses of radiation directly. And if there is a big reflector surface errors, the antenna gain would decrease, which thereby affecting the radar range and resolution[1-3]. The radar system is installed on the moving platform, such as airborne radar, ship-based radar and vehicle-borne radar, affected by vibration, gravity deformation in the process of movement and other factors, which leading to inevitable deformation[4]. Therefore, to ensure the radar performance is in good condition, high precision measurement of the radar antenna surface must be carried out regularly[5, 6].

The three-dimensional laser scanning is a common method for deformation measurement of large structures[7], but it needs a stable benchmark, which limits the application in airborne, ship-based and vehicle-borne radar antenna deformation measurement. This paper proposes a method for the radar antenna deformation measuring based on videometrics, implementing real-time deformation monitoring of ship-based radar antenna. The proposed method has the advantages of simple equipment, convenient measurement, high precision and automation, and low requirements on environmental conditions.

2. Measurement principle

The basic principle of videometrics is shown in Figure 1, multiple cameras(or a camera in the different position) image the space object point P, point P is the intersection of the rays composed by designated C_i and the corresponding image points p_i. Given the internal and external parameters of the camera, the object three-dimensional position can be calculated via line-line intersection.

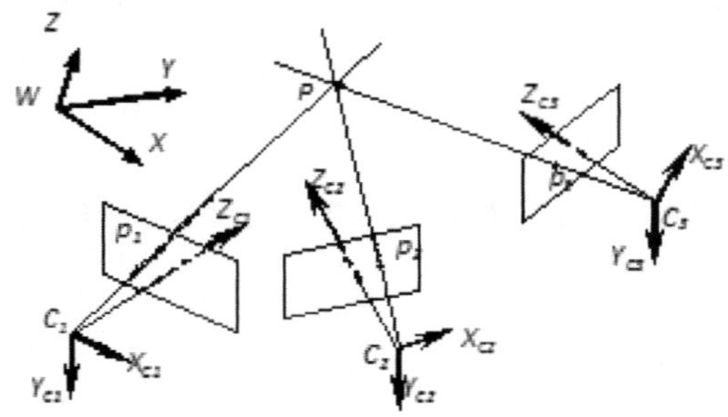

Figure 1. the basic principle of line-line intersection.

Known the internal and external parameters of each camera, defining the corresponding image point of space object point $P(X, Y, Z)$ captured by the camera i is $p_i(x_i, y_i)$, and the ideal image point is $(\tilde{x}_i, \tilde{y}_i)$ after aberration correction. Hence, the collinear equation can be transformed into linear equations with regard to the unknown space point P's coordinate (X, Y, Z).

$$\begin{cases} \left[(\tilde{x}_i - C_x)r_6 - F_x r_0\right]X + \left[(\tilde{x}_i - C_x)r_7 - F_x r_1\right]Y + \left[(\tilde{x}_i - C_x)r_8 - F_x r_2\right]Z + \left[(\tilde{x}_i - C_x)T_Z - F_x T_X\right] = 0 \\ \left[(\tilde{y}_i - C_y)r_6 - F_y r_3\right]X + \left[(\tilde{y}_i - C_y)r_7 - F_y r_4\right]Y + \left[(\tilde{y}_i - C_y)r_8 - F_y r_5\right]Z + \left[(\tilde{y}_i - C_y)T_Z - F_y T_Y\right] = 0 \end{cases} \quad (1)$$

For a single camera, there shall be two equations while the unknown parameters (X, Y, Z) are three. Hence, equation (1) is a Diophantine Equations which therefore has no unique solution. In fact, eq.(1) identifies a spatial line through the camera optical center and the image point, namely a line in collinear equation. When the $n(n \geqslant 2)$ cameras are used in line-line intersection, the equations number like eq.(1) increases to 2n while the unknown parameters are still three. So it's convenient to solve the linear equations by least square method.

3. Measurement Scheme

The measurement scheme is shown in Figure 2, two cameras were installed above the antenna for binocular intersection measurement; Set the diagonal reflecting mark points as shown in Figure 3 in the antenna and other parts to measure the deformation during the antenna motion. Ignoring the deformation of high stiffness and high strength parts of the antenna base, and with three reference points in the high stiffness and strength parts of the antenna base, the reference coordinate system is established, as shown in Figure 4.

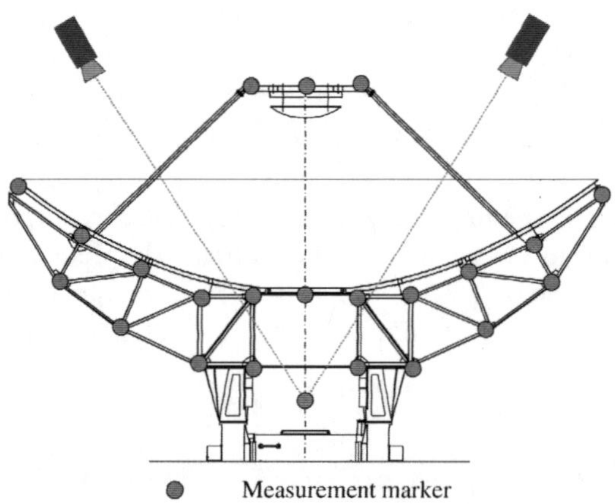

Figure 2. Schematic diagram of measurement scheme

Figure 3. Schematic diagram of diagonal reflecting mark

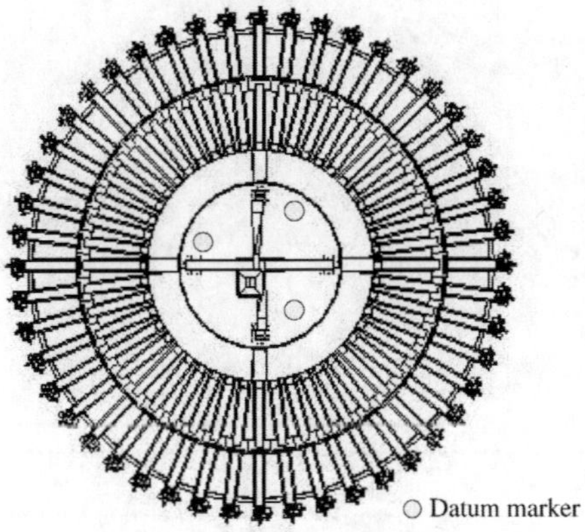

Figure 4. Schematic diagram of the layout of reference diagonal marks

The algorithm process of antenna deformation estimation is presented in Figure5. First, extract the image coordinate of diagonal targets using Harris corner detection algorithm[8]; Second, calculate the initial values of the measuring cameras external parameters via pose estimation algorithm[9] with the above three reference points and corresponding image points; then, according to the camera external parameters, carry out the points matching of other diagonal marks with the epipolar constraint matching method[10]; finally, obtain the optimum spatial coordinates of diagonal marks and measuring cameras external parameters via bundle adjustment algorithm[11].

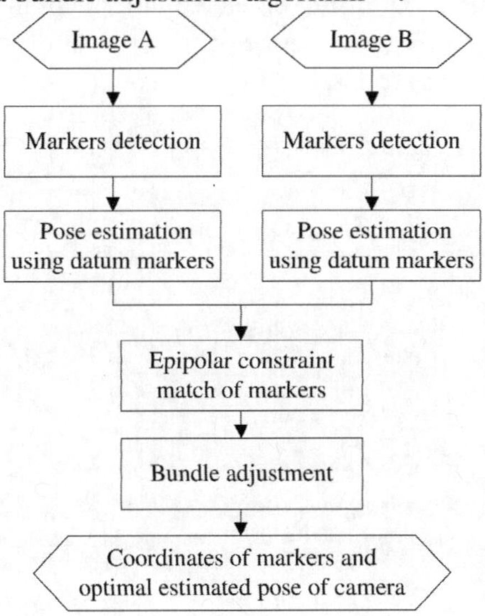

Figure 5. The flow chart of data processing

4. Error Analysis

The position precision of measured antenna is related to physical spatial resolution, image quality, corner detection, calibration of camera parameters and 3D intersection angle. The layout of cameras and the antenna in the static measurement is shown in Figure 6.

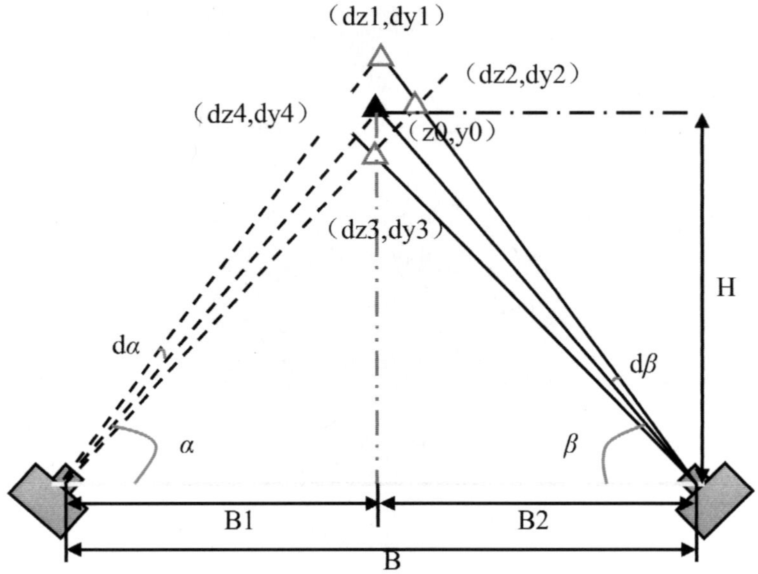

Figure 6. The layout of cameras and antenna

Define the baseline direction between the two cameras as the *Y* axis and the depth direction as the *Z* axis respectively. Letter B composed of B1 and B2 is the baseline while H is the distance between cameras and the antenna. α and β represent the stereo cameras' ideal observation angles. The error of observation angles dα,dβ caused by corner extraction error can be expressed as follows.

$$d\alpha = d\beta = \delta dv \tag{2}$$

Where dv represent the camera angle resolution, which is the ratio of camera angle of view and image resolution. (z0,y0) is the ideal result of stereo intersection, and (dz1,dy1), (dz2,dy2), (dz3,dy3), (dz4,dy4) represent the maximum error of target position error. According to the analytic geometry, equations can be attained as follows.

$$\begin{cases} \dfrac{B_1 - dy_1}{H + dz_1} = \cot(\alpha + d\alpha) \\ \dfrac{B_2 + dy_1}{H + dz_1} = \cot(\beta + d\beta) \end{cases} \quad \begin{cases} \dfrac{B_1 + dy_2}{H + dz_2} = \cot(\alpha - d\alpha) \\ \dfrac{B_2 - dy_2}{H + dz_2} = \cot(\beta + d\beta) \end{cases}$$
$$\begin{cases} \dfrac{B_1 + dy_3}{H - dz_3} = \cot(\alpha - d\alpha) \\ \dfrac{B_2 - dy_3}{H - dz_3} = \cot(\beta - d\beta) \end{cases} \quad \begin{cases} \dfrac{B_1 - dy_4}{H - dz_4} = \cot(\alpha + d\alpha) \\ \dfrac{B_2 + dy_4}{H - dz_4} = \cot(\beta - d\beta) \end{cases} \tag{3}$$

Combine the above equations,

$$\begin{cases} dz_1 = \left|\dfrac{B_1 + B_2}{k1 + k3} - H\right| \\ dy_1 = \left|\dfrac{k3B_1 - k1B_2}{k1 + k3}\right| \end{cases} \quad \begin{cases} dz_2 = \left|\dfrac{B_1 + B_2}{k2 + k3} - H\right| \\ dy_2 = \left|\dfrac{k3B_1 - k2B_2}{k2 + k3}\right| \end{cases}$$
$$\begin{cases} dz_3 = \left|\dfrac{B_1 + B_2}{k2 + k4} - H\right| \\ dy_3 = \left|\dfrac{k4B_1 - k2B_2}{k2 + k4}\right| \end{cases} \quad \begin{cases} dz_4 = \left|\dfrac{B_1 + B_2}{k1 + k4} - H\right| \\ dy_4 = \left|\dfrac{k4B_1 - k1B_2}{k1 + k4}\right| \end{cases} \tag{4}$$

Where $k1=\cot(\alpha+d\alpha), k2=\cot(\alpha-d\alpha), k3=\cot(\beta+d\beta), k4=\cot(\beta-d\beta)$ 。

In the vertical direction, the error of intersection measurement dX is the actual deviation of the target point extraction error, and will not increase the error due to the intersection error, so it is less than the error in the Z axis and the Y axis direction,

$$dX=\max\left(\frac{H}{\sin\alpha}\tan(d\alpha),\frac{H}{\sin\beta}\tan(d\beta)\right) \quad (5)$$

The layout of cameras and the antenna in the static measurement is shown in Figure 6. Two phantom Flex4k cameras are installed 6m in the distance of antenna, its focal length is 75mm and resolution is 4096×2048 pixels with pixel pitch of 6.75μm. The horizontal angle of field of view is about 20.9°,then the view angle of the unit of the camera is 20.9° /4096=5.10×10^{-3}degress per pixel which is also called resolution of view angle of cameras dv. Supposed that H is 6m and baseline of stereo cameras B is 5m(B1=B2=2.5m), the observation angle α、β can be calculated . According to the experience of the algorithm, the image point error of the target is 0.8 pixel, then the observation error is:

$$d\alpha=d\beta=0.8\times5.10\times10^{-3}=0.0041° \quad (6)$$

Combine eq.(6) and eq.(4), the maximum error in the direction of Z axis and Y axis can be calculated, and the maximum error in the direction of X axis can be attained according to eq.(5). The maximum errors of the depth, horizontal and vertical direction are as follows: (0.94mm, 0.51mm, 0.47mm).

Therefore, the position of the system can reach 1.0mm without considering atmospheric disturbance and other factors.

5. Conclusion

This paper proposes a videometrics based method for measuring the deformation of the radar antenna which can implement implementing real-time and dynamic deformation monitoring of ship-based radar antenna. The theoretical analysis indicates that the proposed method precision can reach 1.0mm and has the advantages of high precision, convenient and flexible operation. This method can also be employed in deformation measurement of other large deployable structures.

References

[1] Chen Xingquan Wang Jiexian 2008 Surface Fitting Algorithm of Radar Antenna Profile Measurement *Journal of Geodesy and Geodynamics* **4** 28

[2] Guo Meng He Hanhui Dingbang 2013 A displacement detection method for satellite phased-array antenna based on strain measurement *Spacecraft Environment Engineering* **6** 29

[3] Wang Cong-Si Duan Bao-Yan Yuan-Ying 2005 Divided-fitting method for distorted surface of antenna based on Coons surface *System Engineering and Electronics* **11** 27

[4] Wang Chang-Wu Ping Li-Hao 2006 A Real-time Measuring System of Operational Deformation of Radar Antenna *Modern Radar* **12** 28

[5] Ruixing Zhou, Guanyunxin Shang, Yongwei Gao 2002 Analysis of Wind Load and Deformation Characteristics of Conic and Paraboloid Antennas *Modern Radar* **2** 24

[6] Yu Jiang Jiang Shangping Linhua 2008 Displacement measurement of antennas by means of digital close range photogrammetry *Spacecraft Environment Engineering* **1** 25

[7] Imbriale W A 2001 Distortion compensation techniques for large reflector antennas *Big Sky, MT: IEEE* 799-805

[8] Harris C, Stephens M. 1988 A Combined Corner and Edge Detector 147-151

[9] Lu C P, Hager G D, Mjolsness E 2000 Fast and Globally Convergent Pose Estimation from Video Images *Pattern Analysis and Machine Intelligence* **6** 610-622

[10] Zhang Z, Deriche R, Faugeras O *et al* 1995 A Robust Technique for Matching Two Uncalibrated Images through the Recovery of the Unknown Epipolar Geometry *Artificial*

Journal of Physics: Conference Series **679** (2016) 012052 doi:10.1088/1742-6596/679/1/012052

Intelligence **2** 87-119
[11] Triggs B, Mclauchlan P, Hartley R *et al* 2000 Bundle Adjustment-A Modern Synthesis, Springer Berlin Heidelberg, 298-372

Temperature Measurement of Power Cable Based on Distributed Optical Fiber Sensor

Xiaoqing Shen, Yang Yang, Bo Cong, Fenghai Ding, Bin Qiu, Lingling Ye
China Satellite Maritime Tracking and Control Department, Jiangyin, China,214431

shxq1st@126.com

Abstract. To measure the temperature of the power cable onboard ships efficiently, a design scheme based on distributed optical fiber sensor is proposed. In this paper, its principle and hardware are described in detail and its feasibility and effectiveness is verified through real test.

1. Introduction

Electrical power cable is one of the most important part of the ship's power infrastructure and for the safety of the ship, close inspection and monitoring of the cable is necessary but this can't be achieved by conventional methods and equipments, so there is an urgent need for a temperature monitoring system for this purpose.

The existing cable temperature monitoring systems can be classified into two kinds, i.e., electric signal and optical signal sensor. Electrical signal sensors include traditional thermocouple sensor and thermal resistance sensor, etc. and have the advantage of high precision, fast response, but the disadvantage of low reliability due to complicated structure. The optical fiber temperature sensor have the ability of anti-corrosion, immune to electromagnetic interference immunity and high sensitivity, can realize distributed real-time measurement. At present, the distributed optical fiber temperature measurement technology[1]is mainly based on Raman scattering.

The design scheme proposed in this paper is based on Raman optical time-domain reflectometry[2], and it only needs a piece of sensing optical fiber, which is used as sensing element and the transmission element simultaneously. This temperature measurement system can be deployed in a large area, has simple structure and low cost.

2. Design of distributed optical fiber temperature measurement system

The distributed optical fiber temperature sensing system is shown in Figure 1. The sensing system comprises a semiconductor laser, divider, calibration unit, optical switch, highly sensitive detector and low noise amplifier circuit, laser driver and coding circuit, high speed data acquisition and signal processing module and optical fiber.

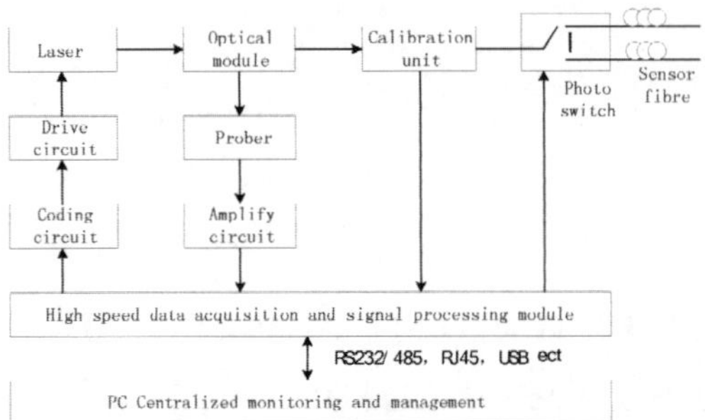

Figure 1. Block diagram of distributed optical fiber temperature sensing system.

The laser diode emits pre-coded pulse train, which is coupled into the fiber through an optical switch. In the fiber, backscattering is produced in three types, i.e., Rayleigh, anti-Stokes and Stokes. The anti-Stokes ray is temperature-sensitive and is used as signal. The Stokes ray is not sensitive to temperature and used as reference.

The two kinds of ray are detected, amplified and the data acquired can be processed to show the temperature distribution.

References

[1] Yanbiao Liao 2003 *Chinese Journal of Physics* vol 9 (China: Institute of Physics CAS Press) pp 323-326

[2] Zhang Zaixuan, Wang Jianfeng and Guo Ning 2001 *Journal of Optoelectronics Laser* vol 6 (China:Science Press) pp 596-600

Journal of Physics: Conference Series **679** (2016) 012054 doi:10.1088/1742-6596/679/1/012054

Principle Assumption of Space Object Detection Using Shipborne Great Aperture Photoelectrical Theodolite

Jia Ouyang, Tong-shuang Zhang and Qian-xue Wang

China Satellite Maritime Tracking and Control Department, China

E-mail: akaerguli@hotmail.com

Abstract. In this paper the use of space object detection is introduced. By analyzing the research actuality of space object detection using photoelectrical equipment, a shipborne great aperture photoelectrical theodolite is designed. The principle assumption of space object detection using shipborne great aperture photoelectrical theodolite is put forward.

1. Introduction

The task of space object detection is accurate detecting and tracking of important objects. By using optical telescopes and detecting radars to composing the land–based observation net, the space object detections of various countries are based on the land-based observation mode presently[1].

Compared with the land-based equipments, the shipborne photoelectrical detection equipments have advantages of portable and watching vary regions. Moreover, the detecting conditions on the sea are greatly exceeding with the land-based because of lower indexes of floating dust and aerosol. So the research of detecting space objects by shipborne photoelectrical detection equipment is very significative.

2. Shipborne great aperture photo-electrical theodolite

The shipborne great aperture photoelectrical theodolite can provide the track, shape, luminosity and spectrum measurement of space objects in surveillance aspects[2].

2.1. Developing direction

In wave band side, the photoelectrical theodolite is developing extended spectrum including visible light, infrared ray, soft X-ray, γ-ray, ultraviolet ray. The functions of photoelectrical theodolite are also developed from singularity to multifunctional integration.

The aperture of photoelectrical detecting equipment is enlarged from Φ230mm to Φ4m. The operation range is increased to 40000km.

2.2. System constitute

(1) Optical system
(2) Tracking and servo system
(3) Angel and range rate measuring system
(4) Track record system

2.3. Technical indexes

When the target image is occupy 15 to 30 picture elements, the shape of object can be differentiated[3]. Then the focus relationship is:

$$R = \frac{l}{l'} f \quad (1)$$

l is the space object size, l' is least target size required, f is the focus of optical system, R is the operation range of detecting target size.

Based on the expression, choosing the focus f = 8m, the optical system of CCD picture elements size is 9μm, then the operation range can reach 1481.5km. According to the design principle of optical system, the relative aperture D/f=1/8~1/12. So the aperture of optics D is 800mm to 1200mm.

2.4. Design indexes of shipborne photo-electrical theodolite

A shipborne great aperture photo-electrical theodolite is designed by analyzing upwards. The indexes of this equipment are:

(1) Aperture of optics $D \geq$ 1000mm

(2) Focus f is 2000mm～10000mm adjustable

(3) Angel of view: ≥20′×20′、≥10′×10′

3. Key technology

(1) Laser ranging

(2) Stable technique of optical axis

(3) Laser communication[4]

4. Conclusion

In conclusion, the shipborne great aperture photo-electrical theodolite can realize the large operation range using long wavelength (10.6μm) and high elevation to detecting the space objects.

The large surface ship installed great aperture photo-electrical theodolite has major coverage area. It can fill the requirements of observing to targets of low earth orbit and geosynchronous orbit.

References

[1] J.X.Zhang 2009 Progress in foreign ground-based optoelectronic detecting system for space target detection *Chinese J. Optics and Applied Optics* **2** 10

[2] L.Pan 2009 *Ship attitude and position measurement technology of space tracking ship* (Beijing: National defense industry press) p289

[3] J.F.Du 2006 Error budget design for photoelectric theodolite *J. Applied Optics* **27** 506

[4] J,Q,Shi, S.C.Wei and X.Y.Han 2008 Practicality calculation method of laser transmitting absorption loss in atmosphere *J. Henan Normal University(Natural Science)* **36** 56

Journal of Physics: Conference Series **679** (2016) 012055 doi:10.1088/1742-6596/679/1/012055

Low-loss, high reflectivity, first-order, pitch-by-pitch fiber Bragg grating fabrication in truly free-standing single-mode fiber

Yicheng Lai, Yijing Chen, Marcus W.O. Cheong

Data Storage Institute (DSI), Agency for Science, Technology and Research (A*STAR), 5 Engineering Drive 1, Singapore (117608).

Email: LAI_yicheng@dsi.a-star.edu.sg

Abstract. Low-loss, high reflectivity, first-order fiber Bragg gratings are achieved by pitch-by-pitch fabrication methodology in a truly free-standing fiber. Slit beam-profiling adopted attain transversely-extended index modification for each grating period defined by single laser pulse. Fiber Bragg grating of 0.52μm period with >27dB transmission stopband and <0.2dB/cm insertion loss are demonstrated.

1. Introduction

The use of intense femtosecond (fs) pulse laser beam to achieve refractive index modifications in optical fibers has enabled formation of fiber Bragg grating (FBG) in a pitch-by-pitch (PbP) process without the need for phase mask or interferometric setup. Indeed, such laser-induced PbP grating inscription methodology has yield FBGs of high spectral quality. However, low-order gratings are typically achieved only by using oil-immersion objectives or require compensating substrate in-proximity in order to overcome distortion effects due to the intrinsic geometry of the optical fiber [1,2]. More importantly, to address the out-of-band insertion losses associated with light scattering in PbP inscribed FBG, time-consuming laser scanning methodologies such as line-by-line inscription and core-scanning technique, are instead adopted to achieve index modification of increased transverse profile for each grating period [3,4]. One should also note that these proposed schemes do not encompasses the simplicity and flexibility of PbP grating fabrication. In particular, it is desirable to achieve rapid FBG generation in truly free-standing fiber where the grating period is simply the ratio of the translation speed of the fiber to the laser pulse repetition rate. And only such methodology can offer practical large scale or long-grating fabrications implementations.

In this paper, we demonstrate, for the first time to the best of our knowledge, a low-loss FBG, PbP fabrication methodology that truly features a free-standing fiber inscribed by a focused fs pulse laser, without the need for laser-scanning or oil immersion-based compensations. We highlight for the first time slit beam-profiling as a means to increased transverse profile of index modification that enable low-loss, high fidelity FBG formation. Proposed scheme retain merit of PbP FBG generation where each grating pitch are individually defined by single focused fs laser pulse. Uniform, 15mm-long first-order (grating period 0.519μm) FBG with >27dB transmission stopband at Bragg resonance and with

Journal of Physics: Conference Series **679** (2016) 012055 doi:10.1088/1742-6596/679/1/012055

out-of-band insertion loss <0.2dB/cm are demonstrated in a free-standing single-mode fiber. Interesting polarization characteristics of the resultant FBG are further highlighted.

2. Slit beam profiling

The beam distortion experience by a focused inscription beam when incident into a free-standing optical fiber can be understood based on the simple schematic as shown in figure 1a, where the objective lens between the slit and the fiber is not explicitly depicted. Due to the intrinsic cylindrical geometry of the fiber, differential divergence between the axial and radial axes of the incident beam leads to dislocation of the focal region between the orthogonal axes. Laser-induced index modification occurs at a region of overlap with intensity above the inscription threshold, leading to a typically-observed axially-elongated inscribed feature.

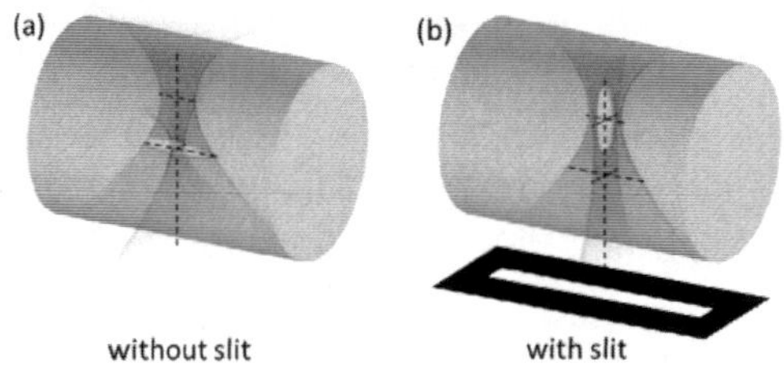

Figure 1. Focusing conditions within a free-standing fiber (a) without and (b) with slit beam-profiling. Index modification at overlap region reshapes transversely due to the slit.

We have previously highlighted the use of slit beam-profiling as a means to reshape the focal energy of a laser within a free-standing fiber [5]. In particular, the inclusion of an axially-oriented slit serves to increases the depth of focus of the radial plane by means of reduction of its effective numerical aperture. With such modification to the field overlap between the two focusing component planes as illustrated in figure 1b, the inscription region reshapes transversely while maintaining a tight axial focusing condition. To visualize these considerations in free-space inscription of grating in a single-mode fiber, grating structure of 1.03μm period are inscribed by a step-wise, pitch-by-pitch methodology where each pitch is defined by a 10msec-interval fs pulse laser exposure at 100kHz. The resultant inscribed grating structures viewed under a microscope at orthogonal axes are as shown in figure 2.

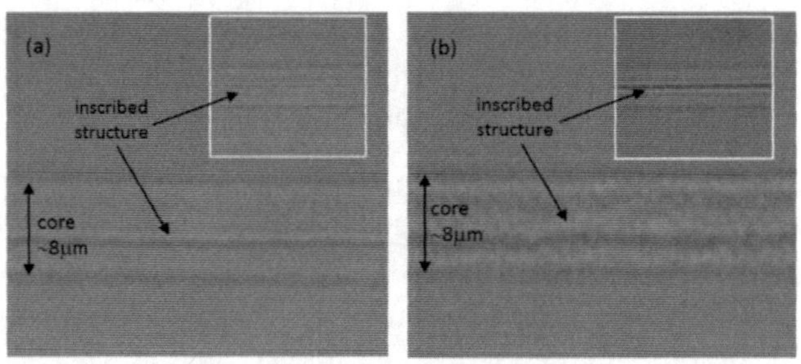

Figure 2. Microscope images (orthogonal view in inset) showing inscribed grating structures in a free-standing single-mode fiber (a) without and (b) with slit beam-profiling.

It is evident that without slit beam-profiling, high fidelity grating inscription cannot be achieved since the focusing condition lead to poor axial spatial resolution. The 1.03μm-pitch inscription resulted in a quasi-continuous structure consisting of an array of laser-modified features that overlap spatially. In comparison to figure 2b, the high grating fringe visibility achieved highlights the effectiveness of slit-profiled inscription. Focal energy redistribution at inscription point is evident

Journal of Physics: Conference Series **679** (2016) 012055 doi:10.1088/1742-6596/679/1/012055

leading to the increased aspect ratio of the inscribed feature. More importantly, the inscription yields the required transversely-extended index-modification that enables low-loss FBG pitch formations.

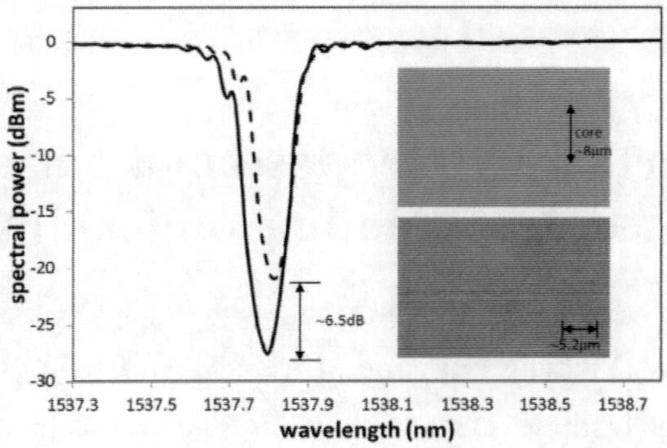

Figure 3. Superimposed measured transmission spectra along orthogonal polarizations of the FBG. Inset show microscope images of the FBG with 1st-order period of ~0.52μm.

To exemplify the merit of the proposed slit beam-profiled inscription methodology in free-standing single-mode optical fiber, a 15mm-long FBG was fabricated using pulse energy of 160nJ. The fiber was translated at a velocity of 51.6mm/sec with the fs pulse laser at 100kHz repetition rate. The inscription process took <300msec and the transmission spectral of the resultant FBG obtained is as shown in figure 3.

The uniform FBG exhibits a transmission stopband of >27dB at Bragg resonance of 1537.8nm. The out-of-band insertion loss of the FBG, monitored throughout the inscription process, measured <0.3dB leading to effective insertion loss of the FBG at <0.2dB/cm. The FBG observed under microscope reveals high grating fidelity of period ~0.52μm as shown in figure 3 inset. It is worthwhile to note that the grating exhibits very low birefringence of ~4x10-5 in comparison to ~1x10-4 [1,3] while the polarization-dependent Bragg response was >6.5dB. These suggest that the index modulation along the polarization axes differs while the average indices remain close. Polarization characteristics of such FBGs are under further investigations.

3. Summary

Low-loss, high reflectivity, first-order FBG are achieved by PbP fabrication methodology in a truly free-standing fiber for the first time. Results exemplify the scheme encompasses fully the simplicity and flexibility of laser-induced PbP grating fabrication.

Acknowledgments

This work is supported by the A*STAR Thematic Strategic Research Programme (122 360 0010).

References

[1] Y. Lai et al., "Point-by-point inscription of first order fiber Bragg grating for C-band applications", *Opt. Exp.*, 15(26), 18318-18325 (2007).

[2] G.D Marshall et al., "Point-by-point written fiber Bragg gratings and their applications in complex grating designs", *Opt. Exp.*, 18(19), 19844-19859 (2010).

[3] K. Zhou et al., "Line-by-line fiber Bragg grating made by femtosecond laser", *IEEE Photon. Tech. Lett.*, 22(16), 1190-1192 (2010).

[4] R.J. Williams et al., "Femtosecond direct-writing of low-loss fiber Bragg gratings using a continuous core-scanning technique", *Opt. Lett.*, 38(11), 1918-1920 (2013).

[5] Y. Lai et al., "High fidelity, pitch-by-pitch grating inscription in free-standing single-mode optical fiber", *IEEE 7th ICAIT*, 166-171 (2014).

Journal of Physics: Conference Series **679** (2016) 012056 doi:10.1088/1742-6596/679/1/012056

Low-energy distortion-free femtosecond pulse laser inscription in coated, free-standing optical fiber

Yijing Chen[1], Yicheng Lai[1], and Marcus W.O. Cheong[1]

[1]Data Storage Institute (DSI), Agency for Science, Technology and Research (A*STAR), 5 Engineering Drive 1, Singapore (117608)

E-mail: cheny@dsi.a-star.edu.sg

Abstract. Low-energy, distortion-free femtosecond pulse laser inscription in coated free-standing fiber is demonstrated for the first time. Methodology features radial plane beam-shaping to achieve optimal focusing and spot-like inscription within the fiber. Strong, first-order fiber Bragg grating is achieved in coated single-mode fiber using only 50nJ pulse energy.

1. Introduction

Femtosecond (fs) pulse laser inscription in fibers has been of interest for realizing monolithic fiber devices as it offers, amongst all, spatial control flexibility. Such inscriptions have since been adopted for fiber Bragg grating (FBG) fabrications with different optical setups [1-4]. It is worthwhile to note that though FBG inscriptions have been widely shown, the techniques are generally restrictive and, suited only for realizing fiber gratings. While these schemes do achieve high axial spatial resolution, the transverse profile, a function of the Rayleigh depth of the focusing lens, of the laser-induced feature remains significant [3,4]. Consequently, these methods do not achieve true spot-like inscription within the fiber. In addition, particularly for oil-immersion-based techniques, it is inherently impractical to keep the fiber in oil for large-scale production.

In this paper, we demonstrate for the first time to the best of our knowledge, a methodology that simultaneously achieves spot-like, laser-induced inscription in a fiber while alleviating limitations due to distortion on the incident focused beam in a free-standing fiber. The proposed method features radial plane beam-shaping of the incident beam by inclusion of cylindrical lens before the objective lens. Optimal focusing condition is achieved within the fiber to lead to ultra-low energy laser inscription in a free-standing fiber without the need for oil immersion-based compensations. To illustrate the effectiveness of the proposed scheme, first-order FBG with 55dB transmission stopband at Bragg resonance is achieved in a coated, free-standing single-mode fiber using lowest inscription energy to-date of 50nJ

2. Beam shaping

Due to the intrinsic cylindrical geometry of the optical fiber, the focused beam inside the fiber experiences differential divergence between the axial and radial axes that leads to dislocation of the focal region between the orthogonal axes. The laser-induced index modification preferentially occurs at radial focal region whereby the radial focusing exhibits the larger numerical aperture to give the stronger optical confinement, which leads to a typically-observed axially-elongated inscribed feature as depicted in Fig. 1.

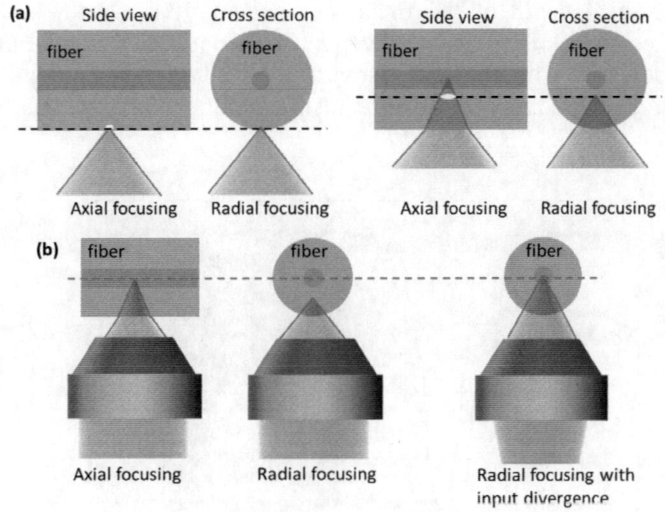

Fig. 1 Focusing conditions in a free-standing fiber (a) without radial plane beam shaping and, (b) with radial plane beam-profiling. Input divergence to radial plane in (b) lead to optimal convergence of focal points between the radial and axial planes.

On the other hand, by introducing beam divergence to the radial plane of the incident beam prior to the objective lens, one can adjust the focal position of the radial component to eliminate the focal point dislocation. When the axial and radial focal points coincide within the fiber, optimal focusing and hence inscriptions, is achieved as illustrated in Fig. 1b

To experimentally illustrate the proposed methodology, we introduce a dual-cylindrical-lens (dual-CL) system (focal length of 75mm and 50mm respectively) prior to the objective lens. By adjusting the separation between the two lenses, the beam divergence can be varied accordingly. Single point inscriptions with 10msec exposure each are carried out for setups with and without the dual-CL system. The divergence of the dual-CL is adjusted to effectuate optimal inscription at the fiber core. The pulse energy required for visible inscribed feature was 40nJ for inscription with dual-CL as compared to 150nJ without the dual-CL setup.

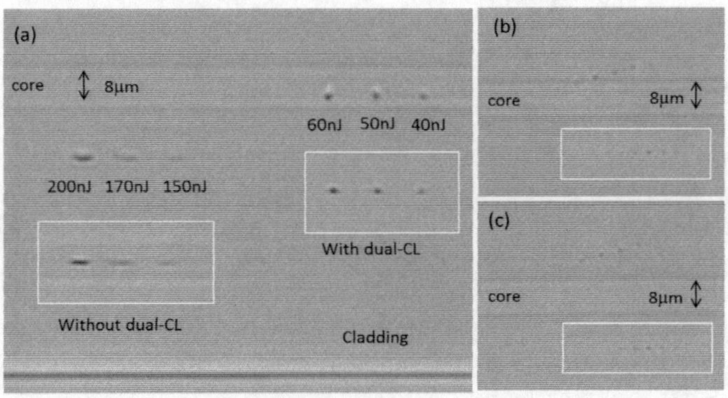

Fig. 2 (a) Comparing point inscriptions in free-standing fiber with and without dual-CL at different pulse energy. (b) and (c) illustrates changes in optimal inscription depth with adjusted input radial plane beam divergence. Beam is incident from the bottom of the pictures. All insets show the orthogonal views.

As shown in Fig. 2(a), the characteristics of the resultant inscribed features are consistent with Fig. 1b. With the dual-CL system, the inscribed features are within the core and exhibits spherical geometry. Without the dual-CL, the axially-elongated inscribed features appear in the cladding

Journal of Physics: Conference Series **679** (2016) 012056 doi:10.1088/1742-6596/679/1/012056

indicating the shorter focal depth of the radial component. It is worthwhile to note that optimal inscription position increases with the beam divergence introduced. Since the dislocation of axial and radial focal points increases with the inscription depth, a larger beam divergence is required for achieving optimal inscription at a deeper position inside the fiber. This is consistent with experimental results obtained as shown in Fig. 2b and 2c.

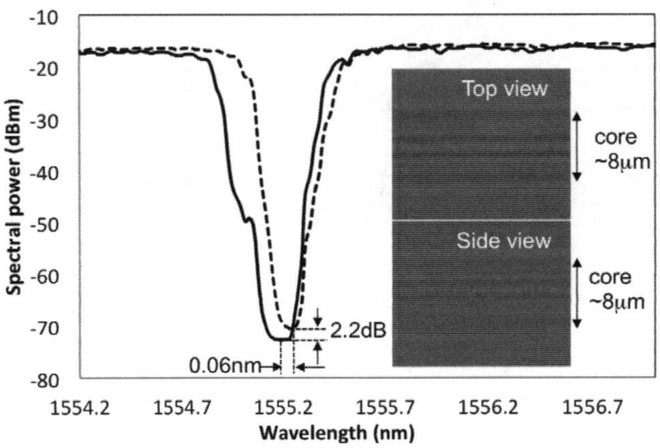

Fig. 3 Superimposed measured transmission spectra along orthogonal polarizations of the FBG. Inset show microscope images of the FBG with 1st-order period of ~0.52μm.

As an exemplary application, the proposed methodology is adopted for point-by-point (PBP) fabrication of a 10mm-long, first-order FBG in a coated, free-standing single-mode fiber. The fs pulse laser repetition rate operates at 100 kHz while the fiber translates at 52.2mm/s, resulting in a grating period of 522nm. The measured incident fs pulse energy was 50nJ. The inscription took only 190msec and the obtained transmission spectral shows a 55dB stopband at Bragg resonance (~1555.2nm) as shown in Fig. 3.

Due to the symmetry of the inscribed features, the polarization dependency of the spectral response is less significant as compared to [3]. The birefringence measured $\sim 5\times10^{-5}$ while the polarization-dependent Bragg response is ~2.2dB. The measured insertion loss of the fabricated FBG is ~1.2dB/cm, which is comparable to PBP FBG fabricated with oil-immersion [3]. It is important to note that the past demonstration of PBP FBG fabrication in a coated single-mode fiber only achieved 2nd-order grating with inscription energy at hundreds of nJ [5].

3. Conclusion

Ultra-low energy, distortion-free, spot-like fs pulse laser inscription in a coated, free-standing fiber is achieved for the first time based on beam-shaping by inclusion of cylindrical lens before the objective lens. Proposed scheme potentially lay the foundation for a truly versatile laser inscription suitable for large-scale implementation.

References

[1] G.D Marshall et al., "Point-by-point written fiber Bragg gratings and their applications in complex grating designs", Optics Express **18,** 19844 (2010).

[2] J. Thomas et al., "Femtosecond pulse written fiber gratings: a new avenue to integrated fiber technology", Laser Photonics Rev. **6**, 709 (2012).

[3] Y. Lai et al., "Point-by-point inscription of first order fiber Bragg grating for C-band applications", Optics Express **15**, 18318 (2007).

[4] Y. Lai et al., "High fidelity, pitch-by-pitch grating inscription in free-standing single-mode optical fiber", IEEE 7th ICAIT, 166 (2014).

Journal of Physics: Conference Series **679** (2016) 012056 doi:10.1088/1742-6596/679/1/012056

[5] A. Martinez et al., "Direct inscription of Bragg gratings in coated fibers by an infrared femtosecond laser", Opt. Lett. **31**, 1603 (2006).

Journal of Physics: Conference Series **679** (2016) 012057 doi:10.1088/1742-6596/679/1/012057

Spot-like inscription with large inscription depth in a free-standing single-mode fiber

Yijing Chen[1], Yicheng Lai[1], and Marcus W. O. Cheong[1]

[1]Data Storage Institute (DSI), Agency for Science, Technology and Research (A*STAR), 5 Engineering Drive 1, Singapore (117608)

E-mail: chenyi@dsi.a-star.edu.sg

Abstract. Spot-like femtosecond pulse laser inscription with large inscription depth in a free-standing fiber is demonstrated for the first time. Methodology uses axial plane slit beam-profiling to improve inscription uniformity near the core to enable spatially-large in-fiber structural inscription.

1. Introduction

There have been substantial interests in using a focused femtosecond (fs) pulse beam to inscribe arbitrary device structures in the optical fiber. However, the inscription flexibility can be restricted by the short inscription depth and the large inscription distortion due to the fiber curvature geometry. Although oil-immersion-based techniques can effectively remove the fiber curvature effect, it is inherently impractical to keep the fiber in the oil especially for large-scale fiber device production [1]. More recently, high-fidelity fs laser inscription is demonstrated in a free-standing fiber based on a radial-plane slit-beam profiling technique. However, the inscription feature is significantly elongated transversely and the short inscription depth remains [2].

In this paper, we report the first demonstration of spot-like inscription with large inscription depth in a free-standing fiber by utilizing an axial plane slit beam-profiling technique. With the improved inscription uniformity near the core, the inscription flexibility is enhanced, laying the foundation for arbitrary device inscription in the fiber. Spatially-large in-fiber structural inscriptions are further highlighted.

2. Inscription distortion in fiber

Due to the intrinsic cylindrical geometry of the optical fiber, the focused beam inside the fiber experiences differential divergence between the axial and radial axes that leads to dislocation of the focal region between the orthogonal axes, as shown in Fig. 1. The inscription preferentially takes place at the radial focal point due to the larger effective numerical aperture in the radial plane, which leads to a typically-observed axially-elongated inscribed feature. Increasing the focusing depth will increase the inscribed feature elongation, causing the drastic increment of the inscription energy requirement, thus limiting the inscription depth.

Experimentally, the inscription distortion in a free-standing single-mode fiber is as shown in Fig. 2, where each point is inscribed by a 10msec-interval fs pulse laser (830nm, 160fs) exposure at 100kHz.

At least twice higher inscription energy up to 215nJ is required to effectuate visually inscription near the core. The inscription feature is significantly elongated along the fiber axis in the vicinity of the core.

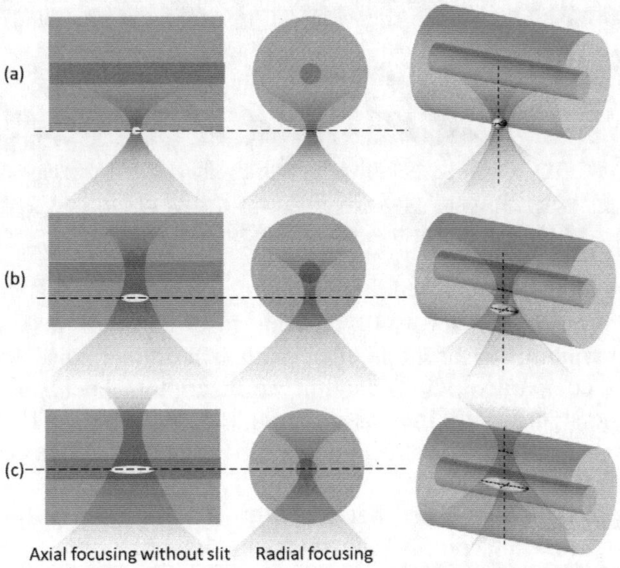

Fig. 1: Schematic illustration of inscription distortion inside a free-standing fiber.

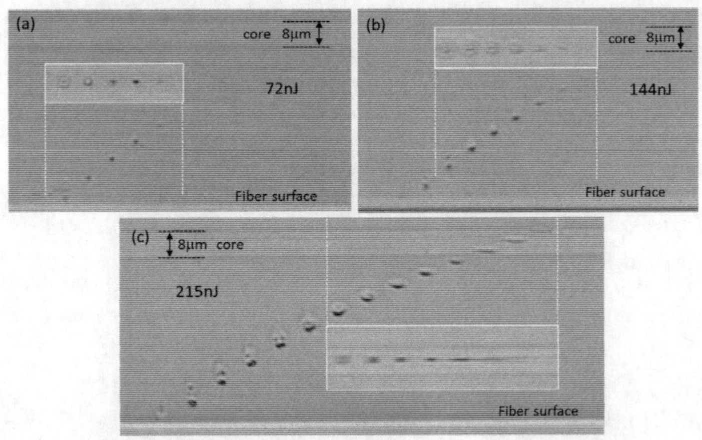

Fig. 2 Fs laser inscribed features in a free-standing fiber with pulse energy of (a) 72nJ (b) 144nJ (c) 215nJ. Beam incident from the bottom. Inserted figures show the orthogonal view.

3. Perpendicular slit-beam profiling

Here we propose a simple free-space approach to achieve large inscription depth inside a free-standing fiber without compromising the inscription energy efficiency and the inscription feature size. The proposed technique is based on the insertion of a perpendicularly-oriented slit before the objective lens to increase the depth of focus (DOF) of the axial focal plane, by means of reduction of its effective numerical aperture. With such modification to the field confinement at the radial focal point (inscription point), as illustrated in Fig. 3, the axial inscription elongation is significantly suppressed, resulting in spot-like inscription feature. Meanwhile, the long DOF of axial focusing effectively compensates the focal-point dislocation and thus improves the inscription uniformity near the core.

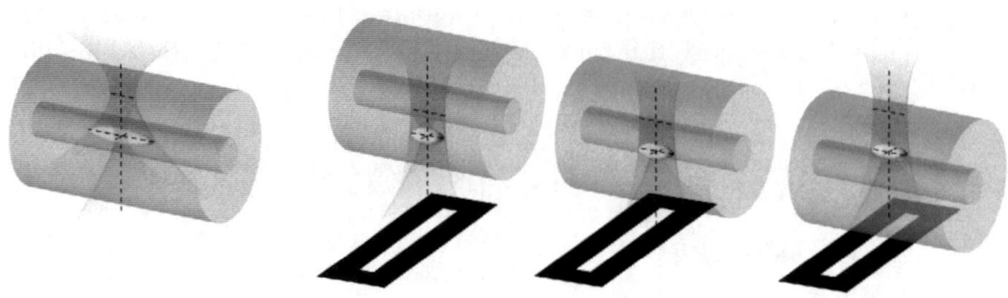

Fig. 3: Schematic illustration of free-space inscription inside the fiber without the slit and with the perpendicular slit beam-profiling at different inscription depth. Objective lens between the slit and the fiber is not explicitly shown.

To experimentally illustrate the proposed methodology, we introduce a 1.5mm-width perpendicular slit prior to the objective lens. First of all, in-core inscription of a 2μm-period dot chain is demonstrated in a free-standing single-mode fiber with inscription pulse energy of only 76nJ, as shown in Fig. 4. This is comparable with near-surface inscription energy in Fig. 2a for the set up without slit, indicating the significantly increased inscription depth here. The resultant dot chain also features both high transverse symmetry and large co-axial fidelity, as viewed from two orthogonal directions.

Next, dot chains are inscribed at varied depths with the fixed inscription energy to demonstrate high inscription uniformity near the core. As shown in Fig. 5a, with adoption of the perpendicular slit, high-fidelity dot chains are obtained over a wide depth of at least 10μm from the core at the inscription energy of 81nJ. In contrast, the near-core inscription without the slit (Fig. 5b) requires the higher inscription energy of 175nJ and the dot chain fidelity is drastically varied over the 10μm depth change, implying the poor uniformity. The zig-zag dot chain is axially broadened and got significantly merged.

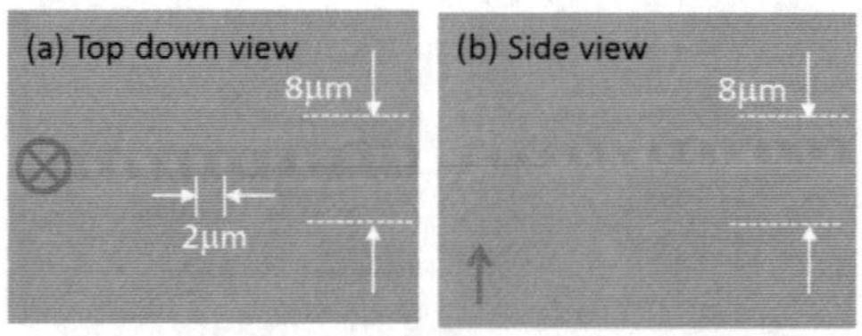

Fig. 4: Dot-by-dot inscription in the core of a free-standing single-mode fiber. (a) Top-down view and (b) side view.

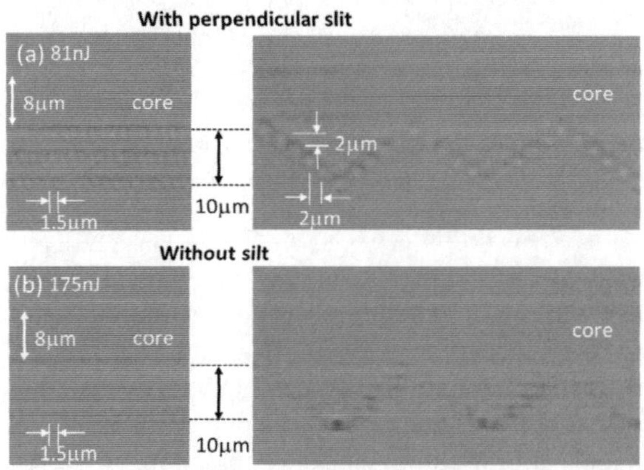

Fig. 5: Dot-by-dot inscription in the single-mode fiber (a) with perpendicular slit at fixed inscription energy of 81nJ and (b) without the slit at fixed inscription energy of 175nJ. Beam incident from the bottom of the picture.

4. Conclusion

We report the first experimental demonstration of large inscription depth in the free-standing single mode fiber, featuring spot-like, high-uniformity inscription in the vicinity of the core. Results obtained highlight the methodology potentially laid the foundation for spatially-large, arbitrary in-fiber device structure inscription.

References

[1] Y. Lai et al., "Point-by-point inscription of first order fiber Bragg grating for C-band applications", Optics Express **15**, 18318 (2007).

[2] Y. Lai et al., "High fidelity, pitch-by-pitch grating inscription in free-standing single-mode optical fiber", ICAIT IEEE 7th international conference, 166 (2014).

[3] H.-S. Chon et al., "Dependence of transverse and longitudinal resolution on incident Gaussian beam widths in the illumination part of optical scanning microscopy", J. Opt. Soc. Am. A **24**, 60 (2007).

Opto-ultrasound imaging in vivo in deep tissue

Ke Si[1,2], YanXu[1], Yao Zheng[1], Xinpei Zhu[2], and Wei Gong[2,*]

College of Optical Science and Engineering, Zhejiang University, Hangzhou, 310027
Institute of Neuroscience, Zhejiang University, Hangzhou, 310058

Correspondingauthor:weigong@zju.edu.cn

Abstract. It is of keen importance of deep tissue imaging with high resolution in vivo. Here we present an opto-ultrasound imaging method which utilizes an ultrasound to confine the laser pulse in a very tiny spot as a guide star. The results show that the imaging depth is 2mm with a resolution of 10um. Meanwhile, the excitation power we used is less than 2mW, which indicates that our methods can be applied in vivo without optical toxicity and optical bleaching due to the excitation power.

1. Introduction

Microscopy imaging technology has been confirmed as the key to the "gate of micro-world" since Leeuwenhoek invented the first microscope in 1665 and no one would deny its contribution to myriads of fields up to now. However, since the biological tissue usually has strong scattering properties, it is difficult for the light to pass through and therefore it is difficult for the general microscope to "look" through and present out. To explore the secret inside brain, which is the most mysterious part in biological tissue, a section technology called MOST[1] is used to image every thin slice first and then reconstructed to 3D images. This method can get the structure information of the whole brain in vitro with high resolution of about 300nm. However, since the tissue is first sliced preparing for imaging, it cannot be use in vivo. Therefore, to explore the function of special biological tissue in living body, it is necessary to solve the problem of imaging in deep tissue. This request promotes another approach called two-photon fluorescence microscopy (TPM), to be widely used in biological science. The advantages of TPM lies on the long wavelength beam employed has much deeper penetration depth, meanwhile only the fluorescence on the focus can be excited. The general penetration depth with TPM can reach at 600um. To further increase the depth penetration, the most common strategy used is increasing the laser power exponentially to maintain adequate ballistic excitation power at relatively larger focal depths. However, this neglects the fact that femtosecond-pulsed laser are power limited. Moreover, increasing the excitation power can lead to saturation, photobleaching and photodamage.

Here we introduce an opto-ultrasound imaging system, which can expend the penetration depth to 2mm with a high resolution of 10um [2,3]. Meanwhile, the excitation power is restricted to 2mw, which is a safe energy and allows the applications in most biological tissue in vivo.

2. Theory and Methods

The opto-ultrasound imaging system combines single-cycle pulsed ultrasound modulation and digital optical phase conjugation.

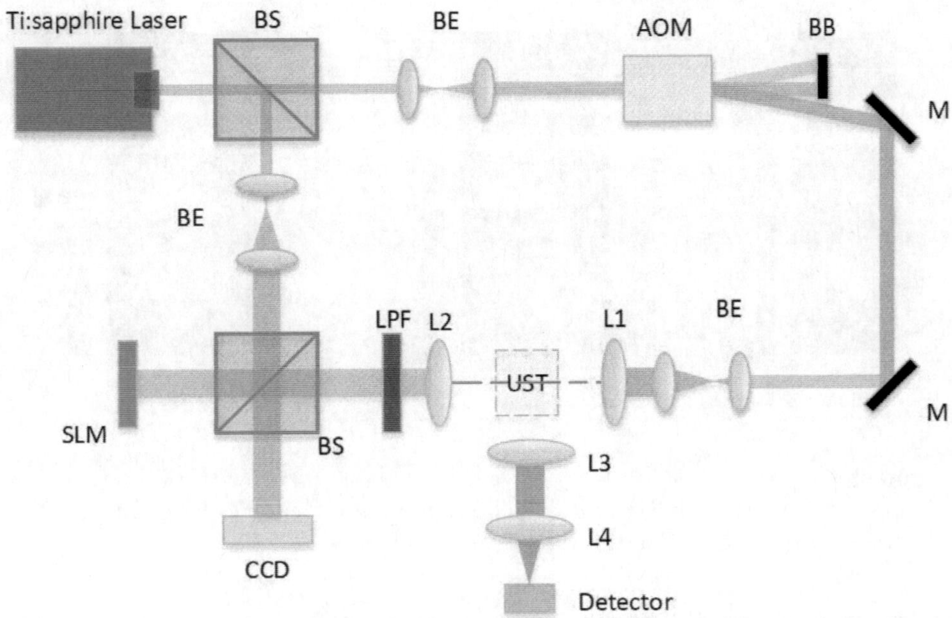

Figure 1. Experiment set-up. BS, non-polarizing beamsplitter; BE, beam expander; M, mirror; BP, bandpass filter; LP, long-pass filter; L1, f=35 mm lens; L2, f=50 mm lens; D, fluorescence detector; stage, three-axis motorized translation stage.) The pixel size of both the SLM and the CMOS camera is 8um. Distance from sound focus to SLM=305 mm.

The system set-up is demonstrated in Fig. 1. We use single-cycle focused ultrasound pulses and tightly synchronized near-infrared laser pulses to achieve a near-isotropic three-dimensional confined interaction volume. The pulsed light and pulsed sound waves are precisely synchronized so that the light wave illuminates the sample only when the single-cycle ultrasound pulse propagates through its spatial focus. Accordingly, the sound modulation zone is confined to 40um in the transverse direction by the sound focusing element, and to 40um in the axial direction by the temporal profile of the single-cycle sound pulse convolved with the temporal profile of the laser pulse. To provide sufficient and also durable optical power for fluorescence excitation, we used DOPC to perform phase conjugation.

3. Results

The fluorescence imaging of the opto-ultrasound imaging system is provided in Fig. 2.Figure 2a shows the fluorescence pattern (hole array)which will be sandwiched by two 2 mm thick high scattering media. In such a condition, as a result of random scatteringthe structural information of the fluorescence pattern was completely lost using direct wide-field fluorescence image. Figure 2b presents an original fluorescence image of the hole array with tissue phantoms around it using our methods.Figure 3c shows the image of Fig 2b using bicubic interpolation.

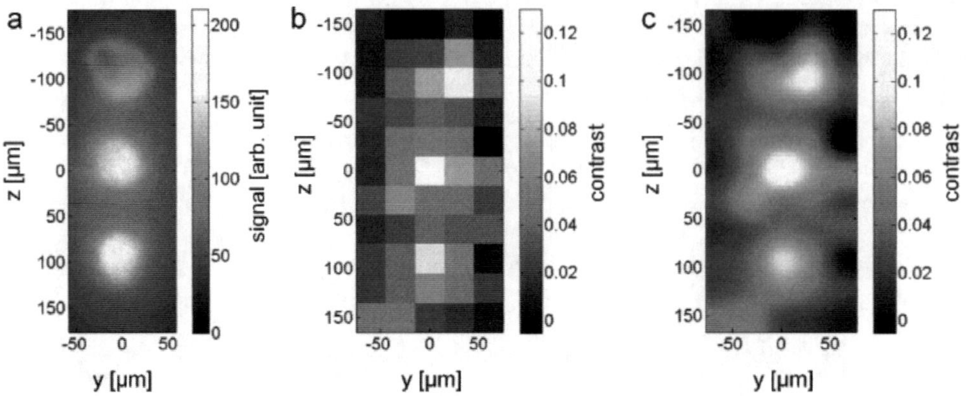

Acknowledgment
This work is supported by National Basic Research Program of China (973 Program) (2015CB352005) and by the Fundamental Research Funds for the Central Universities.

References

[1] Li, Anan, et al. "Micro-optical sectioning tomography to obtain a high-resolution atlas of the mouse brain." Science 330.6009 (2010): 1404-1408.
[2] Si, Ke, et al. "Fluorescence imaging beyond the ballistic regime by ultrasound-pulse-guided digital phase conjugation." Nature photonics 6.10 (2012): 657-661.
[3] Si, Ke, et al. "Breaking the spatial resolution barrier via iterative sound-light interaction in deep tissue microscopy." Scientific reports 2 (2012).

A compact wideband monopole antenna fed by coplanar waveguide

Guoping Pan, Wenhua Zhao, Jiang Zhou, Yan Su, Qifu Li

China Satellite Maritime Tracking and Control Department, Jiangyin, China 214431

panlei812@126.com

Abstract. A compact wideband printed antenna fed by coplanar waveguide (CPW) is presented in this paper. The wideband characteristic is achieved by coupling three monopole resonant modes of the proposed antenna. The proposed antenna is built and simulated to verify the design strategy. A wide impedance bandwidth of 100.5% (1.7866-5.3953 GHz) with S11< -10 dB is achieved. The proposed antenna has stable performance, including bi-directional radiation pattern and gain. It also shows the merits of low cost and simple structure.

1. Introduction

Recently, with the rapid development of modern wireless communication technology, the design of wideband antennas is more important. Printed planar antennas have the advantages of light weight, low profile, easy fabrication, low cost. The wideband antennas have the basic demand of good impedance matching and stable radiation patterns in the operation band.

In order to attain wider bandwidths, many impedance matching techniques have been used. A reconfigurable wideband and multi-band patch antenna with dual-patch elements and C-Slots was reported in [1], and an impedance bandwidth of 33.52% could be obtained in the wideband mode. Some printed wide slot antennas were extensively researched, like wide rectangular slot [2-4], bow-tie slots [5].

In this paper, a compact wideband printed antenna fed by coplanar waveguide for wideband wireless communication is proposed. To analyze the proposed antenna, High Frequency Structure Simulator (HFSS) software is used. A parametric analysis has been carried out, and the current distributions at three different resonant frequencies have been analyzed. The proposed antenna has wide impedance bandwidth of 100.5% (S11 < -10 dB) ranging from 1.7866 GHz to 5.3953 GHz, and bi-directional radiation patterns in the whole band.

2. Antenna configuration and design

The configuration of the proposed antenna is shown in Fig. 1. The size of 1.6-mm-thick FR4 substrate is $L \times W$, with a relative permittivity of 4.4 and a loss tangent of 0.02. The width of both U-shape monopoles is uniform and is defined as W_L. The length of upper branch of the left U-shape monopole is W_a. The width of the slot between two upper branches of both U-shape monopoles is W_g. CPW-Fed microstrip line has width of W_f, and the microstrip line length is L_f. The distance between co-planar ground and the microstrip line is denoted by S.

Journal of Physics: Conference Series **679** (2016) 012059 doi:10.1088/1742-6596/679/1/012059

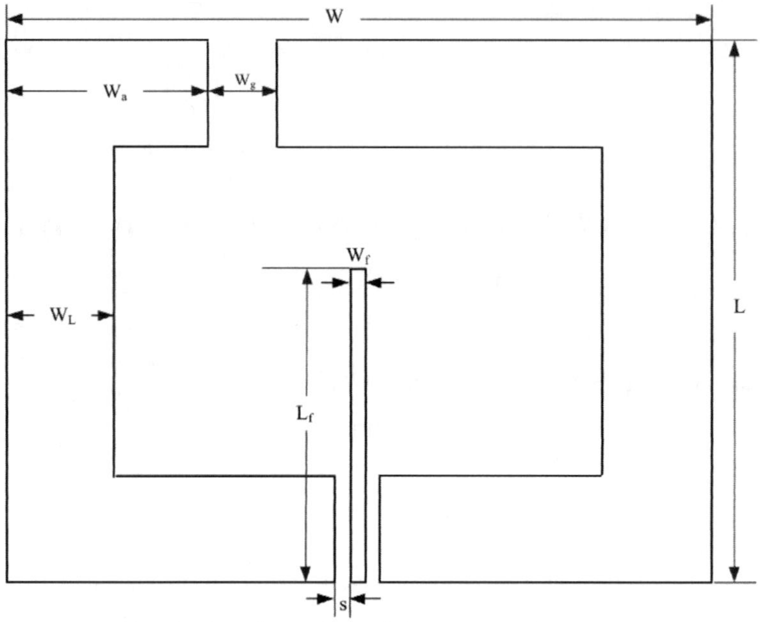

Fig. 1. Configuration of the proposed antenna.

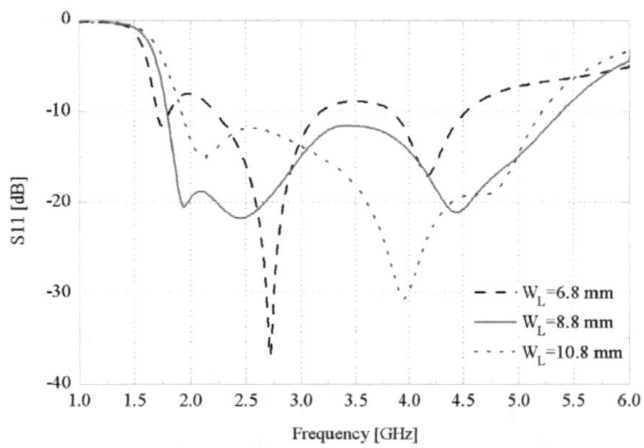

Fig. 2. Simulated reflection coefficient of the antenna with different W_L.

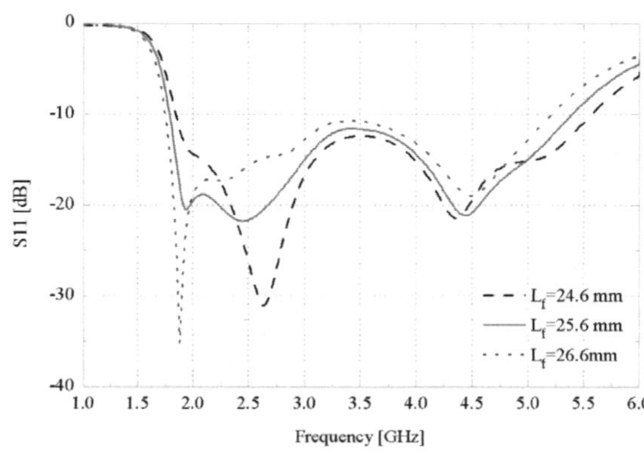

Fig. 3. Simulated reflection coefficient of the antenna with different L_f.

HFSS is used for the parameter optimization. Fig. 2 presents the simulated reflection coefficient of the proposed antenna with different W_L. The simulated results show that the lowest resonant frequency becomes higher as the width W_L increases. When the width of W_L is 8.8 mm, the proposed antenna has the best impedance matching properties. Fig.3 indicates simulated reflection coefficient of the antenna with different L_f. As we all know, the highest resonant frequency is observed to become lower as the length L_f increases. When the length of L_f is 25.6 mm, the impedance matching properties show the best. The microstrip line not only feeds to excite the compact wideband antenna but also works as a radiator. Other parameters can also be optimized for best matching performance with HFSS. The final design parameter values of the proposed antenna are shown in Table 1.

Journal of Physics: Conference Series **679** (2016) 012059 doi:10.1088/1742-6596/679/1/012059

Table 1. final design parameters

Parameter	*L*	*W*	W_a	W_g	W_L	L_f	W_f	*S*
Value (mm)	45	57.4	16.7	6	8.8	25.6	1.2	1

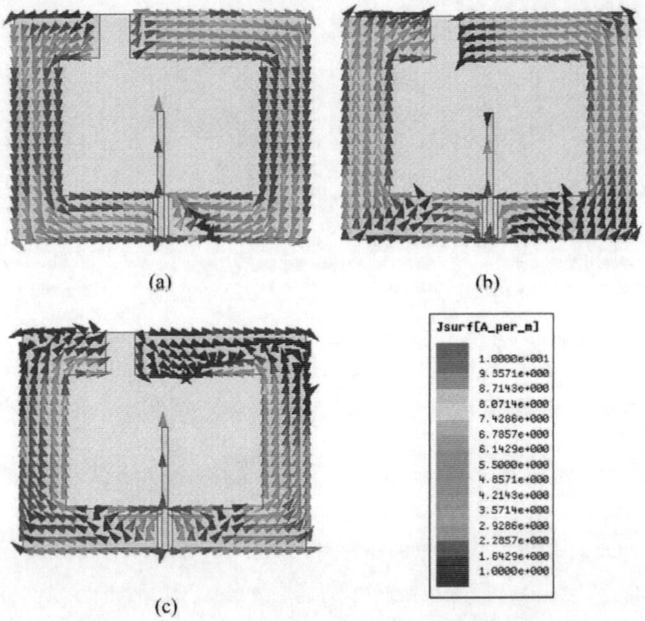

Fig. 4. Current distributions of the proposed antenna at: (a) 2 GHz; (b) 3 GHz; (c) 4.44 GHz.

Fig. 4 shows the current distributions of the proposed antenna at 2 GHz, 3 GHz and 4.44 GHz to analyze the operation modes. Fig. 4(a) shows the first resonance mode. The current distribution is mainly concentrated at both U-shape monopoles. It's a monopole mode, and the resonant frequency is mainly decided by the total length of both U-shape monopoles. Fig. 4(b) shows the second resonance mode. The current distribution is mainly concentrated at the left U-shape monopoles. It's also monopole mode, and the resonant frequency is mainly due to the length of the left U-shape monopoles. Fig. 4(c) shows the third resonance mode. It's also monopole mode. The length of the microstrip line above co-planar ground is $\lambda/4$ (λ is the resonance wavelength), and the current distribution is mainly concentrated at the microstrip line. Therefore, wide impedance bandwidth can be achieved by coupling these three monopole modes. For the lower band, both U-shape monopoles are the main radiators; for the middle band, the left U-shape monopole is the main radiator; for the higher band, the microstrip line is the main radiator.

3. Simulation results and analysis

Fig. 5 illustrates the simulated reflection coefficients of the proposed antenna. The simulated -10-dB bandwidth of reflection coefficients is 3.6087 GHz (1.7866–5.3953 GHz, 100.5%).

Fig. 6(a) to 6(c) show the simulated normalized radiation patterns in the X-Y and the Y-Z planes of the proposed antenna at 2 GHz, 3 GHz and 4.44 GHz, respectively. As we can see, at all these given frequencies, radiation patterns are stably bi-directional. The 3-dB beamwidths at different frequencies are defined to be the sums of 3-dB beamwidths in two different directions (+Y and -Y directions).The 3-dB beamwidths are 150° in the H-plane (X-Y plane) and 110° in the E-plane (Y-Z plane) at 2 GHz. The 3-dB beamwidths are 150° and 155° in the H-plane (X-Y plane) and the E-plane (Y-Z plane) at 3

Journal of Physics: Conference Series **679** (2016) 012059 doi:10.1088/1742-6596/679/1/012059

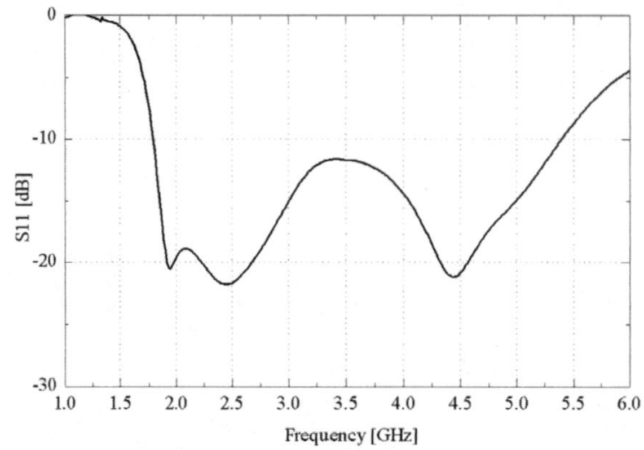

Fig. 5. Simulated reflection coefficient of the proposed antenna

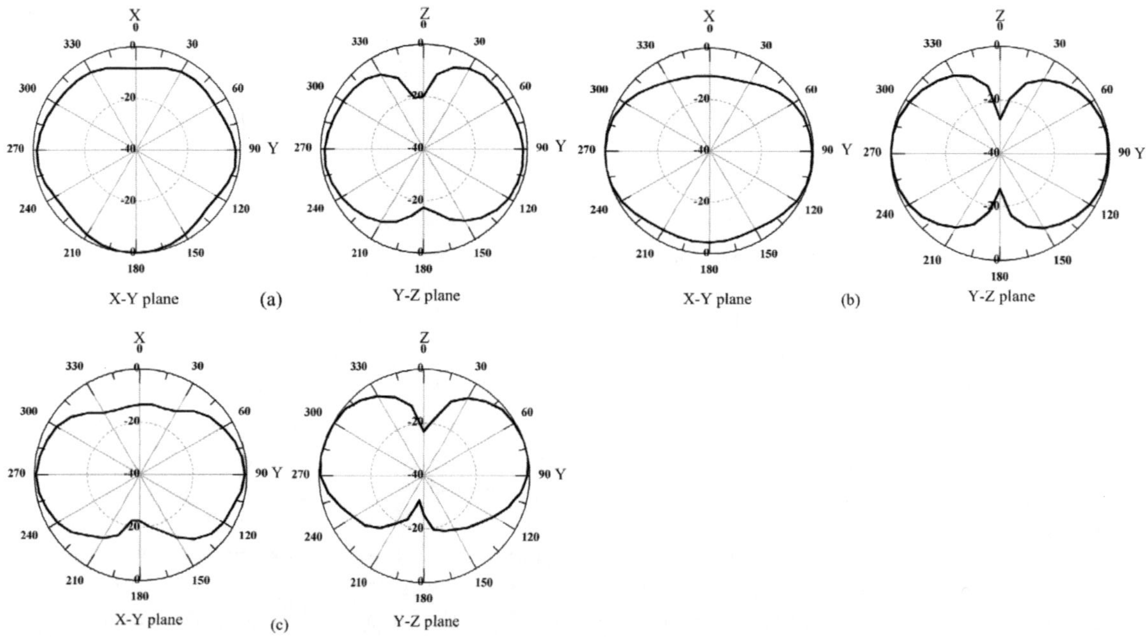

Fig. 6. The simulated normalized radiation patterns of the proposed antenna: (a) 2 GHz; (b) 3 GHz; (c) 4.44 GHz.

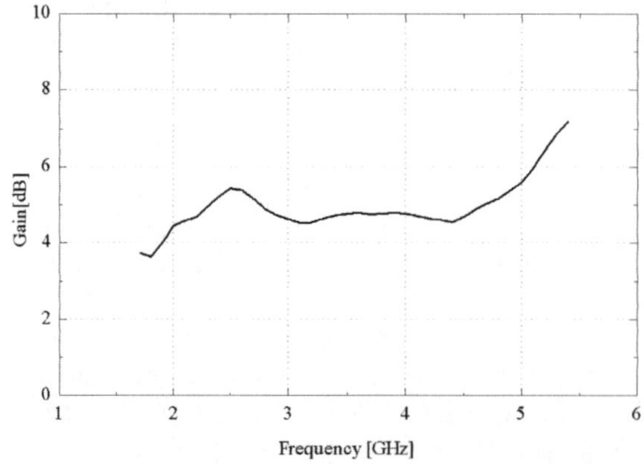

Fig. 7. The simulated gain of the proposed antenna

GHz. The 3-dB beamwidths are 100° and 120° in the H-plane (X-Y plane) and the E-plane (Y-Z plane) at 4.44 GHz.

Fig. 7 shows the realized gain variation with frequency of the proposed antenna. In the band between 1.7866 and 5.3953 GHz, the minimum gain is at 1.8 GHz with a value 3.65 dBi and the maximum gain is at 5.3953 GHz with a value 7.18 dBi.

4. Conclusion

In this paper, a compact wideband printed antenna fed by coplanar waveguide for wireless communication is presented. The overall dimensions of the proposed antenna are $45 \times 57.4 \times 1.6$ mm^3, and the structure is simple and compact. By exciting different radiators at different operation bands, the wideband characteristic is achieved. The simulated -10-dB bandwidth of reflection coefficients is 100.5% (1.7866–5.3953 GHz). The radiation patterns are stably bi-directional in the whole band. The proposed antenna has good application value for a wideband wireless communication system.

Reference

[1] H. F. AbuTarboush, R. Nilavalan, S. W. Cheung, K. M. Nasr, T. Peter, D. Budimir, and H. Al-Raweshidy, "A reconfigurable wideband and multiband antenna using dual-patch elements for compact wireless devices," IEEE Trans. Antennas Propag., vol. 60, no. 1, pp. 36–43, Jan. 2012.

[2] J. F. Huang and C. W. Kuo, "CPW-fed slot antenna with CPW tuning stub loading", Microwave Opt. Technol. Letters, vol. 19, pp. 257-258, 1998.

[3] M. Miao, B. L. Ooi, and P. S. and P. S. Kooi, " Broadband CPW-fed Wide slot Antenna", Microwave Opt. Technology Letters, vol. 25, pp. 206-211, 2000.

[4] P. L. Chin, A. Z. Elsherbeni, and C. E. Smith, "Characteristics of Coplanar Bow-Tie Patch Antennas", Antennas and Propagation Society International Symposium, vol. 1, pp. 398-401, 2002.

[5] P. L. Chin, A. Z. Elsherbeni, and C. E. Smith, "A Coplanar Waveguide Bow-Tie Aperture Antennas", Antennas and Propagation Society International Symposium, vol. 1, pp. 564-567, 2002.

Institute of Physics
Dirac House, Temple Back
Bristol BS1 6BE UK

CURRAN ASSOCIATES INC.
proceedings.com

ISSN: 1742-6588
ISBN 978-1-5108-2234-4

9 781510 822344